# Current Topics in Membranes and Transport

## Volume 20
## Molecular Approaches to Epithelial Transport

**Contributors**

*William P. Alles*
*Daniel Biemesderfer*
*Chris Clausen*
*Malcolm Cox*
*M. E. M. Da Cruz*
*Troy E. Dixon*
*Darrell D. Fanestil*
*Bliss Forbush III*
*E. Frömter*
*Michael Geheb*
*David B. P. Goodman*
*Victoria Guckian*
*Michael C. Gustin*
*Doris A. Herzlinger*
*T. Hoshiko*
*Ivan Emanuilov Ivanov*
*L. Kampmann*
*Michael Kashgarian*
*Ralph J. Kessler*
*R. Kinne*
*Ingeborg Koeppen*
*G. Kottra*
*Simon A. Lewis*
*J. J. Lim*
*J. T. Lin*
*Alicia A. McDonough*
*George K. Ojakian*
*Lawrence G. Palmer*
*Chun Sik Park*
*Heide Plesken*
*Uzi Reiss*
*Michael J. Rindler*
*B. C. Rossier*
*David D. Sabatini*
*Bertram Sacktor*
*James B. Wade*
*N. K. Wills*

# Current Topics in Membranes and Transport

*Edited by*
**Arnost Kleinzeller**
*Department of Physiology*
*University of Pennsylvania School of Medicine*
*Philadelphia, Pennsylvania*

**Felix Bronner**
*Department of Oral Biology*
*University of Connecticut Health Center*
*Farmington, Connecticut*

## VOLUME 20

## Molecular Approaches to Epithelial Transport

*Guest Editors*
**James B. Wade**
*Department of Physiology*
*Yale University School of Medicine*
*New Haven, Connecticut*

**Simon A. Lewis**
*Department of Physiology*
*Yale University School of Medicine*
*New Haven, Connecticut*

Volume 20 is part of the series (p. xix) from the Yale Department of Physiology under the editorial supervision of:

**Joseph F. Hoffman**
*Department of Physiology*
*Yale University School of Medicine*
*New Haven, Connecticut*

**Gerhard Giebisch**
*Department of Physiology*
*Yale University School of Medicine*
*New Haven, Connecticut*

1984

**ACADEMIC PRESS, INC.**
*(Harcourt Brace Jovanovich, Publishers)*

**Orlando San Diego San Francisco New York London**
**Toronto Montreal Sydney Tokyo São Paulo**

ACADEMIC PRESS, INC.
Orlando, Florida 32887

*United Kingdom Edition published by*
ACADEMIC PRESS, INC. (LONDON) LTD.
24/28 Oval Road, London NW1 7DX

LIBRARY OF CONGRESS CATALOG CARD NUMBER: 70-117091

ISBN 0-12-153320-4

PRINTED IN THE UNITED STATES OF AMERICA

84 85 86 87 9 8 7 6 5 4 3 2 1

*The editors wish to dedicate this volume to Dr. J. J. Lim, author of Chapter 2. The tragedy of his death (September 1, 1983) on Korean Air Lines flight 007 will be felt as a personal and professional loss to all of us.*

# Contents

## PART II. USE OF ANTIBODIES TO EPITHELIAL MEMBRANE PROTEINS

# Contributors

Numbers in parentheses indicate the pages on which the authors' contributions begin.

**William P. Alles,** Department of Physiology, Yale University School of Medicine, New Haven, Connecticut 06510 (87)

**Daniel Biemesderfer,** Department of Physiology, Yale University School of Medicine, New Haven, Connecticut 06510 (161)

**Chris Clausen,** Department of Physiology and Biophysics, Health Sciences Center, State University of New York, Stony Brook, New York 11794 (47)

**Malcolm Cox,** Renal Electrolyte Section, Medical Service, Philadelphia Veterans Administration Medical Center, and Department of Medicine, University of Pennsylvania School of Medicine, Philadelphia, Pennsylvania 19104 (271)

**M. E. M. Da Cruz,** Laboratório Nacional de Engenheria e Technologia Industrial, Lisbon, Portugal (245)

**Troy E. Dixon,** Department of Medicine, Northport Veterans Administration Hospital, Northport, New York 11768 (47)

**Darrell D. Fanestil,** Division of Nephrology, Department of Medicine, University of California San Diego, La Jolla, California 92093 (259)

**Bliss Forbush III,** Department of Physiology, Yale University School of Medicine, New Haven, Connecticut 06510 (161)

**E. Frömter,** Max-Planck-Institut für Biophysik, 6000 Frankfurt 70, Federal Republic of Germany (27)

**Michael Geheb,**[1] Renal Electrolyte Section, Medical Service, Philadelphia Veterans Administration Medical Center, and Department of Medicine, University of Pennsylvania School of Medicine, Philadelphia, Pennsylvania 19104 (271)

**David B. P. Goodman,** Department of Pathology and Laboratory Medicine, Hospital of the University of Pennsylvania, University of Pennsylvania School of Medicine, Philadelphia, Pennsylvania 19104 (295)

**Victoria Guckian,**[2] Department of Physiology, Yale University School of Medicine, New Haven, Connecticut 06520 (217)

**Michael C. Gustin,** Department of Biochemistry, University of Wisconsin, Madison, Wisconsin 53706 (295)

**Doris A. Herzlinger,** Department of Anatomy and Cell Biology, Downstate Medical Center, State University of New York, Brooklyn, New York 11203 (181)

**T. Hoshiko,** Department of Physiology, Case Western Reserve University School of Medicine, Cleveland, Ohio 44106 (3)

[1]Present address: Department of Medicine, Wayne State University, Harper-Grace Hospital, Detroit, Michigan 48201.

[2]Present address: Department of Physiology, University of Maryland School of Medicine, Baltimore, Maryland 21201.

**Ivan Emanuilov Ivanov,** Department of Cell Biology, New York University School of Medicine, New York, New York 10016 (199)

**L. Kampmann,** Max-Planck-Institut für Biophysik, 6000 Frankfurt 70, Federal Republic of Germany (27)

**Michael Kashgarian,** Department of Pathology, Yale University School of Medicine, New Haven, Connecticut 06510 (161)

**Ralph J. Kessler,** Division of Nephrology, Department of Medicine, University of California San Diego, La Jolla, California 92093 (259)

**R. Kinne,**[3] Department of Physiology and Biophysics, Albert Einstein College of Medicine, Bronx, New York 10461 (245)

**Ingeborg Koeppen,**[4] Department of Physiology, Yale University School of Medicine, New Haven, Connecticut 06510 (217)

**G. Kottra,**[5] Max-Planck-Institut für Biophysik, 6000 Frankfurt 70, Federal Republic of Germany (27)

**Simon A. Lewis,** Department of Physiology, Yale University School of Medicine, New Haven, Connecticut 06510 (87)

**J. J. Lim,**[6] Max-Planck-Institut für Biophysik, 6000 Frankfurt 70, Federal Republic of Germany (27)

**J. T. Lin,** Department of Physiology and Biophysics, Albert Einstein College of Medicine, Bronx, New York 10461 (245)

**Alicia A. McDonough,** Department of Physiology and Biophysics, University of Southern California School of Medicine, Los Angeles, California 90033 (147)

**George K. Ojakian,** Department of Anatomy and Cell Biology, Downstate Medical Center, State University of New York, Brooklyn, New York 11203 (181)

**Lawrence G. Palmer,** Department of Physiology, Cornell University Medical College, New York, New York 10021 (105)

**Chun Sik Park,** Division of Nephrology, Department of Medicine, University of California San Diego, La Jolla, California 92093 (259)

**Heide Plesken,** Department of Cell Biology, New York University School of Medicine, New York, New York 10016 (199)

**Uzi Reiss,** Laboratory of Molecular Aging, National Institute on Aging, National Institutes of Health, Gerontology Research Center, Baltimore City Hospitals, Baltimore, Maryland 21224 (235)

**Michael J. Rindler,** Department of Cell Biology, New York University School of Medicine, New York, New York 10016 (199)

**B. C. Rossier,** Institut de Pharmacologie de l'Université de Lausanne, CH-1011 Lausanne, Switzerland (125)

**David D. Sabatini,** Department of Cell Biology, New York University School of Medicine, New York, New York 10016 (199)

**Bertram Sacktor,** Laboratory of Molecular Aging, National Institute on Aging, National Institutes of Health, Gerontology Research Center, Baltimore City Hospitals, Baltimore, Maryland 21224 (235)

[3]Present address: Max-Planck-Institut für Systemphysiologie, D-4600 Dortmund 1, Federal Republic of Germany.

[4]Present address: Anatomisches Institut der Universität Heidelberg, D-6900 Heidelberg, Federal Republic of Germany.

[5]Present address: Department of Internal Medicine, Semmelweis Medical University, 1088 Budapest, Hungary.

[6]Deceased.

**James B. Wade,**[7] Department of Physiology, Yale University School of Medicine, New Haven, Connecticut 06510 (217)

**N. K. Wills,** Department of Physiology, Yale University School of Medicine, New Haven, Connecticut 06510 (61)

[7]Present address: Department of Physiology, University of Maryland School of Medicine, Baltimore, Maryland 21201.

# Preface

In recent years the issues in epithelial transport research have gradually evolved toward the molecular level. As recently as ten years ago, the predominant problem was to determine exactly what electrolytes and nonelectrolytes particular epithelia absorb or secrete. Once armed with an understanding of the macroscopic transport processes, researchers asked whether the movement of substances was through the cells (i.e., active) or between the cells (i.e., passive) flowing along favorable electrical and/or chemical gradients. In turn, using electrical or chemical methods, the question was raised at which step in transcellular transport was energy in the form of ATP required. As an example, in transepithelial sodium transport, sodium enters the cell passively down a net electrochemical gradient and is actively extruded into the blood via an ATP-requiring transport protein (the $Na^+,K^+$-ATPase). One of the most recent topics being addressed is the mechanism involved in the regulation of these electrolyte and nonelectrolyte transport proteins (synthesis and/or activation), and whether such proteins move substances by a channel-type configuration or a carrier configuration.

It is obvious that the more classic approaches to studying epithelial transport are inadequate by themselves to address these questions fully. It is the purpose of this book to outline and illustrate, by example, some recently developed approaches that can provide important new insight into epithelial transport mechanisms.

Part I of this volume is devoted to the electrical methodology used to address questions such as the following: Does sodium entry across the apical membrane of tight epithelia occur by a channel or carrier mechanism? Do hormones increase the number of transport proteins or the ability of a single protein to carry more ions per unit time? During stimulation of ion transport, are quiescent channels activated, or are cytoplasmic vesicles containing the transport protein mobilized on a certain signal and inserted into the membrane? The methods to be used consist of impedance analysis to measure changes in membrane area associated with stimulation of transport and fluctuation analysis to evaluate alterations in channel density.

Part II provides a wide range of examples of how antibodies to epithelial membrane proteins can be useful. Antibodies are clearly a powerful tool for evaluating the biosynthesis of transport proteins such as $Na^+,K^+$-ATPase and provide a means whereby membrane proteins can be identified and localized in epithelia.

Part III of the volume describes biochemical approaches to characterizing epithelial transport systems. These chapters illustrate approaches being taken for isolation and identification of transport proteins. In addition, these studies show how substrate protection can be utilized to identify chemical groups associated with important sites of a transport system.

We would like to acknowledge the generous financial support of Abbott Laboratories, North Chicago, Illinois; Hoffmann-LaRoche Inc., Nutley, New Jersey; ICI Americas, Inc., Wilmington, Delaware; Miles Laboratories, Inc., Elkhart, Indiana; C. F. Searle & Co., Chicago, Illinois; and the Upjohn Company, Kalamazoo, Michigan, for the Eighth Conference on Membrane Transport Processes sponsored by the Department of Physiology at Yale University School of Medicine which provided a basis for this volume. We also wish to thank Marie Santore for her invaluable assistance in organizing that meeting.

James B. Wade
Simon A. Lewis

# Yale Membrane Transport Processes Volumes

Joseph F. Hoffman (ed.). (1978). "Membrane Transport Processes," Vol. 1. Raven, New York.
Daniel C. Tosteson, Yu. A. Ovchinnikov, and Ramon Latorre (eds.). (1978). "Membrane Transport Processes," Vol. 2. Raven, New York.
Charles F. Stevens and Richard W. Tsien (eds.). (1979). "Membrane Transport Processes," Vol. 3: Ion Permeation through Membrane Channels. Raven, New York.
Emile L. Boulpaep (ed.). (1980). "Cellular Mechanisms of Renal Tubular Ion Transport": Volume 13 of *Current Topics in Membranes and Transport* (F. Bronner and A. Kleinzeller, eds.). Academic Press, New York.
William H. Miller (ed.). (1981). "Molecular Mechanisms of Photoreceptor Transduction": Volume 15 of *Current Topics in Membranes and Transport* (F. Bronner and A. Kleinzeller, eds.). Academic Press, New York.
Clifford L. Slayman (ed.). (1982). "Electrogenic Ion Pumps": Volume 16 of *Current Topics in Membranes and Transport* (A. Kleinzeller and F. Bronner, eds.). Academic Press, New York.
Joseph F. Hoffman and Bliss Forbush III (eds.). (1983). "Structure, Mechanism, and Function of the Na/K Pump": Volume 19 of *Current Topics in Membranes and Transport* (F. Bronner and A. Kleinzeller, eds.). Academic Press, New York.
James B. Wade and Simon A. Lewis (eds.). (1984). "Molecular Approaches to Epithelial Transport": Volume 20 of *Current Topics in Membranes and Transport* (A. Kleinzeller and F. Bronner, eds.). Academic Press, New York.

# Part I

# Frequency Domain Analysis of Ion Transport

CURRENT TOPICS IN MEMBRANES AND TRANSPORT, VOLUME 20

# Chapter 1

# Fluctuation Analysis of Apical Sodium Transport

*T. HOSHIKO*

*Department of Physiology*
*Case Western Reserve University School of Medicine*
*Cleveland, Ohio*



## I. INTRODUCTION

Fluctuation analysis has been used in the study of nerve and black lipid membranes for several years now, but it is only fairly recently that much has been accomplished in epithelial membranes using this technique. The complexity of epithelial membranes has been perhaps the primary deterrent. Four years ago I described some of the very early results (Hoshiko and Moore, 1978) as did Lindemann and Van Driessche (1978). Lindemann has subsequently reviewed work to 1980 (Lindemann, 1980). Since then two types of models for interpreta-




ISBN 0-12-153320-4

tion of the fluctuations have been analyzed: (1) as electrical events modified by the electrical properties of epithelial membrane structures, and (2) as current fluctuations due to interruption of ion flow through a conducting channel.

One of the great appeals of fluctuation analysis is that it offers a way of studying an undisturbed system. Instead of applying a pulse or stepwise stimulus, the natural, spontaneous fluctuations can be observed, and from this apparent chaos, the kinetic behavior of the system can be obtained. There is something of a magical appeal in this. I suppose this also appeals to the parsimonious bent drummed into us as scientists, looking for the simplest explanations for the least complicated experiment. Another appeal is the promise of being able to detect elementary processes—the hope that the fluctuations reflect random thermal motion of the molecules involved. Then we might model the actual molecular events underlying transport. The situation as it turns out is rather more complicated. The fluctuations do reflect the individual molecular events but the elementary processes may be masked, because the electrical events so initiated could be distorted or altered by the electrical effects of the multiple membranes present in epithelia.

## II. MECHANICS OF FLUCTUATION ANALYSIS

### A. Sources of Fluctuations

The fluctuations in current and voltage are so minute that great care must be taken to avoid extraneous disturbances. These of course include the usual things electrophysiologists have long been concerned about (Neumcke, 1982; Fishman, 1982): EMR pickup, power-line transients, battery-pack noise, vibrations that cause lead-in wires to change input capacitance, and amplifier noise in general. Epitheliologists used to worrying about disturbances in the millivolt and microamp scale now have to be concerned with these same disturbances at a much lower level—microvolts and subnanoamperes. In studying current fluctuations, the input voltage noise of the voltage clamp amplifier appears to be the major culprit as the source of instrumentation noise (Poussart, 1971). This assumes that the electrodes and various connectors are clean and of good quality. Any sort of high resistance in electrodes, salt bridges, or electrical connections to the preparation can be a problem. Today, clamp amplifiers with extremely low input voltage noise can be built and some are available commercially (Van Driessche and Lindemann, 1978). The configuration of the chamber and the way the membrane is mounted is critical in achieving a low-noise preparation. One of the biggest problems in studying the apical sodium-selective channel is that its kinetics are very slow and many of the artifacts arising from chamber design and membrane mounting give rise to large-amplitude, low-frequency noise. Such low-frequency noise can obscure the fluctuations due to the sodium channels themselves.

What are the types of noise we might observe in epithelia? Stevens (1972) outlined them for nerve and muscle membranes some 10 years ago and Louis DeFelice (1981) has described them in some detail in his book. These include the basic, irreducible sources of noise grounded in the physics of the structures involved. There is thermal or Johnson noise due to thermal agitation of charge carriers. This is an equilibrium process which is usually drowned out in functioning epithelial membranes. In nonequilibrium processes, excess noise appears, meaning that some current is flowing even if not in the external circuit. Part of this excess noise is true shot noise, due only to the graininess of the charge carriers (charge is transported as quantal units). For shot noise, the noise power is directly proportional to the dc current. Both Johnson noise and shot noise occur without any regular pattern and given rise to slow and fast events. They have no characteristic frequencies (at least in the time scale of interest) and the power is distributed uniformly in frequency, so they are termed "white noise." Both are drowned out by other noise. Some of the excess noise is the infamous $1/f$ noise. It has been said to arise from diffusive processes (cf. Neumcke, 1982), from anisotropic constraints to ion flow (Dorset and Fishman, 1975), from turbulence in convective processes associated with ionic migration in a field (Lifson *et al.*, 1978), and from all sorts of unexpected events like earthquakes and slamming doors (Machlup, 1981). Certainly it seems to increase if the bathing fluid is leaking (Hoshiko and Van Driessche, unpublished). Leaks could mean small variations in the effective area being exposed, and hence fluctuations in the current. Baseline drift is also associated with $1/f$ noise (Conti *et al.*, 1980). One possibility is that $1/f$ noise is the result of a series of Lorentzian processes distributed over a wide range of frequencies (Van der Ziel, 1959). In other words, unrelated events could sum up to give that type of noise.

What about conductance fluctuations? The kinetics of a blocking–unblocking process of a conductive pore is a first-order process. It is analogous to the regeneration–recombination process in semiconductors, whose kinetics were modeled by Machlup (1954). A first-order process is a single exponential with time; in the frequency domain, it has constant power density up to its characteristic frequency. Above that, the power falls off rapidly with the square of the frequency, producing a shape of curve called Lorentzian and defined by two parameters, the corner frequency and the low-frequency plateau. This is the type of response seen in the fluctuations in frog skin short-circuit current when amiloride (Lindemann and Van Driessche, 1977) is applied to the apical surface, or in the case of spontaneous potassium noise described by Van Driessche and Zeiske (1980a).

## B. Spectral Properties of Fluctuations

The computation of the spectral density function is usually done with the fast Fourier transform (FFT) algorithm on a computer. The validity of the input data can be checked crudely by observing the time record, by monitoring each spectrum in real time, or by examining the histogram of the amplitudes. The dc level must be subtracted before the FFT is computed. The input voltage should be as large as possible since a small dynamic range leads to errors in quantization of the data. But one can go too far and saturate the amplifiers which could result in an apparent leveling off of the power at the low frequencies, since it is usually the low frequencies that give rise to large excursions. The point is that it is unwise to let the machine blindly produce spectra, and the spectrum analysis program should check the input data for saturation of the amplifiers. If such an overflow happens only once or twice it will probably wash out in the overall average. However, it is easy to miss overflows while observing the time record during data collection. To see how the preparation is behaving, it is useful to use a dedicated real-time spectrum analyzer.

Another possibility is to record fluctuations onto FM tape, monitoring the spectrum with a real-time spectrum analyzer. Then later it is possible to choose those records that are acceptable for further processing on the computer. With recorded data one can use several overlapping frequency bands to compute the power spectrum over a wider frequency range than with just one band. The FFT takes current or potential values measured at equal time intervals and converts them into half the number of power density points at equally spaced frequency intervals. The lowest frequency present in the record depends on the total duration of the sampling period, whereas the sampling rate determines the highest frequency. Actually the highest usable frequency is about one-half the sampling frequency because of the sampling artifact called aliasing (depending on the low pass filter used at the input). Aliasing is analogous to the effects seen in silent films in which an automobile wheel seems to go forward too slowly, or even backward. Because of this, starting with 1024 points means ending up with as few as 256 frequency points. The 1024 points result in less than three decades in frequency, but to get another decade would require over 10,000 points. Rather, it is more economical to take two overlapping bands of a thousand points each to get a wider frequency band than to increase the number of points in one very wide spectrum.

Another route is to use a dedicated computer to take several FFTs simultaneously. This is the approach used by Van Driessche and Gullentops (see Hoshiko and Van Driessche, 1981; Zeiske *et al.*, 1982). Up to four overlapping frequency bands can be computed in real time in a dual processor system, one processor taking in the data and the second carrying out the FFTs. Successive averages of these overlapping bands are taken almost in real time with subsequent storage and data analysis done on another host computer system. This is certainly more efficient than replaying the data tape several times over, but it is more expensive since it requires special equipment. However, once the frequency bands have been chosen, it is no longer possible to rerun the data to get at another frequency band. Only having the analog data on FM tape can allow that.

The error inherent in a single estimate of a spectral density in a given frequency band is astonishingly large, as large as the value of the spectral density itself (Bendat and Piersol, 1971). Thus, it is necessary to average a number of spectra and/or to smooth the spectral curve by averaging over some number of adjacent frequency intervals. The usual practice is to average some 20–60 spectra and then to smooth the curve by averaging over frequency. If we reduce the density of points at the higher frequencies, the spectra look less cluttered on the log–log plot. This procedure however makes the curve fitting somewhat ambiguous since proper weights should be assigned to the power densities at the different frequencies. The point is that a proper nonlinear curve-fitting procedure is essential in order to arrive at an objective estimate of model parameters, especially since records are often mixed with low-frequency noise.

## III. EPITHELIAL SOURCES OF FLUCTUATIONS

### A. Noise Equivalent Circuit of Epithelium

Let us now consider where fluctuations may arise in the epithelium itself. Although current fluctuations have not been studied in split skin, resistance, spontaneous potential, and short-circuit current are the same as in the intact skin, and there seems no reason to believe fluctuations would be any different. It seems reasonable to ignore the corium as a significant source of noise. Also, the unstimulated skin gland should not contribute to the fluctuations in current and it seems safe to consider only the epithelium proper. Since the epithelial layers seem to constitute a syncytium, common practice has been to lump all the electrical parameters together and look at a noise equivalent circuit of the epithelium under voltage clamp as in Fig. 1. Let us assume that all structures could have at least one kind of noise source—thermal or Johnson noise. The main structures are the apical and basolateral membranes of the epithelial cell, shunted by a paracellular pathway and accessed via a series resistance made up of the bathing solution, stratum corneum, corium, and current electrode pathway all lumped into one. We have four possible noise sources. The paracellular and series paths are taken to be purely resistive, but the cell membranes each have a parallel capacitance. I showed (Hoshiko, 1961) that over 80% of the total resistance of the cellular pathway is accounted for by the apical membrane. The resistance of the paracellular pathway is roughly twice that of the cellular path-

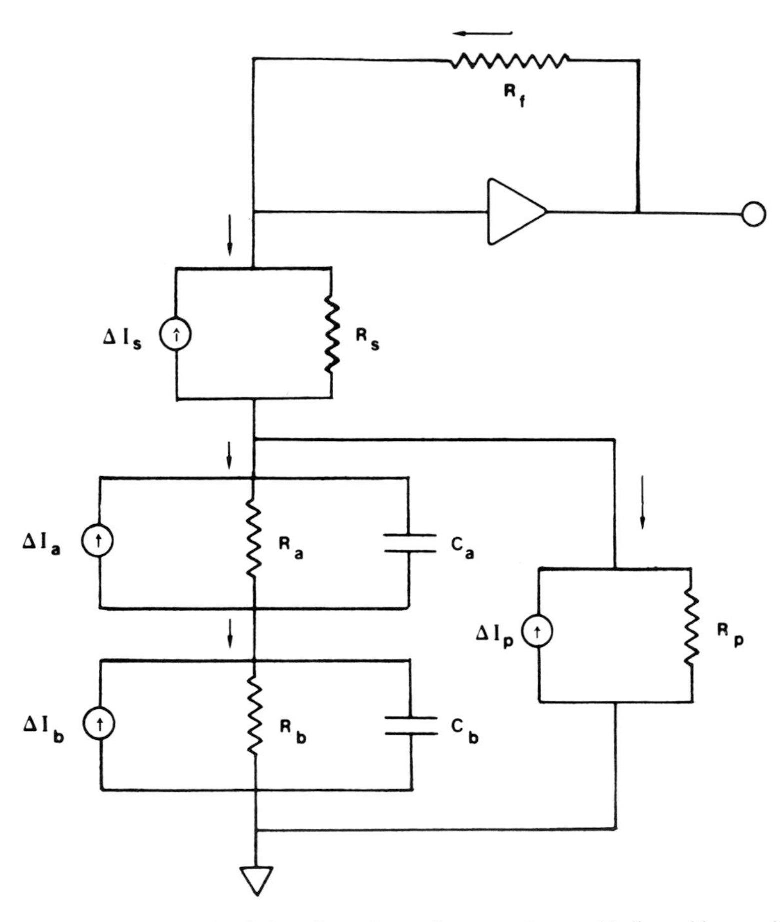


FIG. 1. Noise equivalent circuit for voltage clamp of two-membrane epithelium with paracellular shunt and series resistance. A voltage clamp amplifier supplies clamping current through a feedback resistance, $R_f$, to clamp via a series access resistance, $R_s$, an epithelium consisting of an apical membrane, a, and a basolateral membrane, b, together in series with a paracellular resistive shunt, $R_p$. Thermal noise sources associated with these purely resistive components are $\Delta I_s$ and $\Delta I_p$, respectively. Active membrane noise sources are $\Delta I_a$ and $\Delta I_b$.

way (Ussing and Windhager, 1964; Nagel, 1977). During short-circuiting, most of the current flows through the cellular pathway and that current flow is controlled primarily by the resistance of the apical membrane. Assuming that the four postulated noise sources are uncorrelated, and that the series and paracellular noise sources contribute only Johnson noise, it turns out that the major noise source must be in the cellular membranes. Moreover, since the apical resistance is so high, any noise source in the basolateral membrane is severely attenuated, and the main noise source must reside in the apical membrane. This situation is further improved in studying amiloride- or triamterene-induced noise since they further increase apical resistance. Lindemann and his group (Fuchs *et al.*, 1977; Lindemann and Van Driessche, 1977; Li *et al.*, 1982; Palmer *et al.*, 1982) have chosen to "depolarize" the basolateral membrane by immersing the basolateral side in a high-potassium solution which acts to reduce the potential and resistance of the basolateral membrane. The appearance of amiloride-induced fluctuations adds additional support to the idea that the major noise source is the apical membrane, since it is well accepted that amiloride is a specific blocker of the sodium-selective apical pathway (for a recent review, see Benos, 1982). Removal of sodium from the apical bathing solution greatly diminishes the fluctuations.

## B. Distortion of Apical Current Fluctuations

Granting that current fluctuations in the epithelium appear to arise in the apical membrane, to what extent might the electrical properties of the other membrane structures distort those current fluctuations? This question has been studied by Lindemann and DeFelice (1981) and Van Driessche and Gögelein (1980). Lindemann and DeFelice assumed that a random sequence of unit pulses of current at the apical membrane drove a passive network, resulting in current and voltage fluctuations in the voltage clamp circuit. Van Driessche and Gögelein assumed that fluctuations in conductance at the apical membrane give rise to the fluctuations current or potential. The results of the two approaches are similar except that the Van Driessche and Gögelein model involves the electromotive forces (emfs) across the apical and basolateral membranes. This happens because current pulses coming from conductance fluctuations should be dependent upon the driving force. On the other hand, the Lindemann and DeFelice model lumps such effects in with the magnitude of the current pulse. One advantage claimed for their method is the possibility of procedures that might allow "correction" of the measured spectra. More importantly the two analyses agree upon the general effects of the basolateral and paracellular electrical properties. These are that given equal apical and basolateral resistances and capacitances, the current fluctuations are attenuated more at the lower frequencies. When this is coupled with a Lorentzian component with a corner frequency near the crossover frequency, peaking of the measured spectrum may occur. This effect is accentuated by the presence of a large series resistance or a large positive basolateral emf. It is possible that some combination of those effects may provide an explanation of the peaking I observed (Hoshiko, 1978) in a patch clamp of the apical membrane, since the patch pipets constituted a rather large series resistance. In frog skin under the usual conditions of measurement, the resistance of the basolateral membrane exceeds that of the apical membrane by severalfold and the capacitance by perhaps 50-fold (Smith, 1971). In amiloride the apical resistance is increased even more (Helman and Fisher, 1977). These factors would appear to reduce distortion to negligible levels. Nevertheless, impedance spectra of the epithelium and its component structures will be needed to confirm these theoretical expectations. In addition, the relationship between the current and admittance (or impedance) spectra in epithelia will be helpful in parameter estimation. Newer methods requiring a very short time for estimation of the admittance function have been developed for use with axon electrophysiology (Fishman *et al.*, 1981; Fishman, 1982) and will no doubt prove useful with epithelia.

### C. Parameter Estimation

After computing the power spectrum, the problem is to extract the parameters that describe the kinetic process. As previously mentioned, nonlinear fitting programs are necessary to fit such data. Various programs are available, and most appear to be based on Marquard's modification of the Gauss–Newton method (cf. Van Driessche and Zeiske, 1980b). Often the spectrum to be fitted consists of a low-frequency component superimposed upon a Lorentzian curve. Then the fitting function used is a low-frequency component proportional to $1/f$ raised to some arbitrary power $\alpha$, added to a Lorentzian function. The value of $\alpha$ may range from 0.3 to 2.0 or even higher. If a great deal of low-frequency noise is present or if the noise levels are very low as in the case when high amiloride concentrations are used, the fitting program may not converge. In order to reduce the low-frequency component, one method has been to subtract a "control" spectrum obtained from an untreated preparation from that obtained from the treated spectrum (Christensen and Bindslev, 1982). This has been tried in nerve, but severe conditions must be met in order for the procedure to be valid (Neumcke, 1982). If more than one Lorentzian shape contributes to the spectrum the curve-fitting problem may become difficult. However, if the two Lorentzian functions have widely differing parameters, clearly separated from each other, then parameter estimation is assured. The interpretation of the Lorentzian parameters is the crucial process and depends upon the particular model hypothesized to generate the fluctuations. These model constructs then can be tested by varying blocker concentrations, clamp voltage, etc., and comparing their effects with predictions from the model. Another approach has been to integrate the computed spectrum to achieve a cumulative spectrum as a function of frequency (Rice, 1954). Christensen and Bindslev (1982) claim this method allows more reliable estimation of the corner frequency.

# IV. FLUCTUATIONS IN APICAL SODIUM CONDUCTANCE

## A. Time Constant of the Apical Sodium Channel

Lindemann and Gebhardt (1973) reported that apical sodium permeability in frog skin decreases after a sudden increase in sodium concentration along a multiexponential time course, a process they call a "recline." The major component has a characteristic time of 2–5 seconds, which seems too slow to be due to diffusion of sodium to the sodium-selective channels and too fast for intracellular accumulation. Rather it appears to involve self-inhibition of the sodium permeability, a "clogging" of the sodium channel. A 3-second time constant translates into a corner frequency of 0.053 Hz. This means a period of 19 seconds to get down to the corner frequency, and a sample period at least 10 times as long (i.e., one decade lower) would be needed to really establish the plateau. Such a low-frequency Lorentzian curve has not been observed. For example, Lindemann and Van Driessche (1978) reported that even down to 0.015 Hz, they were unable to observe a reliable flattening of the curve. Lindemann and Gebhardt (1973) also looked at the time constant for the recline in frog bladder. They found a time constant of about 50 msec, almost 100 times faster than in the skin. This translates into a corner frequency of about 3.2 Hz, well within the range of what can be detected. Li *et al.* (1982) and Palmer *et al.* (1982) have published spectra from toad bladder going down to 0.1 Hz. There is only the slightest hint of a corner at the very lowest frequencies. The absence of a low-frequency sodium corner may be due to the fact that these spectra are carried out at a high sodium concentration, when the sodium permeability has been saturated. Fluctuations in the sodium activity would have negligible effects on sodium permeability (or conductance) since it is already saturated. However, current fluctuations at low sodium concentration are yet to be studied in frog bladder. Another possibility is the use of agents which stimulate short-circuit current such as novobiocin (Johnston and Hoshiko, 1971), lanthanide ion (see Goudeau *et al.*, 1979 for recent references), guanidinobenzimidazole (Zeiske and Lindemann, 1974), etc.

## B. Blocker-Induced Conductance Fluctuations

The inability to detect a sodium-induced Lorentzian response led Lindemann and Van Driessche (1977) to study the kinetics of fluctuations induced by amiloride. This has been a most productive tool in the study of the apical sodium channel. They postulated a simple two-state model of amiloride-induced block-

TABLE I
TWO-STATE MODEL[a]

$$R_o \underset{k_-}{\overset{k[A]}{\rightleftharpoons}} R_A$$
(Probability open) (Probability blocked)

1. In the steady state

$R_o = (1 + [A]/K_A)^{-1}$ where $K_A = k_-/k$

$I = \bar{\imath} \cdot N \cdot R_o$ where $\bar{\imath}$ = mean single-channel current; $N$ = channel density

2. Corner frequency

$$\omega_c = 2\pi f_c = k_- + k \cdot [A]$$

3. Plateau

$$S_o = \frac{N \cdot \bar{\imath}^2}{\pi} \frac{R_o (1 - R_o)}{\omega_c}$$

4. Power spectrum

$$S(\omega) = S_o/[1 + (\omega/\omega_c)^2]$$

5. Values in frog skin

$\bar{\imath}$ = 0.3–0.5 pA
$N = 0.7\text{–}2.0 \times 10^8 \text{ cm}^{-2}$

[a] Lindemann and Van Driessche (1977).

age of the sodium-selective channel of the apical membrane (see Table I). The probability the channel is open ($R_o$) and the probability the channel is closed ($R_A$) sum to unity. The rate of closing ($k$[A]) is proportional to the blocker (amiloride) concentration [A], and the opening rate constant is $k_-$. The characteristic rate for this process is the sum of the forward and backward rates, which determines the corner frequency $f_c$. The power spectrum as a function of frequency can be represented in terms of the association rate constant $K_A$, which is in turn the ratio of the backward to forward rate constants. The mean current is simply the product of the average current per channel and the number of open channels. These three basic relations can be used to compute the single-channel current $i$, and the total number of channels $N$. Lindemann and Van Driessche have estimated in *Rana esculenta* that the single-channel current ranged from 0.3 to 0.5 pA and that channel density ranged from 0.7 to 2 × $10^8$ pores per square centimeter of skin area. The magnitude of the single-channel current was used to argue for a pore mechanism. The value 0.3 pA is equivalent to over a million ions transported per second at each transport site. This large number is much more than could be expected of a carrier mechanism—no known carrier turns over at such a high rate. Thus a carrier mechanism seems ruled out and a pore mechanism is likely.

Estimates of channel density from fluctuation and amiloride binding studies are summarized in Table II. Channel density was estimated by Cuthbert and Shum (1974) from the density of amiloride binding sites at the apical surface. They report that the measured binding constants are approximately the same as the inhibition constant (measured as the concentration required for 50% inhibition of the short-circuit current), which gave them confidence that their binding capacity estimates might be functionally related to transport inhibition. However, their values for the density of transport sites range up to a thousand times larger than those estimated by fluctuation analysis. When stimulated by aldosterone (ALDO), Palmer *et al.* (1982) report channel density doubled but both single-channel current and the amiloride rate constants were unchanged. Cuthbert and Shum (1975) report that amiloride binding also increased by almost a factor of 2, with no change in binding constant. On the other hand, Li *et al.* (1982) report that ADH (oxytocin) increased channel density while both single-channel current and the amiloride rate constants remained unchanged. In contrast, Cuthbert and Shum (1975) found no change in amiloride binding capacity but found amiloride binding affinity was increased twofold. Both groups (see Cuthbert, 1981 for review) have suggested ways of accounting for these differences, but further results are needed to resolve the differences. Zeiske *et al.*

TABLE II
CHANNEL DENSITY (*N*)

| Membrane and reference | | Channel density |
|---|---|---|
| Frog skin | | |
| Lindemann and Van Driessche (1978) | | $0.7–2.0 \times 10^8$ cm$^{-2}$ |
| Cuthbert and Shum (1974) | | $86–378 \times 10^8$ cm$^{-2}$ |
| | Control (sites/cell) | Treatment change |
| Toad bladder | | |
| | | ADH |
| Li *et al.* (1982) | 600 | 4.7× |
| Cuthbert and Shum (1974) | 360,000 | No change |
| | | ALDO |
| Palmer *et al.* (1982) | 600 | 2.6× |
| Cuthbert and Shum (1974) | 30,000[a] | 2.× |
| | $(400 \times 10^8)$[b] | No change |
| Rabbit colon (37°C) | | |
| Zeiske *et al.* (1982) | | $0.25–10.8 \times 10^8$ cm$^{-2}$ |

[a] As isolated cells.
[b] As membrane, number of channels per square centimeter.

(1982) have studied amiloride-induced fluctuations in the descending colon of rabbit. They report a channel density ranging from 2.5 to 10.8 × $10^8$ at 37°C which is almost an order of magnitude higher than in amphibian epithelia thus far reported. They found single-channel current to range from 0.12 to 1.07 pA, which may be somewhat higher than values reported in amphibia. At 27°C both values were reduced by factors of 2 and 4, respectively.

## V. MODEL PREDICTIONS

### A. Two-State Model

The simple two-state model, representing a first-order process, predicts a Lorentzian spectrum as noted before, as well as the classical steady state inhibition curve. When the amiloride concentration is altered, a number of additional

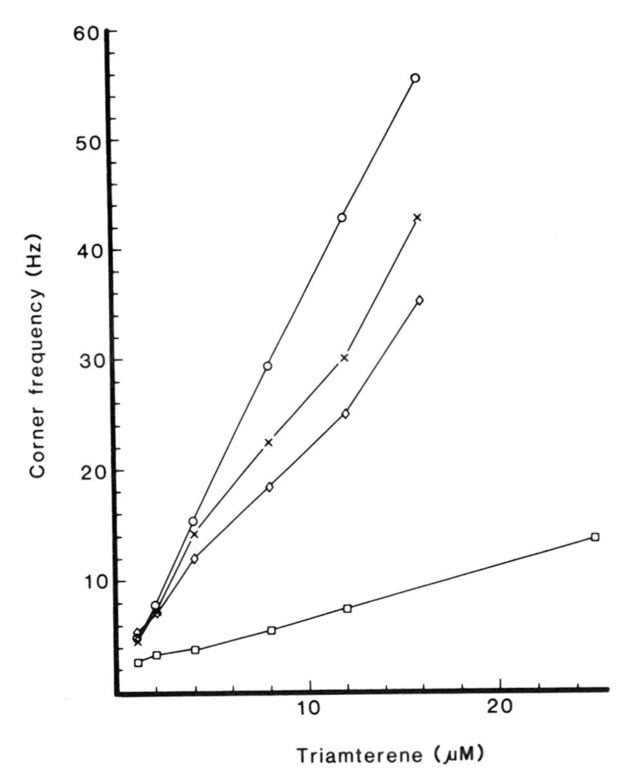


FIG. 2. Corner frequency versus triamterene concentration. The lowest points (□) were obtained when the apical solution contained 120 mEq/liter NaCl, at pH 7.8. The upper three sets of lines were obtained at pH 6.0. Of these, the lower set of points (◇) was obtained in 120 mEq/liter NaCl, (×) in 40 mEq/liter NaCl, and (○) in 20 mEq/liter NaCl in skins from *R. temporaria*.

TABLE III
TWO-STATE MODEL PREDICTIONS

A. When blocker concentration [A] is varied

1. $\omega_c = 2\pi f_c = k_- + k[A]$
2. $I = k_- \cdot I_{max}/\omega_c$
3. $S_o = \dfrac{N\bar{i}^2}{\pi k_-} \cdot \dfrac{[A]/K_A}{(1 + [A]/K_A)^3}$
4. $S_o = \dfrac{N\bar{i}^2}{\pi k_-} \left[ \left(\dfrac{k_A}{\omega_c}\right)^2 - \left(\dfrac{k_-}{\omega_c}\right)^3 \right]$

B. When clamp voltage is varied

1. $f_c$ is constant
2. $S_o = \left[ \dfrac{[A]/K_A}{k_- N \pi (1 + [A]/K_A)} \right] \cdot I^2$

relationships are predicted [Table III (A)] which can serve as additional tests of the model (Hoshiko and Van Driessche, 1981, 1982). The first prediction [Table III (A,1)], that the corner frequency is a linear function of the amiloride concentration, was first demonstrated by Lindemann and Van Driessche (1977) as mentioned above. This relationship also holds when triamterene is used as the

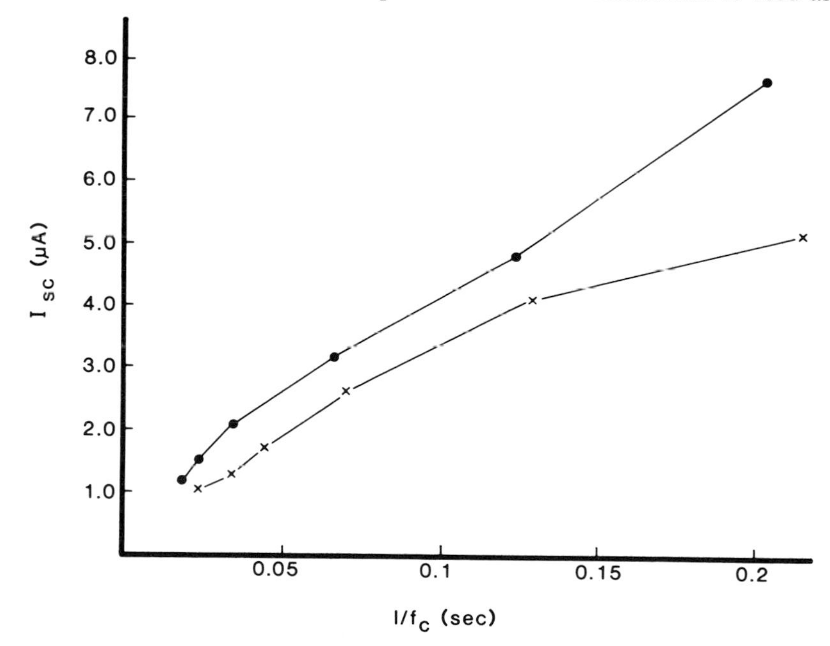


FIG. 3. Short-circuit current $I_{sc}$ versus inverse corner frequency $f_c$. Data are from the same experiments shown in the upper two curves in Fig. 2: (●) at 20 mEq/liter NaCl, pH 6.0; (×) at 40 mEq/liter NaCl, pH 6.0. The triamterene concentrations were progressively from left to right 1, 2, 4, 8, 12, 16 μ*M*.

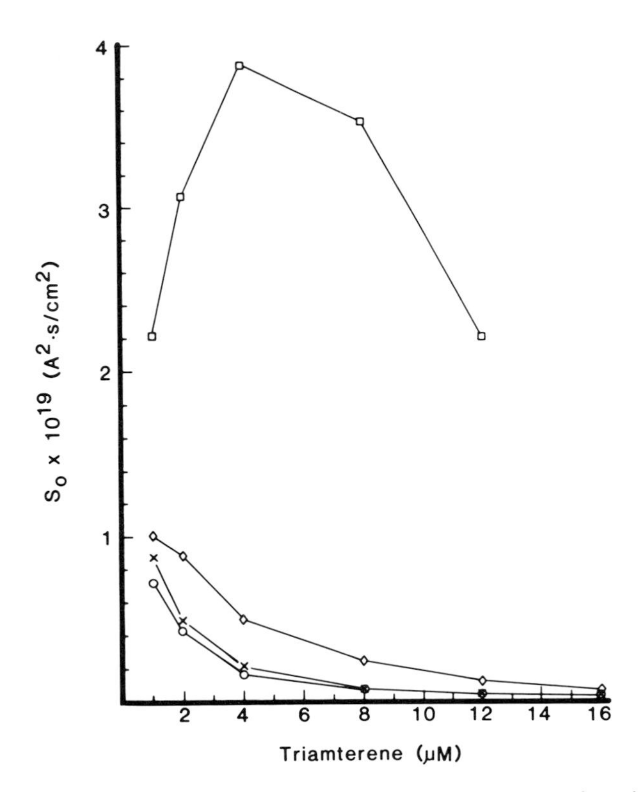


FIG. 4. Low-frequency plateaus versus triamterene concentration. Data are from the same experiment as in Fig. 2. Upper curve (□) is at 120 mEq/liter NaCl, pH 7.8. The lower three curves at pH 6.0; NaCl concentrations were (○) 20 mEq/liter, (×) 40 mEq/liter, and (◇) 120 mEq/liter. From Hoshiko and Van Driessche (1981) with permission.

blocking agent (Fig. 2). Since the probability that a channel is open is the ratio of the dissociation rate constant to the sum of the rate constants (or the corner frequency), the short-circuit current is predicted [Table III (A,2)] to be proportional to the inverse of the corner frequency. This prediction, as shown in Fig. 3, is borne out fairly well in one case but seems to falter in the second.

A third prediction [Table III (A,3)] arises when we express the plateau in terms of the amiloride (or triamterene) concentration. The plateau will be low at low amiloride concentrations, rise to some value, and then decrease again as the cubic term in the denominator takes over. The maximum should occur when the blocker concentration is one-half its equilibrium constant. This relationship was first predicted and observed by Van Driessche and Zeiske (1980b) for barium blockage of the apical potassium in *Rana temporaria*. It also appears in triamterene blockage at pH 7.8, but not at pH 5.5, as shown in Fig. 4. The difference is that the triamterene equilibrium constant is very low at the acid pH and

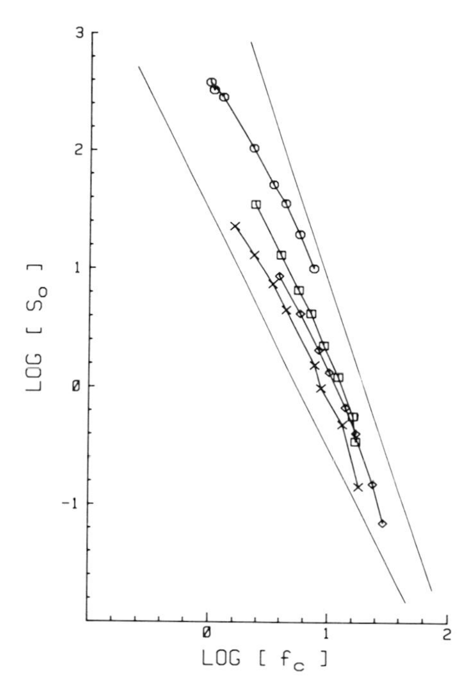


FIG. 5. The plateau ($S_o$) versus corner frequency ($f_c$) on a log–log plot, with increasing amiloride concentrations in skin from *R. temporaria*. Four sets of spectra were obtained at 120, 40, 20, and 10 mEq/liter NaCl, pH 7.8, with amiloride at 0.5, 1.0, 1.5, 2.0, 4.0, 6.0, 8.0 $\mu M$. The two straight lines show slopes of 2 (lower) and 3 (upper).

therefore the peak occurs at too low a triamterene concentration to be observed. The same is true for amiloride in the usual experimental conditions. The final prediction [Table III (A,4)] is a little more nebulous but consistent. In this we express the plateau as a function of the corner frequency. Now the plateau should be proportional to the inverse square or to the inverse cube of the corner frequency. If we make a log–log plot of the plateau versus the corner frequency, the slope should fall somewhere between $-2$ and $-3$. A representative experiment is shown in Fig. 5. Each line represents spectra at varying amiloride concentrations repeated at four different sodium concentrations. The prediction seems to hold for this whole set of curves.

A second set of predictions is possible concerning the behavior according to the two-state model when the clamp voltage is altered [Table III (B)]. First of all the corner frequency should be constant. The half-inhibition concentration $K_I$ for amiloride has been reported by Cuthbert and Shum (1976) to be independent of the clamp potential. In this experiment (Fig. 6) the clamp potential was varied over a relatively wide range of values. The corner frequencies and plateaus were

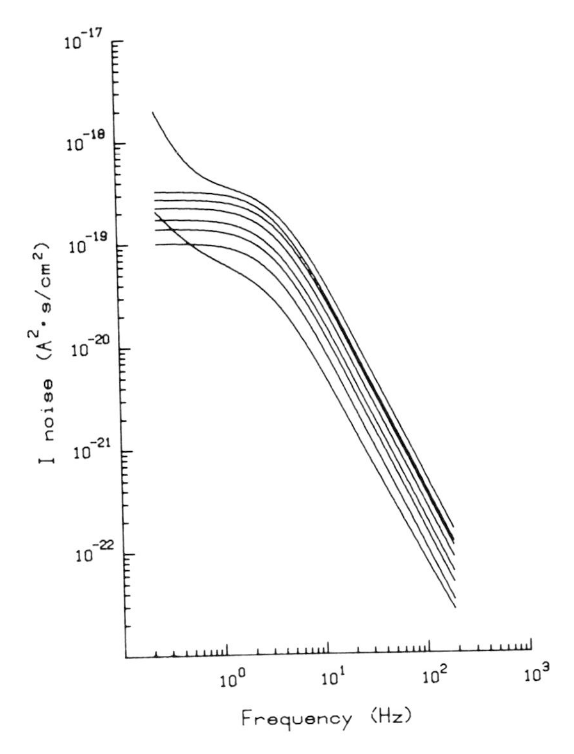


FIG. 6. Effect of voltage clamp levels on amiloride current spectra. Spectra shown were obtained in rapid succession in 1 μ*M* amiloride, 40 mEq/liter NaCl, pH 7.8, skin from *R. temporaria*. The clamp levels were from the highest curve to the lowest: 100, 70, 60, 40, 20, 0, −20, −40 mV. Open circuit potential at the start of the experiment was −61 mV. Note that the corner did not change whereas the plateau values decreased.

estimated and are shown in Fig. 7 together with the short-circuit current. It is apparent that the corner frequency is indeed relatively constant. Of course this prediction does not constitute a unique test of the model, but it indicates that the association and dissociation rates of the blocking agent (triamterene) are unaffected by the clamp potential. The reason we chose to use triamterene is that its corner frequency is significantly higher than for amiloride. This means that only a short time segment is necessary to get the data to define the spectrum. Also, the dissociation rate constant for triamterene is larger and therefore more easily measured than that for amiloride. A second prediction arises because at a constant concentration of the blocker, the plateau should be proportional to the square of the current. This should hold true for any kind of channel model in which the single-channel current is driven by the electrochemical potential gradient. However, in testing this expectation, one difficulty is that the current may

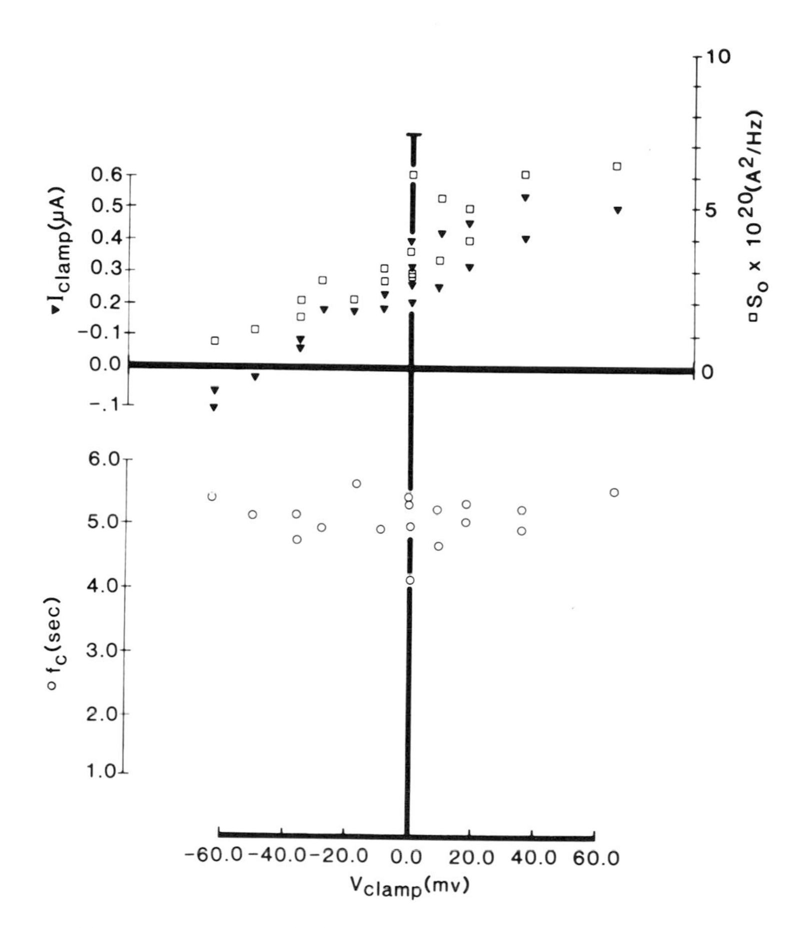


FIG. 7. Effect of voltage clamp levels on corner frequency, short-circuit current, and plateau. Spectra from a 0.1-$cm^2$ area of skin from *R. temporaria* bathed in 20 mEq/liter NaCl, pH 7.8, 1 $\mu M$ amiloride. The spontaneous open circuit potential was −64 mV. The open circles show that the corner frequency was independent of clamp level, whereas the current and plateau values increased with increasing clamp levels.

include an additive amiloride-insensitive component. By examining a log–log plot it should be possible to deduce the power dependence between the plateau and the current. If the noise were true shot noise and the current fluctuations were due movement of quantal units of charge, the plateau should be proportional to the first power of the current, independently of the driving force. The data in Fig. 8 show a slope close to 2, consistent with the predictions from the channel-blocking model. This again shows that a carrier model, even for a big carrier capable of carrying a large number of ions per cycle, is excluded.

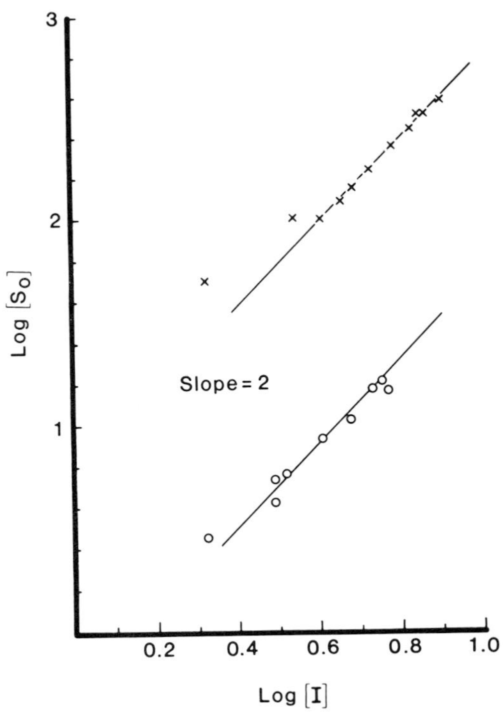


FIG. 8. Log–log plot of relationship between the plateau and the short-circuit current as the clamp level is changed. Two sets of data are shown, both in 1 $\mu M$ amiloride: the upper curve in 40 mEq/liter NaCl and the lower in 20 mEq/liter NaCl. The straight line has a slope of 2.

## B. Three- and Four-State Models

We turn now to some results that cannot be accommodated by the two-state model. One finding is that the equilibrium constant measured by fluctuation analysis and the inhibition constant estimated from the short-circuit current do not in general agree. I have already alluded to the equilibrium constant for the blocker $K_A$ as the ratio of the dissociation to the association rate constants. In the two-state model, this equilibrium constant should be identical to the half-maximal inhibition concentration $K_I$ obtained from the log/dose inhibition curve, as shown in Table IV (A). In fact, as noted by Li and Lindemann (1981), the two may differ by a factor of 6 or more in some cases. A more complicated model is required. The three-state model of Lindemann and Van Driessche (1978) does predict such a discrepancy and has been worked out in a subsequent paper (Li *et al.*, 1982). These results are summarized in Table IV (B).

Another effect that cannot be accommodated by the two-state model is the

TABLE IV
COMPARISON OF $K_A$ AND $K_I$

A. Two-state model predicts $K_A$ from fluctuation analysis and $K_I$ from the steady state inhibition curve to be the same

$$I = \frac{I_{max}}{(1 + [A]/K_A)}$$

B. Measured values of $K_A$ and $K_I$ do not agree (Li and Lindemann, 1981)

| | $K_A$ (μ*M*) | $K_I$ (μ*M*) |
|---|---|---|
| TAP | 300 | 4240. |
| 5-Cl amiloride | 29.6 | 180. |
| Triamterene | 2.78 | 18.2 |
| Amiloride | 0.30 | 1.8 |

influence of changing the apical sodium concentration. In 1979, Van Driessche and Lindemann reported that in *R. esculenta* the corner frequency was unchanged as the apical sodium concentration was altered. Using *R. temporaria,* we have found that the corner frequency was not independent of sodium but decreased with increasing sodium concentration in the apical bathing medium (Hoshiko and Van Driessche, 1981). The slope of the line relating corner frequency to blocker concentration was higher at the lower sodium concentrations. This is shown with triamterene in Fig. 2 and in Table V. The difference is that we did not use a high potassium concentration in the corium bathing medium. In repeating the experiments with high potassium concentrations in the inside or corium bathing solution, the sodium effect was greatly suppressed or absent. The effect of corium potassium is observable when the apical sodium concentration is low.

TABLE V
TRIAMTERENE KINETICS[a,b]

| $[Na]_o$ (mEq/liter) | $k_-$ ($sec^{-1}$) | $k$ ($\mu M^{-1}$ $sec^{-1}$) | $i$ (pA) | $K_I$ (μ*M*) | $N \times 10^{-8}$ ($cm^{-2}$) |
|---|---|---|---|---|---|
| 120 | 19.2 | 12.4[c] | 0.46 | 1.61 | 0.17 |
| 40 | 12.0 | 15.9 | 0.31 | 0.77 | 0.32 |
| 20 | 13.2 | 18.9 | 0.23 | 0.77 | 0.23 |

[a] From Hoshiko and Van Driessche (1982) with permission.
[b] $n = 5$.
[c] Analysis of variance showed a significant sodium effect at the 5% level on the association rate constant $k$.

It seems unlikely that this difference can be attributed to either the depolarization effect or to the decreased basolateral membrane resistance associated with high corium potassium concentration. The point was made earlier that with low apical sodium, apical membrane resistance is increased and the intrinsic noise from the apical membrane is enhanced. If the effect of corium potassium is primarily to enhance the apical/basolateral resistance ratio, the effect should be most prominent with the high-sodium case, since then the apical resistance would need the extra bolstering. If the potassium effect is via the reduction of the intracellular potential, it should again be the high-sodium case that should see the biggest change. It is possible that the explanation lies in another direction. The effect of the high potassium may in fact be the absence of sodium coupled with an indirect effect of potassium.

It is well known that sodium is required in the corium bathing solution to maintain active transport. Taylor and Windhager (1979) have suggested that removal of sodium from the basolateral surface inhibits the extrusion of intracellular calcium. It would appear that sodium removal entailed in the "potassium depolarization" maneuver may have the effect of increasing cytosolic calcium. Chase and Al-Awqati (1981) reported that treatment of the serosal surface with a calcium ionophore or low serosal sodium reduced the uptake of sodium across the apical membrane in toad bladder. Taylor (1981) reported that removal of serosal sodium in toad bladder diminished and dampened the recline phenomenon. Quinidine treatment, which also is believed to increase cytosolic calcium, gave the same results. Cuthbert and Wilson (1981) mention that in the $K^+$-depolarized preparation they found dramatically increased levels of cAMP. Taken together, these findings suggest that the $K^+$-depolarized preparation may result in alteration of key aspects of the kinetic behavior of the apical membrane. These considerations remain speculative, however, and a more precise study is required.

It seems probable therefore that the increased association rate constant in the presence of low apical sodium is due to an effect on the apical sodium channel proper. In order to account for this effect, Ekhart Frehland, Stefan Machlup, and I (1983) postulated a new three-state model. We were led to this effort since the three-state model of Lindemann and Van Driessche (1978) does not account for the sodium dependence of the association rate constant. In their model the third state is a nonconducting sodium-clogged channel. Although the sodium competes with amiloride for a binding site, the effect is also to impede the current. Their model gives the opposite sign for sodium dependence. Frehland suggested that a sodium-occupied channel might still conduct but prevent the binding of the amiloride. In this way, the presence of the sodium could alter the apparent association rate of the amiloride. In a qualitative way the sodium dependence of the association rate constant is accounted for by this model. However, it does not predict the discrepancy between the apparent equilibrium constant determined by

TABLE VI
FOUR-STATE MODEL

$$R_2 \underset{k_{-2}}{\overset{k_2[\mathrm{Na}]}{\rightleftharpoons}} R_1 \underset{k_{-1}}{\overset{k_1[\mathrm{Na}]}{\rightleftharpoons}} R_o \underset{k_-}{\overset{k[\mathrm{A}]}{\rightleftharpoons}} R_A$$

(very slow) (very fast) (slow)

Equilibrium constants

$K_{ss} = k_{-2}/k_2 \qquad K_s = k_{-1}/k_1 \qquad K_A = k_-/k$

Blocking corner[a]

$\omega_c = 2\pi f_c = k_- + k[\mathrm{A}]/(1 + [\mathrm{Na}]/K_s)$ + second-order correction

Steady state short-circuit current

$I_{sc} = I_{max}(1 + [\mathrm{Na}]/K_s)/(1 + [\mathrm{A}]/K_A + [\mathrm{Na}]/K_s + [\mathrm{Na}]^2/K_s K_{ss})$

Low-frequency plateau

$$S_o = \frac{N\bar{i}^2}{\pi} \cdot \frac{(1 + [\mathrm{Na}]/K_s)^2}{(1 + [\mathrm{A}]/K_A + [\mathrm{Na}]/K_s + [\mathrm{Na}]^2/K_s K_{ss})} \cdot \left[\frac{1}{k_-}\frac{[\mathrm{A}]}{K_A} + \frac{1}{k_{-1}}\frac{[\mathrm{Na}]^2}{K_s K_{ss}}\right]$$

[a] Assume $k_{-2} << k_- << k_1$.

fluctuation analysis and the inhibition constant determined from the dose–response curve. What this three-state model lacks is the sodium-clogged state. A simple possibility is that such a state may result when a second sodium ion entered the channel. This four-state model does account for the self-inhibition and does predict the discrepancy. The form of the relationship to sodium concentration is given in Table VI. At this stage it seems that theory has outstripped experiment, and more experimental results are required to resolve these issues.

## VI. CONCLUSION

Fluctuation analysis as a kinetic tool offers the possibility of obtaining new information not available using traditional methods. The relationship between current power and the steady state current makes it likely that the fluctuations arise from a blocking–unblocking action of a conducting channel. Fluctuation analysis makes it possible to estimate single-channel current and channel density. Although estimates for these quantities are obtained using an exogenously applied blocking agent, their values appear to be relatively independent of the blocker and reflect intrinsic channel properties. The magnitude of the single-channel current makes it probable that the transport mechanism is a channel and not a carrier. Access to such quantities makes it possible to study the mode of action of regulatory agents including hormones which stimulate transport—whether their actions are through an increase in channel density or the magnitude of the single-channel current. Thus far it seems that both ADH and aldosterone act by increasing channel density rather than single-channel current. Additional parameters measured are the association and dissociation rate constants for the blocking agent. Still to be sorted out are the questions involving sodium self-inhibition and sodium interactions with the blocking agent. Nonetheless, the new results to date and the promise of uncovering new properties in the future are encouraging enough to think that the technique will become another useful tool available to all epitheliologists.

## ACKNOWLEDGMENTS

I wish to express heartfelt gratitude to Dr. Willy Van Driessche for the gracious hospitality shown me and my family during our stay in Leuven, Belgium and to the staff of the Laboratorium voor Fysiologie, K.U.L., particularly to Professor Rick Casteels for his support. Much of the new experimental work reported here was done in that laboratory in collaboration with Dr. Van Driessche. I would also like to acknowledge my debt to Dr. Stefan Machlup for his encouragement and collaborative help, and to Dr. Eckart Frehland who together with Dr. Machlup have clarified many phenomena with their theoretical analyses. I am grateful to Professor Bernd Lindemann and Dr. Jack Li for many fruitful discussions and to Dr. Sandy Helman for the gift of one of his low-noise chambers. This work was supported in part by grants from the USPHS (AM05865) and from the Onderzoeks fonds., K.U.L.

## REFERENCES

Bendat, J. C., and Piersol, A. G. (1971). "Random Data: Analysis and Measurement Procedures," p. 191. Wiley (Interscience), New York.

Benos, D. J. (1982). Amiloride: A molecular probe of sodium transport in tissues and cells. *Am. J. Physiol.* **242,** C131–C145.

Chase, H. S., and Al-Awqati, Q. (1981). Regulation of the sodium permeability of the luminal border of toad bladder by intracellular sodium and calcium. *J. Gen. Physiol.* **77,** 693–712.

Christensen, O., and Bindslev, N. (1982). Fluctuation analysis of short-circuit current in a warm-blooded sodium-retaining epithelium: Site current, density and interaction with triamterene. *J. Membr. Biol.* **65,** 19–30.

Conti, F., Neumcke, B., Nonner, W., and Stampfli, R. (1980). Conductance fluctuations from the inactivation process of sodium channels in myelinated nerve fibers. *J. Physiol. (London)* **308,** 217–239.

Cuthbert, A. W. (1981). Sodium entry step in transporting epithelia: Results of ligand binding studies. *In* "Ion Transport by Epithelia" (S. G. Schultz, ed.), pp. 181–195. Raven, New York.

Cuthbert, A. W., and Shum, W. K. (1974). Binding of amiloride to sodium channels in frog skin. *Mol. Pharmacol.* **10,** 880–891.

Cuthbert, A. W., and Shum, W. K. (1975). Effects of vasopressin and aldosterone on amiloride binding in toad bladder epithelial cells. *Proc. R. Soc. London, Ser. B* **189,** 543–575.

Cuthbert, A. W., and Shum, W. K. (1976). Characteristics of the entry process for sodium in transporting epithelia as revealed with amiloride. *J. Physiol. (London)* **255,** 587–604.

Cuthbert, A. W., and Wilson, S. A. (1981). Mechanisms for the effects of acetylcholine on sodium transport in frog skin. *J. Membr. Biol.* **59,** 65–75.

DeFelice, L. (1981). "Introduction to Membrane Noise." Plenum, New York.

Dorset, D. L., and Fishman, H. M. (1975). Excess electrical noise during current flow through porous membranes separating ionic solutions. *J. Membr. Biol.* **21,** 291–309.

Fishman, H. M. (1982). Current and voltage clamp techniques. *In* "Techniques in Cellular Physiology Part II," pp. 1–42. North-Holland Publ., Amsterdam.

Fishman, H. M., Moore, L. E., and Poussart, D. (1981). Squid axon K conduction: Admittance and noise during short- versus long-duration step clamps. *In* "The Biophysical Approach to Excitable Systems" (W. J. Adelman and D. E. Goldman, eds.), pp. 65–95. Plenum, New York.

Frehland, E., Hoshiko, T., and Machlup, S. (1983). *Biochim. Biophys. Acta* **732,** 636–646.

Fuchs, W., Hviid Larsen, E., and Lindemann, B. (1977). Current-voltage curve of sodium channels and concentration dependence of sodium permeability in frog skin. *J. Physiol. (London)* **267,** 137–166.

Goudeau, H., Wietzerbin, J., and Gary-Bobo, C. M. (1979). Effects of mucosal lanthanum on electrical parameters of isolated frog skin. *Pfluegers Arch.* **379,** 71–80.

Helman, S. I., and Fisher, R. S. (1977). Microelectrode studies of the active Na transport pathway of frog skin. *J. Gen. Physiol.* **69,** 571–604.

Hoshiko, T. (1961). Electrogenesis in frog skin. *In* "Biophysics of Physiological and Pharmacological Actions" (A. M. Shanes, ed.), pp. 31–47. Amer. Assoc. Advance. Sci., Washington, D.C.

Hoshiko, T. (1978). Power density spectra of frog skin potential, current and admittance functions during patch clamp. *J. Membr. Biol.* **40,** 121–132.

Hoshiko, T., and Moore, L. E. (1978). Fluctuation analysis of epithelial membrane kinetics. *In* "Membrane Transport Processes" (J. F. Hoffman, ed.), Vol. 1, pp. 179–198. Raven, New York.

Hoshiko, T., and Van Driessche, W. (1981). Triamterene-induced sodium current fluctuations in frog-skin. *Arch. Int. Physiol. Biochim.* **89,** 58–60.

Hoshiko, T., and Van Driessche, W. (1982). Current fluctuations in frog skin are due to apical sodium channels: More evidence. *Biophys. J.* **37,** 281a.

Johnston, K. H., and Hoshiko, T. (1971). Novobiocin stimulation of frog skin current and some metabolic consequences. *Am. J. Physiol.* **220,** 792–798.

Li, J. H.-Y., and Lindemann, B. (1981). Blockage of epithelial Na-channels by organic cations: The relationship of microscopic and macroscopic inhibition constants. *Pfluegers Arch.* **391,** R25.

Li, J. H.-Y., Palmer, L. G., Edelman, I. S., and Lindemann, B. (1982). The role of sodium-channel density in the natriferic response of the toad urinary bladder to an antidiuretic hormone. *J. Membr. Biol.* **64,** 77–89.

Lifson, S., Gavish, B., and Reich, S. (1978). Flicker noise of ion-selective membranes and turbulent convection in the depleted layer. *Biophys. Struct. Mech.* **4,** 53–65.

Lindemann, B. (1980). The beginning of fluctuation analysis of epithelial transport. *J. Membr. Biol.* **54,** 1–11.

Lindemann, B., and DeFelice, L. J. (1981). On the use of generalized network functions in the evaluation of noise spectra obtained from epithelia. *In* "Ion Transport by Epithelia" (S. J. Schultz, ed.), pp. 1–3. Raven, New York.

Lindemann, B., and Gebhardt, U. (1973). Delayed changes in Na-permeability in response to steps of $(Na)_o$ at the outer surface at frog skin and frog bladder. *In* "Transport Mechanisms in Epithelia" (H. H. Ussing and N. Thorn, eds.), pp. 115–130. Munksgaard, Copenhagen.

Lindemann, B., and Van Driessche, W. (1977). Sodium-specific membrane channels of frog skin are pores: Current fluctuations reveal high turnover. *Science* **195,** 292–294.

Lindemann, B., and Van Driessche, W. (1978). The mechanism of Na uptake through Na-selective channels in the epithelium of frog skin. *In* "Membrane Transport Processes" (J. F. Hoffman, ed.), Vol. 1, pp. 155–178. Raven, New York.

Machlup, S. (1954). Noise in semiconductors: Spectrum of a two-parameter random signal. *J. Appl. Phys.* **25,** 341–343.
Machlup, S. (1981). Earthquakes, thunderstorms and other $1/f$ noises. *NBS Spec. Publ. (U.S.)* **614,** 157–160.
Nagel, W. (1977). The dependence of the electrical potentials across the membranes of the frog skin upon the concentration of sodium in the mucosal solution. *J. Physiol. (London)* **269,** 777–796.
Neumcke, B. (1982). Fluctuation of Na and K currents in excitable membranes. *Int. Rev. Neurobiol.* **23,** 35–67.
Palmer, L. G., Li, J. H.-Y., Lindemann, B., and Edelman, I. S. (1982). Aldosterone control of the density of sodium channels in the toad urinary bladder. *J. Membr. Biol.* **64,** 91–102.
Poussart, D. J. M. (1971). Membrane current noise in lobster axon under voltage clamp. *Biophys. J.* **11,** 211–234.
Rice, S. O. (1954). Mathematical analysis of random noise. *In* "Selected Papers on Noise and Stochastic Processes" (N. Wax, ed.), p. 164. Dover, New York.
Smith, P. G. (1971). The low-frequency electrical impedance of the isolated frog skin. *Acta Physiol. Scand.* **81,** 355–366.
Stevens, C. F. (1972). Inferences about membrane properties from electrical noise measurements. *Biophys. J.* **12,** 1028–1047.
Taylor, A. (1981). Role of cytosolic calcium and sodium–calcium exchange in regulation of transepithelial sodium and water absorption. *In* "Ion Transport by Epithelia" (S. G. Schultz, ed.), pp. 233–259. Raven, New York.
Taylor, A., and Windhager, E. E. (1979). Possible role of cytosolic calcium and Na–Ca exchange in regulation of transepithelial sodium transport. *Am. J. Physiol.* **236,** F505–F512.
Ussing, H. H., and Windhager, E. (1964). Nature of shunt path and active sodium transport path through frog skin epithelium. *Acta Physiol. Scand.* **61,** 484–509.
Van der Ziel, A. (1959). "Fluctuation Phenomena in Semiconductors," p. 56. Academic Press, New York.
Van Driessche, W., and Gögelein, J. (1980). Attenuation of current and voltage noise signals recorded from epithelia. *J. Theor. Biol.* **86,** 629–648.
Van Driessche, W., and Lindemann, B. (1978) Low noise amplification of voltage and current fluctuations arising in epithelia. *Rev. Sci. Instrum.* **49,** 52–55.
Van Driessche, W., and Lindemann, B. (1979). Concentration dependence of currents through single sodium-selective pores in frog skin. *Nature (London)* **282,** 519–520.
Van Driessche, W., and Zeiske, W. (1980a). Spontaneous fluctuations of potassium channels in the apical membrane of frog skin. *J. Physiol. (London)* **299,** 101–116.
Van Driessche, W., and Zeiske, W. (1980b). $Ba^{2+}$-induced conductance fluctuations of spontaneously fluctuating $K^+$ channels in the apical membrane of frog skin (*Rana temporaria*). *J. Membr. Biol.* **56,** 31–42.
Zeiske, W., and Lindemann, B. (1974). Chemical stimulation of $Na^+$ current through the outer surface of frog skin epithelium. *Biochim. Biophys. Acta* **352,** 323–326.
Zeiske, W., Wills, N. K., and Van Driessche, W. (1982). $Na^+$ channels and amiloride-induced noise in the mammalian colon epithelium. *Biochim. Biophys. Acta* **688,** 201–210.

# Chapter 2

# Impedance Analysis of *Necturus* Gallbladder Epithelium Using Extra- and Intracellular Microelectrodes

*J. J. LIM,*[1] *G. KOTTRA,*[2] *L. KAMPMANN, AND E. FRÖMTER*

*Max-Planck-Institut für Biophysik*
*Frankfurt, Federal Republic of Germany*



## I. INTRODUCTION

The conductive and capacitive (polarizable) properties of cell membranes are normally described by means of a resistance $R$ and a capacitance $C$, which are interconnected by the time constant $\tau = R \cdot C$ of charging in response to a square wave constant-current pulse. More accurately, $R$ and $C$ can be determined by alternating current (ac) spectroscopy, i.e., by measuring the voltage response to sine wave currents as a function of frequency. In such experiments, the cell

[1]Deceased.

[2]Present address: Department of Internal Medicine, Semmelweis Medical University, Budapest, Hungary.




ISBN 0-12-153320-4

membrane impedance shows a dispersion in the frequency range between 10 Hz and 100 kHz (i.e., the impedance changes from one constant level to another constant level with frequency) (Schwan, 1957). The center frequency $f_c$ of the dispersion is related to the time constant by $f_c = 1/2\tau\pi$. Since the specific conductance of cell membranes is known to vary considerably but the capacitance is approximately constant ($\sim 1\ \mu F/cm^2$), one may expect composite membrane systems such as epithelia, which consist of serial arrays of cell membranes, to exhibit multiple dispersions in their ac spectrum which may be referable to individual membrane elements.

Although many attempts have been made in the past to determine the impedance of epithelia such as frog stomach (Teorell and Wersäll, 1945; Rehm *et al.*, 1973), frog skin (Cuthbert and Painter, 1969; Smith, 1971), rabbit urinary bladder (Clausen *et al.*, 1979), rabbit corneal endothelium (Lim and Fischbarg, 1981), frog lens (Mathias *et al.*, 1981), and *Necturus* gallbladder (Schifferdecker and Frömter, 1978; Gögelein and Van Driessche, 1981), with a few exceptions the approach has not been as fruitful as expected for the following reasons. (1) Technically, the experiments required long measuring times (on the order of 20 minutes) which created problems with regard to the stability of the preparation under investigation and restricted the analysis to steady states. (2) In many cases the dispersions happened to be indistinguishable from nonideal capacitances. (3) After discovery of the paracellular shunt path in leaky epithelia (Frömter and Diamond, 1972), it became apparent that the derivation of individual cell membrane impedances from transepithelial measurements required either knowledge of the magnitude of the shunt conductance or else certain other pieces of information (Schifferdecker and Frömter, 1978).

In an attempt to overcome these limitations, we have extended the concept of epithelial impedance measurements by combining them with simultaneous intracellular recordings from microelectrodes. This improvement became possible after the development of an appropriate, almost completely shielded, microelectrode (Suzuki *et al.*, 1978). In a first series of experiments, we analyzed the time course of the transepithelial and intracellular voltage responses of *Necturus* gallbladder to a square wave constant-current pulse (Suzuki *et al.*, 1982). These experiments demonstrated that such records suffice to determine the magnitude of the resistances and capacitances of the apical and basal cell membranes ($R_a$, $C_a$, $R_b$, $C_b$), and, in addition, the resistance of the paracellular shunt $R_{sh}$ in leaky epithelia, provided the lateral intercellular spaces are extended so that their resistive contribution is negligible.

With the availability of modern data acquisition and computation facilities, we have now developed an experimental setup that allows us to perform transepithelial and intracellular ac spectroscopy on epithelia quasi-automatically by scanning the frequency range between 2.5 Hz and 10 kHz in less than 1 minute. This technique and the first experimental results are described in this chapter. In addition to simplifying the measurements and improving their accuracy, this technique yields data that can be analyzed with more detailed equivalent-circuit models than was hitherto possible. In fact, we are now able to analyze the resistive properties of *Necturus* gallbladder not only on the basis of the lumped-circuit model described and applied previously (Frömter, 1972; Reuss and Finn, 1975; Schifferdecker and Frömter, 1978; Suzuki *et al.*, 1982), but also on the basis of the more realistic distributed-circuit model. The distributed model, which we had proposed long ago (Frömter, 1972; see also Boulpaep and Sackin, 1980, and Clausen *et al.*, 1979), allows resolution of the resistance of the paracellular shunt path into its contributions from the tight junction proper and from the lateral intercellular space and allows more accurate determination of the basal cell membrane resistance.

## II. DATA ANALYSIS TECHNIQUE

### A. Principle

In principle, we determine the small-signal transfer function of the gallbladder for electrical current. This is done by passing a small-amplitude input wave form containing many spectral (frequency) components across the tissue and measuring the transepithelial and intracellular potential response. By using fast Fourier transform (FFT) we then determine the autospectrum of the current wave form and the cross-spectrum between each voltage response wave form and the current wave form. By dividing the cross-spectra through the autospectrum, we obtain the complex transepithelial transfer function as well as an equivalent apparent intracellular transfer function. The transfer function can be represented in various ways. We prefer here to plot the real versus the imaginary component, which describes the so-called impedance locus of the epithelium (Nyquist plot). By using mathematical models which attempt to express the morphological structure of the epithelium in terms of the relative position of membrane structures and interspaces, we try to fit the data with model equations in order to obtain quantitative estimates of the model parameters that are thought to represent individual membrane elements.

### B. The Current Wave Form

The pseudorandom binary sequence (PRBS) noise of a Waveteck 132 noise generator was used as current signal. Its frequency spectrum is virtually flat (see, for example, Clausen and Fernandez, 1981). The PRBS wave form was fed into a PC1 stimulus isolation unit (WPI, New Haven, Connecticut), which converted it into a corresponding time series of constant-current steps that were passed across the preparation. Since the PRBS signal is periodic with a period of $2^n - 1$ but the FFT algorithm requires data blocks of $2^n$ data, it is not possible to repeat the signal continuously and to analyze and average the data directly unless special synchronization circuits are introduced (Poussart and Ganguly, 1977; Clausen and Fernandez, 1981). We have therefore used only one long sequence of $2^{16} - 1$ clock pulses and have cut the sequence of voltage and current data into equal smaller blocks of $2^n$ data for performing FFT under appropriate precautions (see below). Due to limitations in the memory usually a block length of 512 data points was chosen, which reduced the low-frequency resolution to 25 Hz for a clock frequency of 25 kHz (the high-frequency limit of resolution was 10 kHz). To obtain a better low-frequency resolution we have therefore passed through the preparation a second, slower PRBS current wave form with a clock frequency of 2.5 kHz (which covered the frequency range between 1 kHz and 2.5 Hz). The fast sequence took 2.5 seconds and the slow sequence 25 seconds for completion. The current density was 100 $\mu A/cm^2$ (peak to peak); however, the dc component was suppressed to zero. To open or close the lateral spaces in some experiments, an additional direct current of 20 $\mu A/cm^2$ could be passed across the epithelium from a battery circuit.

## C. Data Acquisition

As schematically depicted in Fig. 1, the transepithelial and the intracellular voltages were measured with shielded microelectrodes and slightly modified versions (increased amplification factor) of the feedback amplifier described previously (Suzuki *et al.*, 1978), and were continuously recorded on a chart recorder. In addition, after backing off the dc component, one of the voltages was fed into one input channel of a TDA 5 signal analysis system (Genrad, Munich, Germany), while the other input channel was connected to the current monitor (amplifier 3 in Fig. 1; Keithley Model 604). The TDA 5 system consists of a two-channel analog conditioning element with direct memory access, a PDP 11/34 computer (Digital Equipment Corporation, Maynard, Massachusetts), and PM/DS11 B disk drive. Data acquisation was synchronized with the current signal by triggering the computer from the noise generator, and operating the analog conditioning element by the noise generator's clock. Each channel of the analog conditioning element contained an eight-pole Butterworth antialiasing filter of 48 dB/octave roll-off, which was set to 10 kHz or 1 kHz in the fast and slow current applications, respectively. Phase match was better than 1 degree. Furthermore, the analog conditioning element contained a selectable gain pre-amplifier and a 12-bit analog-to-digital converter. All input parameter settings were software controlled. To be able to test various analysis procedures on the

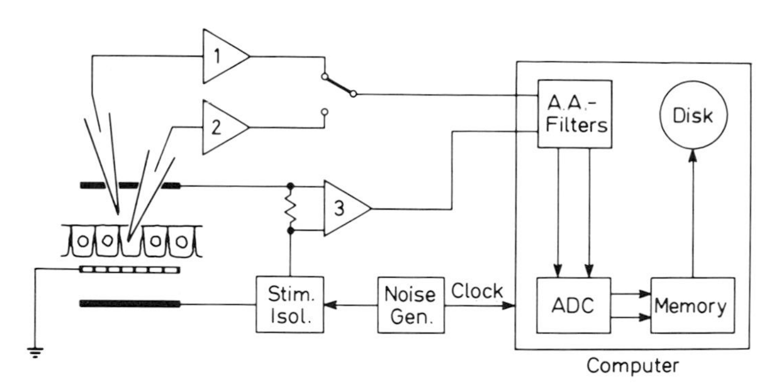


FIG. 1. Schematic of the experimental setup showing the trans- and intraepithelial microelectrodes connected to amplifiers 1 and 2, respectively, the gallbladder epithelium, the sieve-like AgCl/Ag grounding electrode, the pair of Pt electrodes for current injection, and the electrical instrumentation.

original data, all digitized data from the fast and slow runs of each experiment were directly transferred onto the disk and stored for off-line processing.

In order to exclude distortions from nonideal behavior of the measuring equipment, the transfer functions of the microelectrodes plus amplifiers were also determined after each puncture (or rather, the respective data were stored on the disk for later consideration). To measure this data, a 1-k$\Omega$ resistor was interposed between the silver grid and ground (Fig. 1), and a PRB current sequence was passed across it while the electrodes were sitting in the bath and one of the amplifier outputs was sampled. Since all other leads were disconnected from the bath, the bath potential had to follow the current signal, so that any deviation in the voltage recorded from the electrodes and amplifiers could be noticed. This method is justified in the present experiments, because the microelectrodes were completely shielded (except for the ultimate tip, which was inserted into the cell) and therefore capacitive coupling between electrode shaft and bath was excluded. If the resistance of the intracellular microelectrode increased during the impalement (as it often did, typically from 20 to 25 M$\Omega$), its transfer function was recorded in a separate beaker filled with Ringer's solution, diluted to establish the same electrode resistance that had been observed inside the cell.

## D. Computations

To analyze the stored data, each long block of $2^{16}-1$ data points was cut into 127 blocks of 512 data (plus 1 block of 511 data) from which the cross- and autospectra as well as the transfer functions were calculated. Since the signal was not periodic in the time interval of the short blocks (and hence the excitation level of the preparation was not identical at the beginning and end of each small block), the contents of each block were multiplied by a "window" function before FFT was performed. We tested different functions, which all gave identical results, and finally chose the Hanning window. The 127 transfer functions of each long block were then averaged, yielding 256 equally spaced frequency points of the real and imaginary component of the transfer function, which were then divided by the respective transfer functions of the electrodes plus amplifiers to obtain the final data (see, for example, Fig. 3). To save time during the fitting procedure, the number of data points was further reduced by selecting according to a fixed sequence only 20 of the 256 complex data points of each averaged transfer function. These 20 points spanned the entire frequency range and were equally spaced on a logarithmic scale rather than on the linear scale of the FFT routine.

To test the validity of the circuit models and determine the magnitude of model parameters, least squares computer fits were obtained with the model equations described in the next paragraph, using the derivative-free fitting routine VA05M of the Harwell subroutine library (Atomic Energy Research Establishment, Harwell, Berkshire, England) to minimize the square sum (in absolute $\Omega^2$) of the deviations of the model equations from the data points of all four data sets (trans- and intraepithelial high- and low-frequency data) jointly. However, obviously aberrant data points, as defined by inspection, were discarded to improve the significance of the fit in the better defined regions of the curves. As a result, instead of 80, only 53 data points on the average entered into the statistical analysis (this figure is low because of missing or inappropriate intracellular high-frequency data in many experiments).

To correct for inadequate positioning of the transepithelial microelectrode tip during data aquisition, a dc correction was also applied to some data sets in order to equalize the high-frequency end of the trans- and intraepithelial measurements. It corrects for unequal series resistances in both types of measurements.

While fitting the six parameters of the lumped model ($R_a$, $R_b$, $R_{sh}$, $C_a$, $C_b$, and $R_s$; see next paragraph) did not present great problems, the seven parameters of the distributed model (see below) required a preselection of the start parameters in order for the program to converge and to yield meaningful results. These start parameters were either taken from the result of the lumped model fit on the same data (using $R_j = 0.75\ R_{sh}$ and $R_{lis} = 0.25\ R_{sh}$) or were obtained from the results of a prefit which utilized only six data points to speed the calculations up.

## III. EQUIVALENT-CIRCUIT MODELS OF *NECTURUS* GALLBLADDER EPITHELIUM

The simplest equivalent circuit that can be used to represent the ac properties of *Necturus* gallbladder is the lumped model depicted in Fig. 2a (Schifferdecker

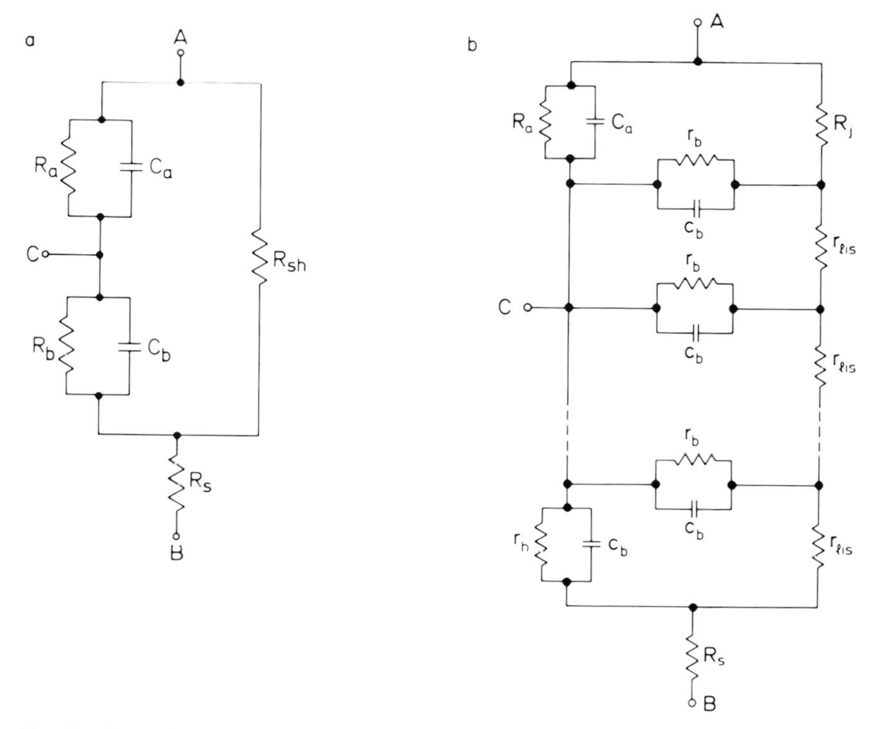


FIG. 2. Lumped (a) and distributed (b) electrical analog models of *Necturus* gallbladder epithelium. A, B, and C represent the mucosal, serosal, and intracellular compartments, respectively.

and Frömter, 1978; Gögelein and Van Driessche, 1981). It consists of one RC unit for the apical and another for the basal cell membrane, and of an ohmic resistor for the paracellular shunt ($R_{sh}$) plus a series resistor ($R_s$) for the bath solution. This model may be expected to describe the epithelium only, if the resistance of the lateral space is negligible. A mathematical formulation of the transepithelial impedance $Z_t$ as defined by the frequency-dependent transepithelial voltage response $\Delta V_t\,(f)$ and the frequency-dependent transepithelial current $I_t\,(f)$,

$$Z_t = \Delta V_t(f)/I_t(f) \tag{1}$$

was given previously in terms of the real and imaginary components (Schifferdecker and Frömter 1978). In complex form the solution is simply

$$Z_t = \frac{1}{(Z_a + Z_b)^{-1} + R_{sh}^{-1}} + R_s \tag{2}$$

and substituting the membrane impedances $Z_a$ and $Z_b$

$$Z_a = \frac{R_a}{1 + j\omega\tau_a} \quad \text{and} \quad Z_b = \frac{R_b}{1 + j\omega\tau_b} \tag{3}$$

we obtain

$$Z_t = \frac{R_{sh}[R_a(1 + j\omega\tau_b) + R_b(1 + j\omega\tau_a)]}{R_{sh}(1 + j\omega\tau_a)(1 + j\omega\tau_b) + R_a(1 + j\omega\tau_b) + R_b(1 + j\omega\tau_a)} + R_s \tag{4}$$

where $j = \sqrt{-1}$, $\omega$ is the angular frequency, and $\tau_a = R_a \cdot C_a$ and $\tau_b = R_b \cdot C_b$ are the intrinsic time constants of the apical and basal cell membranes, respectively.

The measurements of the intracellular voltage response $\Delta V_b(f)$ can be handled in different ways. One possibility is to calculate a frequency-dependent voltage divider fraction, e.g., $\Delta V_b(f)/\Delta V_t(f)$, from the original data and to fit this with the respective mathematical expression derived from the circuit model.

Another possibility is to process $\Delta V_b(f)$ in the same way as $\Delta V_t(f)$, i.e., to divide it by the transepithelial current $I_t(f)$. The result has the dimension of an impedance and may be interpreted as apparent "intracellular" impedance. This possibility was preferred in the present study, because it allows both the transepithelial and intracellular data sets to be presented in the same graph. It should be emphasized, however, that the apparent intracellular impedance

$$Z_b^{app} = \frac{\Delta V_b(f)}{I_t(f)} \tag{5}$$

is not identical with the true impedance of either the basal or the apical cell membrane ($Z_b$ or $Z_a$ respectively). Derivation of the true membrane impedances would require knowing which part of the transepithelial current flows across the cell membranes. This current, $I_b(f)$, cannot be directly measured, however.

To determine the mathematical equivalent of $Z_b^{app}$ we derive first the relation between $I_b$ and $I_t$

$$I_b = \frac{Z_t - R_s}{Z_a + Z_b} I_t \tag{6}$$

from which $\Delta V_b(f)$ can be calculated:

$$\Delta V_b(f) = \left[ \frac{Z_b(Z_t - R_s)}{Z_a + Z_b} + R_s \right] I_t \tag{7}$$

By inserting Eq. (3) we obtain

$$Z_b^{app} = \frac{R_{sh}\, R_b(1 + j\omega\tau_a)}{R_{sh}(1 + j\omega\tau_a)(1 + j\omega\tau_b) + R_a(1 + j\omega\tau_b) + R_b(1 + j\omega\tau_a)} + R_s \tag{8}$$

If one separates the real and imaginary parts of $Z_b^{app}$, one notices that the numerator of the expression for the imaginary component contains negative and positive additive terms, which may result in the existence of three roots (i.e., three crossing points with the real axis, one at $\omega = 0$, one at $\omega = \infty$, and another one at an intermediate frequency $\omega = \omega_0$), and that the imaginary component may be positive between $\omega = 0$ and $\omega_0$ ("apparent inductive semicircle"; see Fig. 3) but negative between $\omega_0$ and $\omega = \infty$. Quantitatively speaking, $\omega_0$ is defined as

$$\omega_0^2 = \frac{R_a^2 C_a - R_b C_b (R_a + R_{sh})}{R_a^2 C_a^2 R_b C_b R_{sh}} \tag{9}$$

This equation indicates that a real value of $\omega_0$ exists and hence the apparent intracellular impedance locus crosses the real axis at $\omega = \omega_0$ if $R_a > R_{sh}$ and if $\tau_a > \tau_b$. [In case $R_a > R_s$ but $\tau_b > \tau_a$, the crossover could be observed in the apparent intracellular impedance locus calculated from $\Delta V_a(f)$.] The apparent inductive semicircle of $Z_b^{app}$ is thus equivalent to the recline of the intracellular voltage response from its transient overshoot which we have recently observed in response to transepithelial constant-current pulses (compare Suzuki *et al.*, 1982).

In addition to the lumped model, we also used the more adequate distributed model to fit the impedance data of the present experiments. This model (see Fig. 2b) allows current–voltage drops to be generated along the lateral space because it represents the space as a cable-like structure with a finite longitudinal resistance. The model was originally proposed by Frömter (1972), later analyzed for the dc case by Boulpaep and Sackin (1980), and recently applied to ac measurements on urinary bladders by Clausen *et al.* (1979). The latter measurements, however, were performed on a tight epithelium, which justified the assumption that the tight-junction resistance was infinite and thus simplified the analysis considerably. In addition, as shown in Fig. 2b, in its present form the distributed model differs from Clausen's model in that we believe the properties of the lateral cell membrane and the true basal cell membrane are indistinguishable. We assume, therefore, a fixed value of $n = 20$ for the ratio of the lateral over the basal cell membrane area, which is taken as an approximation from electron microscopic observations (Schifferdecker and Frömter, 1978). Because of space limitations, we dispense with the lengthy derivation here and explain only the definitions and the final result.

The additional or differently defined circuit parameters in Fig. 2b are the resistance of the tight junction, $R_j$, the longidutinal resistance for current flow along the entire lateral space, $R_{lis} = n \cdot r_{lis}$, and the resistance and capacitance of the basolateral cell membrane, $R_b = r_b/(n + 1)$ and $C_b = c_b \cdot (n + 1)$. Assuming the latter three to be continuously distributed along the lateral space, we obtain the following analytical solutions for the transepithelial impedance

$$Z_t = \frac{\det A - (a_{21}a_{32}R_s + a_{32} - a_{12}a_{21} + a_{22})\, B + (a_{22}a_{33}R_s - a_{13}a_{22} - B)Y_a}{Y_a(\det A) - (a_{32} + Y_a)a_{21}B} \tag{10}$$

and for the apparent intracellular impedance

$$Z_b^{app} = \frac{a_{22}a_{33}R_s - a_{13}a_{22} - B + a_{21}BZ_t}{\det A} \tag{11}$$

where the $a_{ij}$ and det $A$ are the elements and the determinant of the matrix

$$\begin{matrix} 1 & -\cosh\sqrt{nR_{lis}\, y_b} & -\sinh\sqrt{nR_{lis}\, y_b} \\ 1/R_j + Y_a & -1/R_j & 0 \\ 0 & -a_{12}y_b - a_{13}B & -a_{13}y_b - a_{12}B \end{matrix}$$

where $Y_a$ and $y_b$ are the admittances of the apical cell membrane and of the true basal cell membrane (facing the serosal surface),

$$Y_a = \frac{1}{Z_a} \quad \text{and} \quad y_b = \frac{1}{z_b} = \frac{(1 + j\omega r_b c_b)}{r_b}$$

and $B$ is defined as

$$B \equiv \sqrt{ny_b/R_{lis}}$$

## IV. TRANSEPITHELIAL AND APPARENT INTRACELLULAR TRANSFER FUNCTIONS UNDER CONTROL CONDITIONS

The results of a representative experiment on a *Necturus* gallbladder under control conditions are shown in Fig. 3. In this experiment the bladder was bathed on both surfaces with oxygenated $HCO_3$–Ringer's solution of the composition described previously (Suzuki *et al.*, 1982). The figure depicts the original data of the transepithelial and the apparent intracellular transfer functions in the form of an impedance plot (Nyquist or Cole plot), as obtained after correction for the transfer functions of the microelectrodes plus amplifiers. It can be seen that the transepithelial impedance describes in good approximation a semicircle with a small depression of the center below the real axis and a small shift of the high-frequency end along the real axis (indicating an ohmic series resistance), in agreement with previous publications (Schifferdecker and Frömter, 1978; Gögelein and Van Driessche, 1981). The slight curling of the high-frequency end is attributable to technical and/or computational problems and was disregarded in the quantitative analysis described below. In contrast to the transepithelial impedance locus, the apparent intracellular impedance locus describes two semicircles: a smaller one with a positive imaginary argument (plotted downward) in the low-frequency range and a larger one with a negative imaginary argument (plotted upward) in the high-frequency range.

By comparing the positions of the transepithelial locus and the apparent intra-

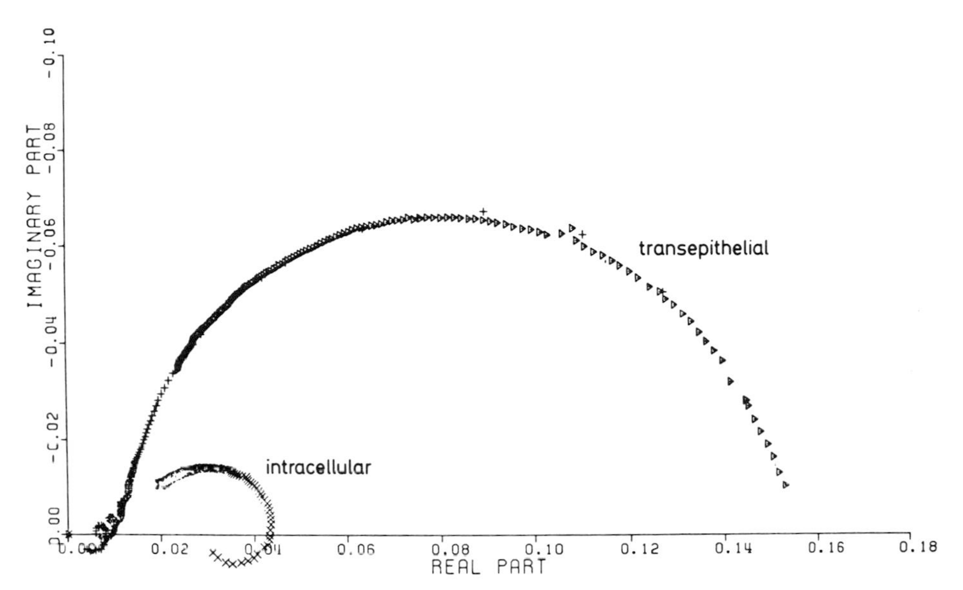


FIG. 3. Transepithelial and apparent intracellular impedance locus of a gallbladder under control conditions. Abscissa, real part; and ordinate, imaginary part (negative plotted upward) of the impedance in k$\Omega$ cm$^2$. The symbols $\triangle$, $+$, and $\times$ represent data from a slow and fast transepithelial, and a slow intracellular measurement run. Because of the linear distribution of the frequency points in the FFT routine, the data points become more crowded at the high-frequency (left) end of each data set. Note the ''apparent inductive semicircle'' at the low-frequency end of the apparent intracellular impedance locus.

cellular locus and extrapolating the left end of the latter toward higher frequencies, it can be seen that both loci seem to start from approximately the same high-frequency point on the real axis. This suggests that both electrodes see the same series resistance in this experiment, as would be expected if the intracellular resistivity was negligible and the series resistance arose only from the underlying connective tissue. Figure 4a and b shows least squares fits of the lumped and distributed model, respectively, using the reduced data of an experiment in which both high- and low-frequency measurements were available for the transepithelial and the apparent intracellular impedance locus. This experiment was also performed under control conditions, so that we assume that the lateral spaces were not collapsed. Both fits appear equally good, as also evidenced by the mean deviation of 1.1% in both cases (expressed as the average distance of a data point from the fitted model curve in percentage of the transepithelial dc resistance). Also, the best-fit parameters from both models compare quite well. For the lumped model we obtain $R_a = 1515$, $R_b = 95$, $R_{sh} = 157$, $R_s = 11.9$, $C_a = 8.1$, and $C_b = 23.8$, and for the distributed model $R_a = 1900$, $R_b = 91$, $R_j = 148$, $R_{lis} = 9.0$, $R_s = 9.4$, $C_a = 8.2$, $C_b = 25.5$; all resistances are in $\Omega$ cm$^2$ and all capacitances in $\mu$F/cm$^2$. This result is not surprising, since the spaces were

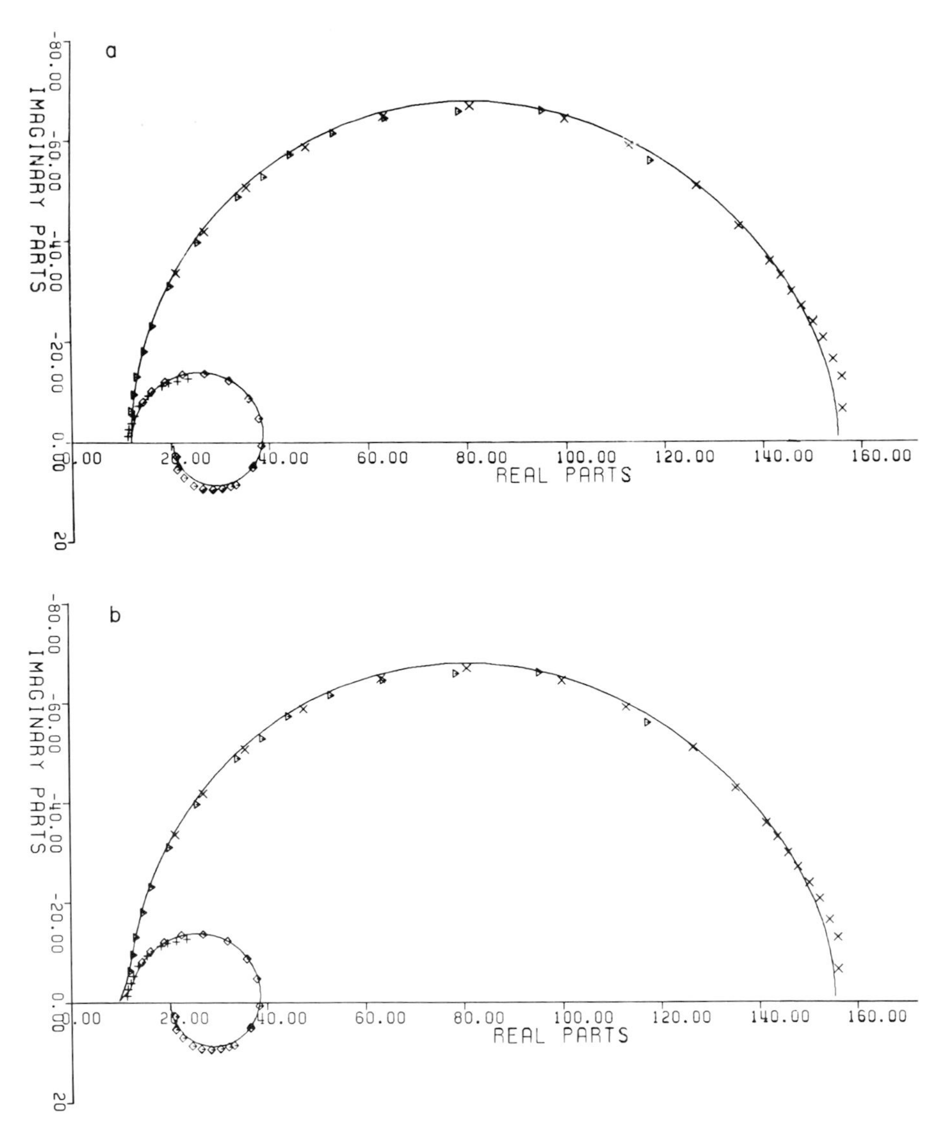


FIG. 4. Lumped (a) and distributed (b) model fit to the reduced impedance data of a gallbladder under control conditions. Details are as in Fig. 3, but scale is in Ω $cm^2$ and symbols do not exactly correspond. For fit parameters see text.

probably open in this experiment and the space resistance too small to distort the impedance locus significantly.

The results of 11 such experiments on five gallbladders are summarized in Table I. These data comprise all data sets available from control conditions,

TABLE I

RESULTS OF LUMPED AND DISTRIBUTED MODEL FITS TO IMPEDANCE DATA UNDER CONTROL CONDITIONS[a]

| Model | $R_a$ | $C_a$ | $R_b$ | $C_b$ | $R_{sh}$ | $R_j$ | $R_{lis}$ | $R_s$ | $s$ |
|---|---|---|---|---|---|---|---|---|---|
| Lumped | | | | | | | | | |
| Mean | 2650 | 6.8 | 118 | 29.7 | 151 | — | — | 13.7 | 1.37 |
| SD | ±1302 | ±2.2 | ±49 | ±6.0 | ±36 | — | — | ±2.0 | ±0.35 |
| Distributed | | | | | | | | | |
| Mean | 2310 | 7.0 | 87 | 32.4 | — | 145 | 9.1 | 11.2 | 1.62 |
| SD | ±770 | ±2.3 | ±37 | ±8.6 | — | ±38 | ±5.8 | ±1.8 | ±0.25 |

[a] All resistances are in $\Omega$ cm$^2$ and all capacitances in $\mu$F/cm$^2$. $s$ is the mean deviation of the data from the theoretical curve in percentage of the transepithelial resistance, which averaged 146 ± 34 $\Omega$ cm$^2$. Eleven cells were investigated, on five gallbladders. The mean cell potential was 84 ± 4 mV.

except for two experiments in which the mean deviation of the fitted curve from the data points was greater than 2% in both the lumped and distributed model fits, and the fits looked correspondingly poor. The distribution of the individual data from which Table I was produced suggests that the best-fit parameters are statistically significant results which justify the calculation of averages. With regard to $R_a$ and possibly also $R_b$, however, this point is not as certain, because the respective histograms (not shown) seem to exhibit some skewing with—in the case of $R_a$—a peak near 1900 $\Omega$ cm$^2$ and some high values (up to 5900 $\Omega$ cm$^2$ in the case of the lumped model fit). This type of distribution suggests that the least squares fit program could not always distinguish between the "correct" minimum and side minima, and that some caution is called for in the interpretation of these data.

Compared to previous estimates of the equivalent-circuit parameters of *Necturus* gallbladder ($R_a$ = 4500 or 3400 $\Omega$ cm$^2$ and $R_b$ = 2900 or 2700 $\Omega$ cm$^2$, see Frömter, 1972, and Reuss and Finn, 1975), the data of Table I show considerably lower values, but the values are similar to those obtained recently with the square wave pulse technique and with 2D-cable analysis (Suzuki *et al.*, 1982). Despite the qualifications mentioned above, we are convinced, however, that the present data represent the true membrane properties of *Necturus* gallbladder cells more correctly. The difference is most likely due to different incubation conditions (high oxygen supply and constant superfusion of the luminal surface in the present experiments versus a probable moderate oxygen deficiency due to insufficient stirring previously). This difference is also reflected in the present experiments in the higher cell potentials and the lower transepithelial resistance (larger distension of the lateral space in response to greater fluid transport).

The new result of Table I is the estimate of the lateral space resistance ($R_{lis}$), which amounts to 9.1 $\Omega$ cm$^2$ or ~7% of the total paracellular shunt resistance in the present experiments. As already mentioned in Section I, this value agrees with the fact that the spaces were distended so that their resistive contribution would be small. This relatively small magnitude of $R_{lis}$ may also be responsible for the fact that a significant improvement of the distributed model fit over the lumped model fit cannot be demonstrated in the control experiments (except possibly for the smaller scatter in $R_a$), in contrast to what will be shown below.

## V. IMPEDANCE LOCI UNDER CURRENT-INDUCED CHANGES OF LATERAL SPACE WIDTH

Larger differences between the lumped and distributed model fits were observed in measurements obtained after the lateral space was collapsed by passage of transepithelial direct current. Figure 5a and b shows the effect of lumen-

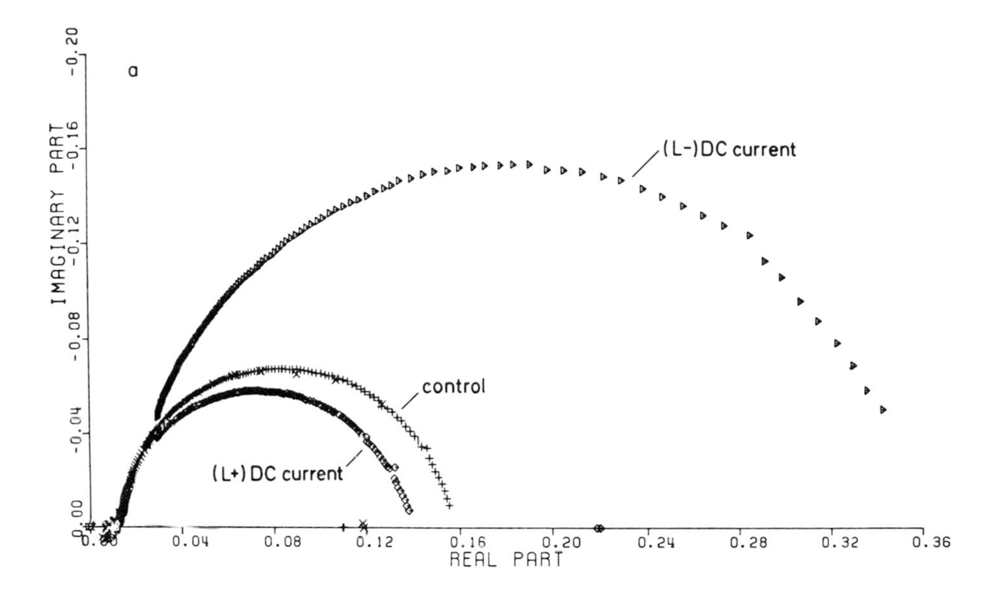


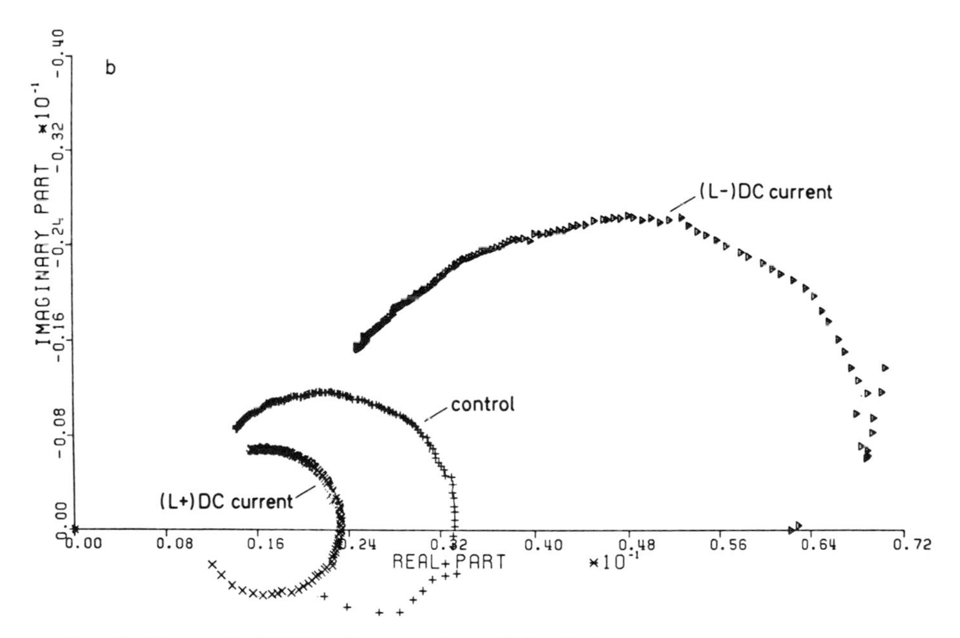


FIG. 5. Transepithelial (a) and apparent intracellular (b) impedance loci of gallbladders under lumen-positive (L+) and lumen-negative (L−) direct current (200 $\mu A/cm^2$). Details as in Fig. 3.

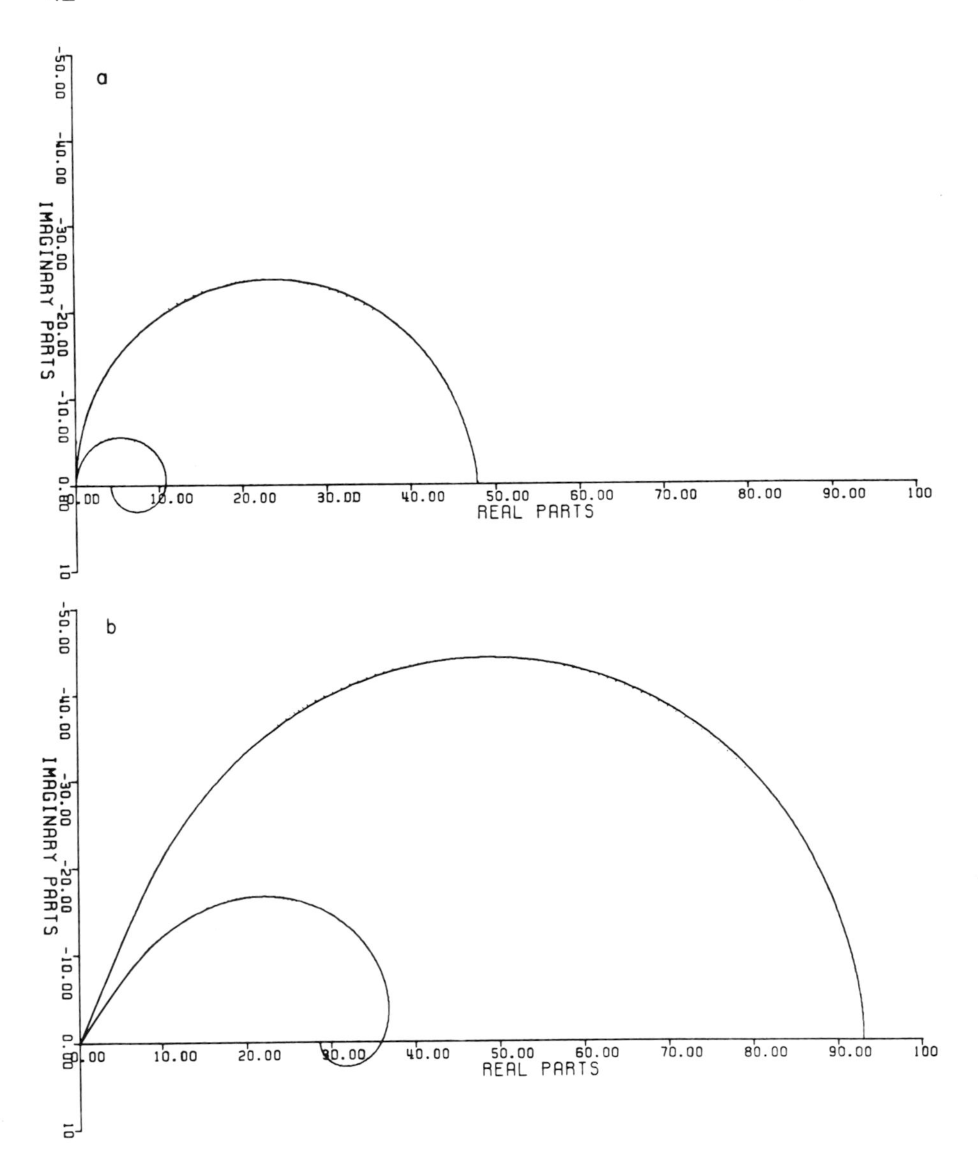


FIG. 6. Model calculations demonstrating the effect of an increase in $R_{lis}$ on the impedance loci. The model parameters which entered Eqs. (10) and (11) were $R_a = 1000$, $C_a = 7.0$, $R_b = 100$, $C_b = 21$, $R_j = 50$, and $R_s = 0$ (resistances in $\Omega$ cm$^2$; capacitances in $\mu$F/cm$^2$). In a, $R_{lis}$ was set to 0 and in b, $R_{lis}$ was set to 50 $\Omega$ cm$^2$. Among other features note particularly the flat angle with which the curves enter the origin at the high-frequency end in b.

positive and lumen-negative currents on the transepithelial and apparent intracellular impedance locus, respectively. As observed previously (Frömter, 1972) and later confirmed and documented by others (Bindslev *et al.*, 1974), lumen-negative current collapses the lateral spaces in gallbladder. As shown in Fig. 5a, this change is associated with a widening of the transepithelial impedance locus (indicative of an increased paracellular path resistance), while lumen-positive current, which dilates the spaces, leads to a contraction of the impedance locus (indicative of a reduction in the paracellular path resistance). Concomitantly, the apparent intracellular impedance locus also widens under lumen-negative current and may even lose its apparent inductive component (see Fig. 5a), while under lumen-positive current it shrinks and the apparent inductive component bulges further to the left. These observations are largely confirmed by the model calculations represented in Fig. 6a and b. In addition, a comparison of these two figures shows a significant flattening of the high-frequency range in the impedance locus of the distributed model, which is not present in the locus of the lumped model. Whereas in the lumped model (Fig. 6a) the high-frequency ends of both the transepithelial and the apparent intracellular impedance locus approach the real axis with an angle of 90° (at the left-hand side of the semicircles), the angle is much smaller for the distributed model impedance locus (Fig. 6b). This feature is a peculiar difference between the models, which relates to the presence of a cable-like structure in the distributed model (compare Falk and Fatt, 1964). Irrespective of the choice of the parameters, such a flattening is never observed in the lumped model.

Finally, Fig. 7a and b shows fits of the lumped and distributed model to data obtained under lumen-negative current flow. It can be seen that the distributed model fits the data much better than the lumped model, particularly in the flattened high-frequency region. Although the number of experimental observations under the flow of lumen-negative or lumen-positive direct current does not suffice at present to justify a statistical evaluation of the changes in all equivalent-circuit parameters, it can now be said with certainty that the calculated lateral space resistance increases during lumen-negative current flow to values on the order of 100–200 $\Omega$ cm$^2$, depending on the duration and density of the current. This result agrees with the observation that the lateral space collapses under lumen-negative direct current (Frömter, 1972; Bindslev *et al.*, 1974) and suggests that the major part of the increase in transepithelial resistance during lumen-negative current flow reflects more or less directly the closing of the lateral space. In addition, this conclusion agrees with unpublished observations of Simon, Ikonomov, and Frömter, that lumen-negative current changes the voltage divider ratio of an electrode inserted into the lateral space in such a way that the voltage drop between serosal fluid and electrode tip increases relatively more than the voltage drop between electrode tip and mucosal bath.

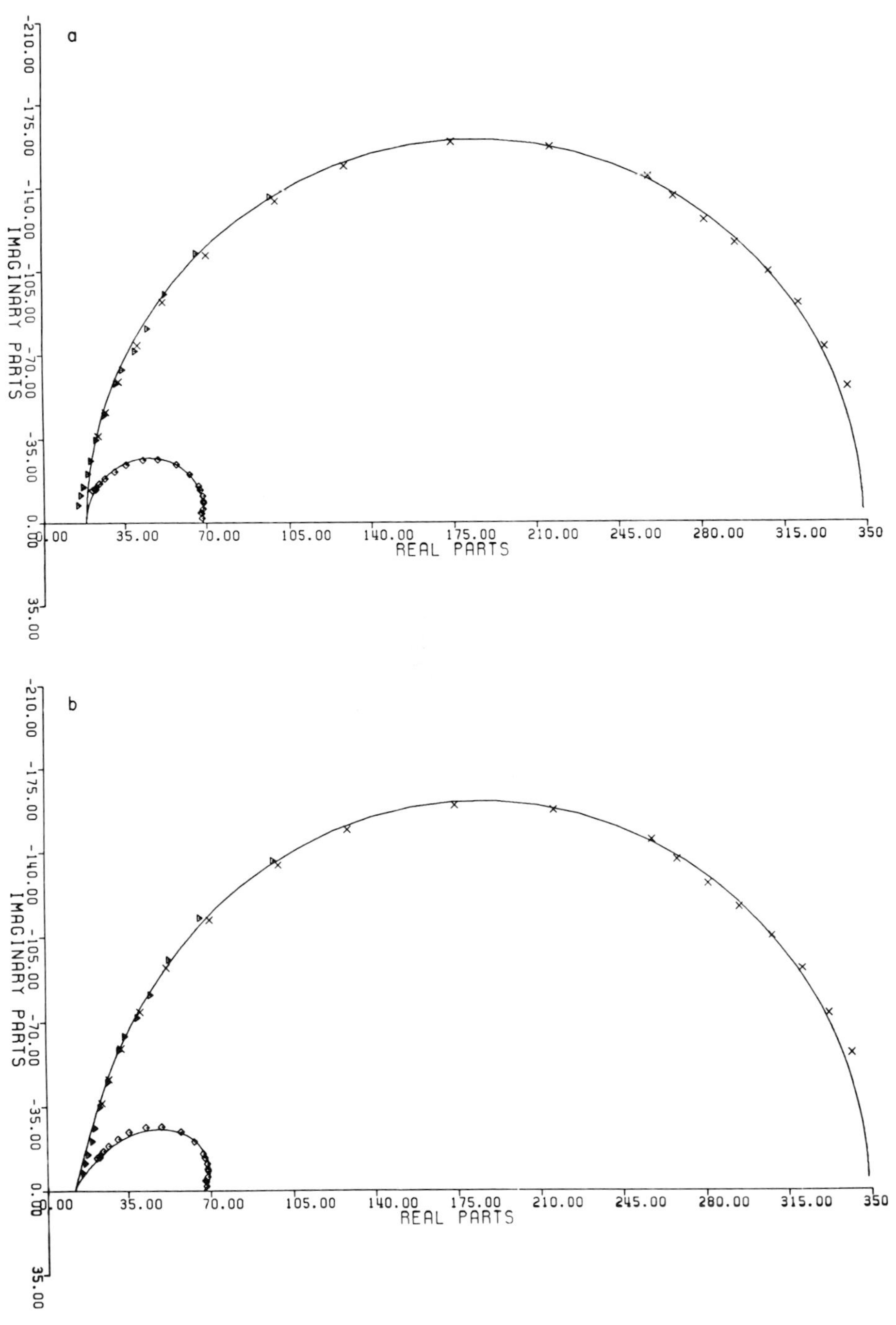


FIG. 7. Lumped (a) and distributed (b) model fits to impedance data of a gallbladder under lumen-negative direct current. Note that the latter model fits the data much better at the high-frequency (left) end. In this experiment the distributed model fit gave values of $R_{\text{lis}} = 110\ \Omega\ \text{cm}^2$ and $R_{\text{j}} = 360\ \Omega\ \text{cm}^2$, while the lumped model fit gave the unrealistic value of $R_{\text{sh}} = 1460\ \Omega\ \text{cm}^2$. The mean deviations were 1.44 and 1.18% in a and b, respectively.

## VI. CONCLUSIONS

Because epithelia consist of serial and parallel arrays of resistive barriers, it is impossible to determine the properties of an individual barrier from transepithelial measurements alone. We have therefore combined transepithelial alternating current spectroscopy with intracellular measurements. The data are analyzed by least squares computer fits with model equations in order to derive the magnitude of the equivalent-circuit parameters of the cell membranes and of the shunt path. Two electrical analog circuits were tested: (1) the lumped model in which the cell membranes and the shunt path are represented by RC elements and by a single resistor, respectively, and (2) the distributed model in which the resistances of the lateral cell membrane and of the lateral space fluid are distributed in the form of a cable-like structure which extends from the tight junction to the serosal fluid compartment. Under control conditions with extended lateral spaces, we find that both models describe the data almost equally as well. This is due to the fact that the lateral space resistance is rather small (on the average ~7% of the total transepithelial resistance). However, when the spaces are collapsed by passage of lumen-negative current the lumped model fits are no longer adequate, particularly in the high-frequency region of the impedance plots. Under those conditions the distributed model fits indicate that the lateral space resistance can increase to 50% or more of the transepithelial resistance. These experiments allow us for the first time to resolve the paracellular shunt resistance into its contributions from the tight junction proper and from the lateral space.

ACKNOWLEDGMENTS

The authors thank Dr. C. Clausen, Department of Physiology, SUNY, Stony Brook, New York, Dr. W. Van Driessche, Lab. voor Fysiologie, Leuven, Belgium, and Dr. F. Sauer, Max-Planck-Institut für Biophysik, Frankfurt, Federal Republic of Germany for helpful discussions during the development of the experimental setup and the methods of data analysis.

REFERENCES

Bindslev, N., Tormey, J. McD., and Wright, E. M. (1974). The effects of electrical and osmotic gradients on lateral intercellular spaces and membrane conductance in a low resistance epithelium. *J. Membr. Biol.* **19,** 357–380.

Boulpaep, E. L., and Sackin, H. (1980). Electrical analysis of intraepithelial barriers. *Curr. Top. Membr. Transp.* **13,** 169–197.

Clausen, C., and Fernandez, J. M. (1981). A low-cost method for rapid transfer function measurements with direct application to biological impedance analysis. *Pfluegers Arch.* **390,** 290–295.

Clausen, C., Lewis, S. A., and Diamond, J. M. (1979). Impedance analysis of a tight epithelium using a distributed resistance model. *Biophys. J.* **26,** 291–318.

Cuthbert, A. W., and Painter, E. (1969). The action of antidiuretic hormone on cell membranes. Voltage transient studies. *Br. J. Pharmacol.* **35,** 29–50.

Falk, G., and Fatt, P. (1964). Linear electrical properties of striated muscle fibres observed with intracellular electrodes. *Proc. R. Soc. London, Ser B* **160,** 69–123.

Frömter, E. (1972). The route of passive ion movement through the epithelium of *Necturus* gallbladder. *J. Membr. Biol.* **8,** 259–301.

Frömter, E., and Diamond, J. M. (1972). Route of passive ion permeation in epithelia. *Nature (London), New Biol.* **235,** 9–13.

Gögelein, H., and Van Driessche, W. (1981). Capacitive and inductive low frequency impedances of *Necturus* gallbladder epithelium. *Pfluegers Arch.* **389,** 105–113.

Lim, J. J., and Fischbarg, J. (1981). Electrical properties of rabbit corneal endothelium as determined from the impedance measurements. *Biophys. J.* **36,** 677–695.

Mathias, R. T., Rae, J. L., and Eisenberg, R. S. (1981). The lens as a nonuniform spherical syncytium. *Biophys. J.* **34,** 61–83.

Poussart, D., and Ganguly, U. S. (1977). Rapid measurement of system kinetics—An instrument for real-time transfer function analysis. *Proc. IEEE* **65,** 741–747.

Rehm, W. S., Shoemaker, R. L., Sanders, S. S., Tarvin, J. T., Wright, J. A., Jr., and Friday, E. A. (1973). Conductance of epithelial tissues with particular reference to the frog's cornea and gastric mucosa. *Exp. Eye Res.* **15,** 533–552.

Reuss, L., and Finn, A. L. (1975). Electrical properties of the cellular transepithelial pathway in *Necturus* gallbladder. I. Circuit analysis and steady-state effects of mucosal solution ionic substitutions. *J. Membr. Biol.* **25,** 115–139.

Schifferdecker, E., and Frömter, E. (1978). The AC impedance of *Necturus* gallbladder epithelium. *Pfluegers Arch.* **377,** 125–133.

Schwan, H. P. (1957). Electrical properties of tissue and cell suspensions. *Adv. Biol. Med. Phys.* **5,** 147–209.

Smith, P. G. (1971). The low-frequency electrical impedance of the isolated frog skin. *Acta Physiol. Scand.* **81,** 355–366.

Suzuki, K., Rohlicek, V., and Frömter, E. (1978). A quasi-totally shielded low-capacitance glass microelectrode with suitable amplifiers for high-frequency intracellular potential and impedance measurements. *Pfluegers Arch.* **378,** 141–148.

Suzuki, K., Kottra, G., Kampmann, L., and Frömter, E. (1982). Square wave pulse analysis of cellular and paracellular conductance pathways in *Necturus* gallbladder epithelium. *Pfluegers Arch.* **394,** 302–312.

Teorell, T., and Wersäll, R. (1945). Electrical impedance properties of surviving gastric mucosa of the frog. *Acta Physiol. Scand.* **10,** 243–257.

# Chapter 3

# Membrane Area Changes Associated with Proton Secretion in Turtle Urinary Bladder Studied Using Impedance Analysis Techniques

*CHRIS CLAUSEN*

*Department of Physiology and Biophysics*
*Health Sciences Center*
*State University of New York*
*Stony Brook, New York*

*AND*

*TROY E. DIXON*

*Department of Medicine*
*Northport Veterans Administration Hospital*
*Northport, New York*






ISBN 0-12-153320-4

## I. INTRODUCTION

In the interpretation of transepithelial electrical measurements (e.g., measurements of spontaneous potential, short-circuit current, transepithelial resistance, etc.), one is faced with a major problem: How does one separate the transepithelial measurements in properties attributable to each of the different membrane structures of the epithelium? Epithelial transport is mediated by selective properties of the apical and basolateral membranes, and by properties of the paracellular pathway via the junctions and lateral spaces.

Another problem that plagues the investigator is how to normalize measured transport processes to a unit area of membrane. Epithelia are highly folded tissues, and one is always faced with the problem that the amount of tissue in a chamber far exceeds *by an unknown factor* the nominal chamber area. In addition, the apical and basolateral membranes possess different areas relative to each other. Moreover, much attention has recently been directed to the investigation of transport-related changes in membrane area, caused by vesicular fusion events. Clearly, changes in the area of a membrane of a given specific ionic conductance will be reflected in changes in transepithelial resistance unrelated to changes in the specific ionic permeability properties of the epithelium.

In this chapter, we briefly discuss the use of impedance analysis techniques in the study of ionic transport processes in epithelia. We discuss how these techniques are used to separate transepithelial measurements into properties of each of the different membrane constituents of the tissue. In addition, we discuss the measurement of membrane capacitance, and how capacitance can be used as an indirect measure of the membrane area. Finally, we present preliminary results from recent studies of turtle bladder, in an attempt to quantify changes in membrane area associated with the proton secretion process in this tissue.

### A. Equivalent-Circuit Analysis

The success of impedance analysis techniques in epithelia relies on the ability to model accurately the epithelial electrical properties, using equivalent electrical circuits composed of resistors and capacitors. The resistors represent the ionic conductance properties of a membrane, and the capacitors represent the electrical capacitance of the membrane. Biological membranes exhibit an unusually high electrical capacitance which is found to be directly proportional to the respective membrane area. The proportionality constant, or so-called membrane specific capacitance, is $\sim 1\ \mu F/cm^2$ and is found to be surprisingly constant in nearly all biological membranes studied (Cole, 1972; Davson, 1964). One can obtain values for the model resistors and capacitors by fitting the model-predicted response to measured epithelial data. The resistor and capacitor values thus obtained can then be related to membrane ionic conductances and areas, respectively.

Four points must be considered when developing an equivalent-circuit model of an epithelium. (1) The model must be composed of a finite number of elements whose values can be obtained uniquely from the data. (2) There must be accurate agreement between the model-predicted results and measured data. (3) Each of the model elements must represent, in a one-to-one fashion, the morphological structures observed in high-resolution micrographs of the tissue. (4) Measurements obtained from the analysis must compare favorably with measurements obtained using other independent methods.

## B. Impedance Analysis

The characterization of an electrical circuit can be accomplished by measuring its electrical impedance. The impedance is simply the ratio of the measured voltage response to a constant applied sinusoidal current at a given frequency. Two quantities are measured at each frequency: the ratio of the amplitudes (impedance magnitude) of the voltage and the current, and the normalized time delay (phase angle) of the voltage signal in reference to the current. The data can then be represented as Bode plots, which plot impedance magnitude and phase angle as a function frequency. By curve fitting measured transepithelial impedance to that predicted by the equivalent-circuit model, one can then determine unique values for the circuit elements (resistors and capacitors), which subsequently can be related to the biological parameters (membrane ionic conductances and areas).

The measurement of transepithelial impedance can be accomplished using several methods. One approach is to apply a sinusoidal current of a given frequency, measure the voltage response using a phase-lock amplifier or phase meter, and then repeat the procedure at several different frequencies (Clausen *et al.*, 1979). Since the Bode plots are mapped out sequentially in frequency, this approach suffers from the disadvantage that roughly 15 minutes is required to collect a sufficient amount of data. More recently, investigators have turned to the utilization of wide-band current signals composed of the sum of many sinusoidal frequencies (Clausen and Wills, 1981; Clausen and Fernandez, 1981). Using Fourier analysis techniques, the composite voltage response to the signal can be separated into the responses at each frequency. This method (see Section II) has the advantage that less than 10 seconds is required to collect the required volume of data.

## C. Vesicle Fusion

Vesicle fusion is known to be a fundamental mechanism for transporting lipid-insoluble substances across the plasma membrane of a variety of cells. Apart from the notable example of neurotransmitter release at the nerve terminal, other examples of this phenomenon can be found in the secretion of norepinephrine by the adrenal gland (Tepperman, 1980), in the secretion of serotonin by platelets (Detweiler *et al.*, 1975), and in the secretion of parathormone by the parathyroid gland (Brown *et al.*, 1978). Also, vesicle fusion has been shown to be an important mechanism for the insertion of transporting proteins into the cell membrane, thereby allowing the transport of substances across that membrane. A notable example of this phenomenon is the insertion of glucose carriers into the plasma membrane of muscle cells under the influence of insulin (Karnieli *et al.*, 1981).

Vesicle fusion is known to be an important mechanism in the regulation of epithelial transport systems as well. In gastric mucosa, the onset of proton secretion in response to secretagogues is known to be dependent on the fusion of intracytosolic vesicles with the apical membrane (see Machen and Forte, 1979, for a recent review). Other examples of this phenomenon in epithelia include antidiuretic hormone-stimulated water transport in the toad urinary bladder, which is associated with an ~20% increase in the apical membrane area due to vesicle fusion (Stetson *et al.*, 1982). In the rabbit urinary bladder, hydrostatic pressure gradients (or simple mechanical stretch) result in changes in sodium transport as well as incorporation of intracellular "plaques" into the apical membrane with a resultant increase in apical surface area (Lewis and de Moura, 1982).

Recent microscopic evidence (see below) suggests that vesicle fusion is an important means of regulating proton secretion by the turtle urinary bladder. In gastric mucosa, toad bladder, and rabbit urinary bladder, the initial observation of apical area change related to the various transport phenomena was obtained by microscopic examination of fixed tissues. Subsequently, impedance analyses were performed which not only corroborated the microscopic findings, but also allowed quantification and an approach to the kinetics of the process in living tissues. In this laboratory, we are currently performing impedance analyses in the turtle urinary bladder.

## D. Transport Processes in Turtle Bladder

The urinary bladder of freshwater turtle (*Pseudymys scripta elegans*) has two major active transport systems. Sodium is reabsorbed utilizing an electrogenic pump located at the basolateral membrane, whereas protons are secreted by a different electrogenic pump located at the apical membrane (Steinmetz, 1974). Microscopic analyses of the tissue demonstrate that there are two distinct predominant cell types, the granular cells and the mitochondria-rich cells (Rosen, 1970). It has been shown by histochemical techniques that carbonic anhydrase is located almost exclusively in the mitochondria-rich cell type (Rosen, 1972).

Moreover, in isolated cells from the turtle bladder, acetazolamide (AZ) and 4-acetamido-4′-isothiocyanostilbene-2,2′-disulfonic acid (SITS), inhibitors of proton transport, reduce $O_2$ consumption in the mitochondria-rich cells, but not the granular cells (Schwartz *et al.*, 1982). On the other hand, amiloride and ouabain, which selectively inhibit sodium transport, reduce $O_2$ consumption in the granular cells, and do not affect $O_2$ consumption in the mitochondria-rich cells (Schwartz *et al.*, 1982). Based on these considerations, the mitochondria-rich cells are thought to be selectively responsible for proton transport, and the granular cells are thought to be the site of the sodium transport system.

Several lines of evidence support the notion that proton transport is, at least in part, regulated by changes in the apical membrane surface areas. Using microscopic techniques, Husted *et al.* (1981) found that inhibition of proton transport with either AZ or SITS led to a loss of microplicae (microvilli-like structures) and a marked decrease in the surface ares of the mitochondria-rich cells. Using the fluorescent pH probe acridine orange, Gluck *et al.* (1982) found intracytosolic vesicles of low pH in these same cells, and evidence that these vesicles contained proton pumps. When proton transport was stimulated with $CO_2$, it was found that these vesicles appeared to fuse with the apical membrane. Also using microscopic techniques, Stetson and Steinmetz (1982) have recently confirmed these results and found that $CO_2$ stimulation leads to a decrease in the number of intracytosolic vesicles and a selective increase in the apical surface area of the mitochondria-rich cells.

## E. Impedance Studies in Turtle Bladder

On initial inspection, two characteristics of the turtle bladder would appear to render impedance analysis as a means of detecting changes in the apical surface area related to proton transport an exercise in futility. First, the mitochondria-rich cells, which are responsible for proton transport, represent only ~10–30% of the total cell population. Hence changes in the apical surface area of these cells might only minimally affect total apical capacitance. Second, the presence of two cell types with known transport differences would presumably preclude the development of a simple, morphologically based equivalent circuit which is needed for quantitative analysis of impedance data. Fortunately, these problems are minimized by two additional pieces of information.

The first problem becomes less limiting when one takes into account that, although the mitochondria-rich cells represent a small percentage of the total cell number, the apical surface area of the mitochondria-rich cells is significantly greater than that of the granular cells. This arises from the fact that mitochondria-rich cells have microplicae which amplify apical membrane area per cross-sectional area. The result is that the mitochondria-rich cells can account for

30–40% of the total apical area present in the tissue. Therefore, it is likely that changes in the area of the mitochondria-rich cell apical membrane could be detected by the analysis.

The problem of the different transport systems present might be addressed by selectively inhibiting the sodium transport of the responsible granular cell population. As mentioned earlier, Schwartz *et al.* (1982) found that ouabain and amiloride selectively inhibited the $O_2$ consumption of granular cells. These authors also found that in the turtle bladder, as in other sodium-transporting epithelia, ouabain and amiloride both dramatically reduce apical membrane ionic conductance and lead to an increase in transepithelial resistance. Based on these findings, these changes would appear to be due to selective inhibition of sodium transport in the granular cell population. Under these conditions, it is reasonable to postulate that changes in the properties of the mitochondria-rich cells could dominate any further changes noted in the transepithelial electrical properties of the tissue.

For these reasons just discussed, we felt optimistic that impedance studies would be capable of quantifying the membrane area changes in this preparation. In this chapter, we report our initial results.

## II. METHODS

Our results arise from experiments performed on turtle bladders that were pretreated with serosal ouabain (1 m*M*) and mucosal amiloride (0.1 m*M*). The proton transport rate was modified by altering $CO_2$ tension in the baths (using bubblers), and by the application of 0.5 m*M* SITS, or 50 μ*M* AZ. SITS is an irreversible inhibitor of proton secretion; AZ at 50 μ*M* concentration was found to be a reversible inhibitor of proton secretion.

The tissues were mounted in a modified Ussing chamber, designed to virtually eliminate edge damage (Lewis and Diamond, 1976). Tissue area was 2 $cm^2$; nominal bath volume was 15 ml. Transepithelial voltage measurements were acquired using a high-speed differential amplifier (Model 113, Princeton Applied Research, Princeton, New Jersey), connected to Ag/AgCl electrodes mounted close to the preparation. A different set of Ag/AgCl electrodes was mounted at both ends of the chamber, and was used to measure short-circuit current ($I_{sc}$) using the voltage clamp technique.

Transepithelial impedance measurements were obtained using the technique of Clausen and Fernandez (1981). Briefly, a wide-band pseudorandom binary signal was generated digitally, and converted to a constant current of 24 μA (peak-to-peak). This current was passed transepithelially using the current electrodes. The voltage response was digitized and acquired by a computer. Total data acquisition time was less than 10 seconds per run. The impedance was computed using standard Fourier analysis techniques (Bendat and Piersol, 1974). Raw data consisted of 1024 frequency measurements acquired at two different bandwidths (2 Hz to 1 kHz, and 20 Hz to 10 kHz). It was subsequently reduced to a manageable number of approximately 100 data points spaced logarithmically in frequency from 2 Hz to 8 kHz. These data were subsequently fitted by an equivalent-circuit model using a nonlinear curve-fitting algorithm, as described in Clausen *et al.* (1979).

# III. RESULTS

## A. Equivalent-Circuit Model

The apical membrane was modeled as a simple parallel resistor–capacitor (RC) circuit, representing the ionic conductance and capacitance of that membrane. In modeling the serosal membrane in a similar fashion, systematic deviations were observed that were similar to those reported by Clausen *et al.* (1979) in studies in rabbit urinary bladder. These investigators could account for the deviations by explicitly considering the path resistance, distributed along the lateral membrane, of the narrow lateral spaces. The serosal membrane was therefore modeled as a distributed impedance shown in Fig. 1. The resulting disagreement between the model-predicted impedance and the data was reduced dramatically and was found to be less than 1.5% in all cases. Figure 2 shows data

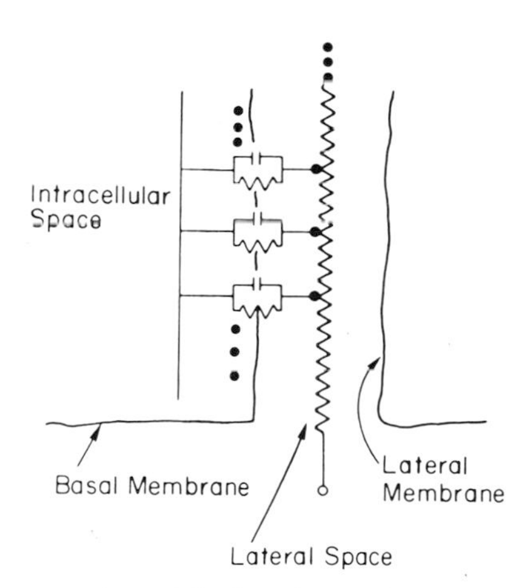


FIG. 1. Schematic representation of the impedance of the basolateral membrane. The path resistance $R_p$ of the narrow lateral spaces is distributed along the lateral membrane producing a distributed membrane impedance. The apical membrane (not shown) is treated as a lumped parallel resistor-capacitor circuit.

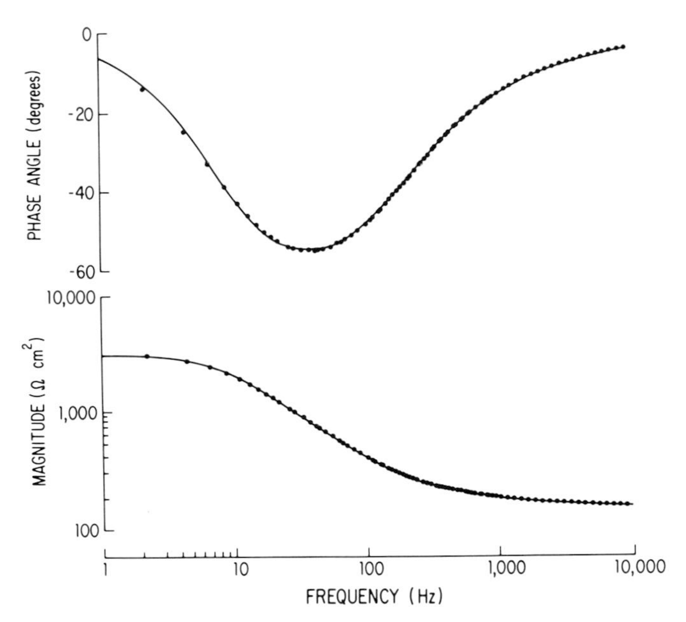


FIG. 2. Measured impedance (symbols) from a randomly chosen run fitted by the distributed equivalent-circuit model (lines drawn through the symbols). The good agreement between the model-predicted and measured impedance is typical.

(symbols) from a randomly chosen impedance run. The lines drawn through the data show the model-predicted response.

The model consists of four membrane parameters corresponding to the apical and basolateral resistances and capacitances ($R_a$, $C_a$, $R_b$, and $C_b$, respectively), the resistance of the lateral space ($R_p$), and a small series resistance ($R_s$) which accounts for the finite resistance of the unstirred layers between the voltage electrodes and the tissue. An underlying assumption in reference to the correspondence between the model parameters and the actual biological parameters is that the epithelium is truly tight, and that it possesses a negligible paracellular conductance via the tight junctions. The validity of this assumption was verified (see below).

The good agreement between the one-cell distributed model and the measured data supports the notion that the electrical properties of the epithelium are dominated by the mitochondria-rich cells, since amiloride and ouabain are known to increase dramatically the apical membrane resistance of sodium-transporting cells (e.g., granular cells). Moreover, the model was unable to accurately fit the measured impedance in tissues where sodium transport was *not* inhibited (Clausen and Dixon, unpublished observation), suggesting that in this case, a more complicated two-cell model might be required. However, we cannot discount the possibility that the two cell types are electrically coupled laterally, and that our measurements represent mean properties of the two types. Although gap junctions have not been found connecting the two cell types, amiloride has been shown to partially inhibit proton transport, possibly by hyperpolarizing granular cells electrically connected to mitochondria-rich cells. Also, microelectrode studies (Nagel *et al.*, 1981) failed to show distinct properties expected from two cell populations, lending additional support for the notion of electrically connected cells.

### B. Resistance Ratio and Paracellular Conductance

If the paracellular conductance was comparable to the membrane conductance, then one can show that the impedance-predicted resistance ratio $R_a/R_b$ would result in an underestimation of the true membrane resistance ratio (see Clausen and Wills, 1981). We can test our assumption of a negligible paracellular conductance by comparing our impedance-predicted resistance ratio measurements with directly measured measurements.

In ouabain-treated bladders, Nagel *et al.* (1981) report an average apical-to-basolateral resistance ratio of ~19, directly measured using microelectrode techniques. Our mean impedance-predicted value, measured under similar experimental conditions, is 20 ± 2 ($n = 9$). The gratifying agreement between the two measurements lends support to the notion that the paracellular conductance pathway is truly negligible when compared with the apical and basolateral conductances, and hence we interpret the parameter values as reflecting true membrane parameters.

Under control conditions (tissues treated only with amiloride and ouabain), the apical resistance was 4160 $\Omega$ cm$^2$ ± 530 ($n = 9$).

### C. Capacitance Values and Capacitance Ratio

Under control conditions, the apical membrane capacitance was 3.7 $\mu$F/cm$^2$ ± 0.7 ($n = 9$), which implies that microvilli and folding of the apical membrane result in an ~3.7-fold increase in area over the nominal chamber area (assuming 1 cm$^2$ ≅ 1 $\mu$F). The capacitance ratio $C_a/C_b$ was 0.46 ± 0.04 ($n = 9$), implying that the basolateral membrane area is 2.2 times greater than the apical area. Micrographs show more basolateral membrane than apical membrane, hence this finding is expected.

### D. Apical Capacitance and Proton Transport

In our initial experiments, we found *what appeared to be* a direct correlation between $C_a$ and proton transport rate. Figure 3 shows data from four different bladders. Transport rate was monitored by measuring short-circuit current (abscissa). The decrease in transport was followed after the application of SITS and removal of $CO_2$ and bicarbonate.

The linearity and similarity of slopes of the responses seen in Fig. 3 were initially interpreted to indicate that there was a one-to-one correlation between $C_a$ and short-circuit current. The difference in zero-transport intercepts seen in data from different bladders was caused simply by different nominal tissue areas in the different bladders, caused by differing degrees of mounting stretch.

Subsequent experiments were then performed in an attempt to investigate this notion further. Namely, if transport is proportional to apical capacitance, then one should be able to obtain the same results using reversible inhibitors: $C_a$ should decrease with the application of the inhibitor, and should increase after stimulation of transport.

Negative results were obtained in all cases. Although reversible inhibitors such as AZ resulted in decreased transport and decreased $C_a$, in no case were we able to observe an increase in $C_a$ correlated with increased transport rate (after removal of AZ and stimulation with $CO_2$).

Observation of $C_a$ measurements taken under control conditions (bathed in bicarbonate/$CO_2$-free solutions, and with stable transport rate) showed that there was with time a consistent decrease in $C_a$ that was independent of transport rate. This is shown in Fig. 4. Here we plot $C_a$ as a function of time. Notice that there is a consistent decrease in $C_a$ prior to the application of a transport inhibitor (in this case, SITS).

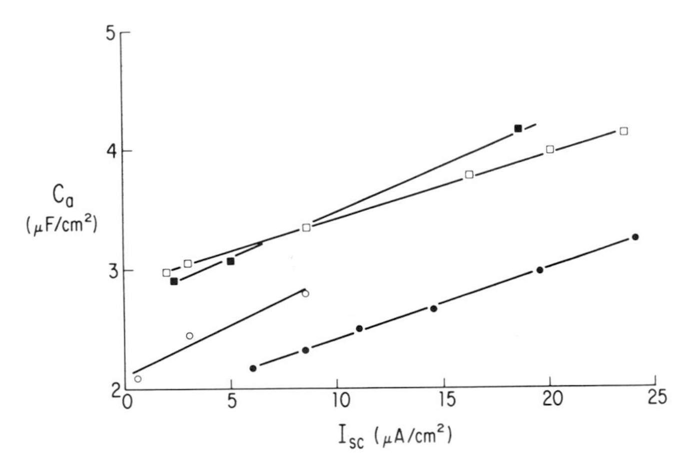


FIG. 3. Apical membrane capacitance plotted against the proton transport rate (measured as the short-circuit current) from four different tissues. Transport rate was inhibited using SITS. These data show *only an apparent correlation* between $C_a$ and $I_{sc}$, since we subsequently found a transport-unrelated decrease in $C_a$ with time.

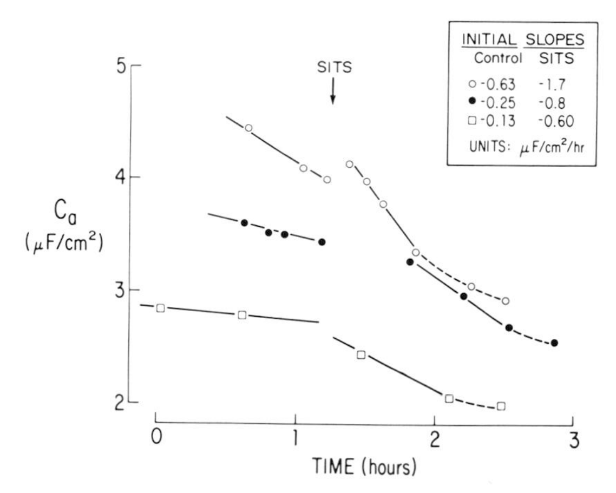


FIG. 4. Apical membrane capacitance, plotted as a function of time, measured in three different tissues bathed initially in bicarbonate/$CO_2$-free solutions. Proton secretion was inhibited by the addition of SITS, which accelerates the rate of decline in $C_a$ (see inset).

Transport inhibitors accelerated the steady decrease in $C_a$. When using SITS, the rate of capacitance decrease changed twofold from $-0.35$ μF/cm² · hour (± 0.11, $n$ = 4) to $-0.73$ μF/cm² · hour (± 0.11, $n$ = 4). A similar observation was also made using AZ to reversibly inhibit the transport (Fig. 5). Here, the change went from $-0.68$ μF/cm² · hour (± 0.25, $n$ = 5) to $-1.04$ μF/cm² · hour (± 0.36, $n$ = 5) after application of the AZ. Finally, after restimulation of

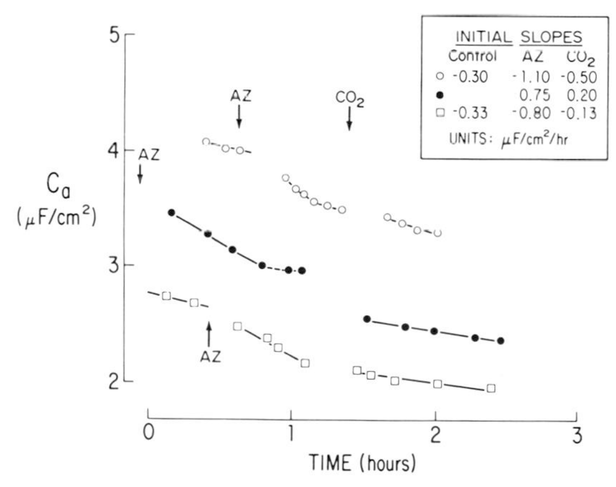


FIG. 5. Apical membrane capacitance, plotted as a function of time, measured in three different tissues bathed initially in bicarbonate/$CO_2$-free solutions. Proton secretion was inhibited by the addition of AZ. Secretion was restored by removing AZ and by bathing the tissues in bicarbonate/$CO_2$-containing solutions (marked $CO_2$). The rate of decrease in $C_a$ is affected by these maneuvers which alter secretion rate (see inset).

transport by washing off the AZ and adding 5% $CO_2$ and 25 m$M$ $HCO_3^-$, the rate of transport increased to a value comparable to the pretreatment state while rate of decline in $C_a$ slowed (to $-0.35$ μF/cm² · hour, $\pm$ 0.15, $n = 5$).

The general result in that changes in proton transport in this tissue are accompanied by changes in $C_a$, and hence apical surface area. However, the tissue also exhibits a transport-unrelated slow decrease in $C_a$ during the course of the experiment.

## E. Effect of SITS on the Membrane Resistances

The effect of SITS on the different membrane resistances was studied in four bladders. Application of 0.5 m$M$ SITS reduced the proton secretion rate, measured as a short-circuit current, from 20 $\pm$ 4 to 2.5 $\pm$ 1.2 μA/cm². This was accompanied by an apical membrane resistance increase from 3.9 $\pm$ 1.1 to 6.7 $\pm$ 1.2 kΩ cm². As stated earlier, this is also accompanied by a decrease in membrane capacitance which itself causes an increase in apical resistance. However, using the membrane capacitance to normalize for unit area of apical membrane, we find that the apical resistance increases from 13 $\pm$ 2 to 17 $\pm$ 2 kΩ · μF, reflecting a modest decrease in apical specific conductance.

SITS had no effect on the basolateral membrane resistance, which exhibited a mean value of 240 Ω cm² $\pm$ 19 ($n = 8$). Using the basolateral capacitance to normalize for membrane area results in a mean normalized resistance of 1.6 kΩ · μF $\pm$ 0.14 ($n = 8$).

# IV. DISCUSSION

The transepithelial electrical properties of turtle bladder, under conditions in which sodium transport is inhibited, are well represented by a simple distributed equivalent-circuit model. Using impedance analysis techniques, we can determine values of the different circuit parameters (resistors and capacitors) which can then be related to the membrane parameters (ionic conductances and areas). The values determined from the impedance analyses are in close agreement with those obtained by direct current microelectrode techniques and micrographic studies of the tissue, under similar experimental conditions.

These studies have yielded the following information.

1. The tissue is truly a tight epithelium, possessing only negligible paracellular conductance via the tight junctions and lateral spaces.
2. The ability of the simple one-cell-type model to fit accurately the data suggests that when sodium transport is inhibited by amiloride and ouabain, the

transepithelial electrical properties of the epithelium are dominated by properties of the mitochondria-rich cells.

3. Inhibition of proton transport by SITS results in a modest decrease in apical ionic conductance, but no change whatsoever in the basolateral membrane conductance properties. Since SITS is known to inhibit bicarbonate exit across the basolateral membrane (Cohen *et al.*, 1978), this suggests that this process involves an electrically silent pathway, possibly an ion-exchange process.

4. Alteration in the rate of proton transport by either SITS or AZ results in a change in the rate of decline of apical capacitance, suggesting that the effect of these agents is mediated through changes in apical membrane area.

5. In addition to the transport-related changes in apical membrane capacitance, we observe a constant transport-unrelated deline with time. This decline does not appear to be related to cell death or removal, since the rate of proton transport remains constant. Another possibility is that under our experimental conditions, cell volume changes are occurring with consequent infolding of apical membrane area. Note that a similar situation is observed in the mammalian bladder, where changes in the hydrostatic pressure are associated with changes in the capacitance of the membrane (Lewis and de Moura, 1982). We have observed that if we include 2% bovine serum albumin in the serosal Ringer's solution, and use only amiloride to inhibit sodium transport (no ouabain), the transport-unrelated changes in apical capacitance disappear (Clausen and Dixon, unpublished observation). We believe that these conditions minimize the cellular swelling that may occur when the tissue is bathed in a hypooncotic solution, where cell volume regulation is hampered by the presence of ouabain. Further studies are currently under way to investigate this.

ACKNOWLEDGMENTS

This work was supported by NIH Grants AM 28074 to Dr. Clausen and AM 30394 to Dr. Dixon. Dr. Dixon is a Research Associate of the Veterans Administration at Northport.

REFERENCES

Bendat, J. S., and Piersol, A. G. (1971). "Random Data: Analysis and Measurement Procedures." Wiley (Interscience), New York.

Brown, E. M., Pazdes, C. J., Cruetz, C. E., Aurbach, G., and Pollard, H. B. (1978). Role of anions in parathyroid hormone release from dispersed bovine parathyroid cells. *Proc. Natl. Acad. Sci. U.S.A.* **75,** 876–880.

Clausen, C., and Fernandez, J. M. (1981). A low-cost method for rapid transfer function measurements with direct application to biological impedance analysis. *Pfluegers Arch.* **390,** 290–295.

Clausen, C., and Wills, N. K. (1981). Impedance analysis in epithelia. *In* "Ion Transport by Epithelia" (S. G. Schultz, ed.), pp. 79–92. Raven, New York.

Clausen, C., Lewis, S. A., and Diamond, J. M. (1979). Impedance analysis of a tight epithelium using a distributed resistance model. *Biophys. J.* **26,** 291–317.

Cohen, L., Mueller, A., and Steinmetz, P. R. (1978). Inhibition of the bicarbonate exit step in urinary acidification by a disulfonic stilbene. *J. Clin. Invest.* **61,** 981–986.

Cole, K. S. (1972). "Membranes, Ions, and Impulses," p. 12. Univ. of California Press, Berkeley.

Davson, H. (1964). "A Textbook of General Physiology," p. 681. Little, Brown, Boston.

Detweiler, T. C., Martin, B. M., and Feinman, R. D. (1975). Stimulus-response coupling in the thrombin-platelet interaction. *Ciba Found. Symp.* **35,** 77–90.

Gluck, S., Cannon, C., and Al-Awqati, Q. (1982). Exocytosis regulates urinary adidification in turtle bladder by rapid insertion of $H^+$ pumps into the luminal membrane. *Proc. Natl. Acad. Sci. U.S.A.* **79,** 4327–4331.

Husted, R. F., Mueller, A. L., Kessel, R. G., and Steinmetz, P. R. (1981). Surface characteristics of carbonic-anyhdrase-rich cells in turtle urinary bladder. *Kidney Int.* **19,** 491–502.

Karnieli, E., Zarnowski, P. J., Simpson, I. A., Sakins, L., and Cushman, S. (1981). Insulin-stimulated translocation of glucose transport systems in the isolated rat adipose cell. *J. Biol. Chem.* **256,** 4772–4777.

Lewis, S. A., and de Moura, J. L. C. (1982). Incorporation of cytoplasmic vesicles into apical membrane of mammalian urinary bladder bladder epithelium. *Nature (London)* **297,** 685–688.

Lewis, S. A., and Diamond, J. M. (1976). $Na^+$ transport by rabbit urinary bladder, a tight epithelium. *J. Membr. Biol.* **28,** 1–35.

Machen, T. E., and Forte, J. G. (1979). Gastric secretion. *In* "Membrane Transport in Biology" (G. Giebish, D. C. Tosteson, and H. H. Ussing, eds.), Vol. 4, pp. 693–748. Springer-Verlag, Berlin and New York.

Nagel, W., Durham, J. H., and Brodsky, W. A. (1981). Electrical characteristics of the apical and basal-lateral membranes in the turtle bladder epithelial cell layer. *Biochim. Biophys. Acta* **646,** 78–87.

Rosen, S. (1970). The turtle bladder, I: Morphological studies under varying conditions of fixation. *Exp. Mol. Pathol.* **12,** 286–296.

Rosen, S. (1972). Localization of carbonic anhydrase activity in turtle and toad urinary bladder mucosa. *J. Histochem. Cytochem.* **20,** 696–702.

Schwartz, J., Bethencourt, D., and Rosen, S. (1982). Specialized function of carbonic-anhydrase-rich and granular cells of turtle bladder. *Am. J. Physiol.* **242,** F627–F633.

Steinmetz, P. R. (1974). Cellular mechanisms of urinary acidification. *Physiol. Rev.* **54,** 980–956.

Stetson, D. L., and Steinmetz, P. R. (1982). $CO_2$ stimulation of vesicle fusion in turtle bladder epithelium. *Fed. Proc., Fed. Am. Soc. Exp. Biol.* **41,** 1263a.

Stetson, D. L., Lewis, S. A., Alles, W., and Wade, J. B. (1982). Evaluation by capacitance measurements of antidiuretic hormone induced membrane area changes in toad bladder. *Biochim. Biophys. Acta* **689,** 267–274.

Tepperman, J. (1980). "Metabolic and Endocrine Physiology," pp. 209–210 Yearbook Publ., Chicago.

# Chapter 4

# Mechanisms of Ion Transport by the Mammalian Colon Revealed by Frequency Domain Analysis Techniques

*N. K. WILLS*

*Department of Physiology*
*Yale University School of Medicine*
*New Haven, Connecticut*



Frequency domain analysis methods have become powerful tools for resolving the mechanisms of ion transport in epithelia. The purpose of this chapter is to describe new information gained from these techniques (particularly impedance analysis and current fluctuation analysis) using the rabbit descending colon. An advantage of this epithelium is that it has been characterized previously by more traditional electrophysiological methods such as microelectrode measurements and radioisotopic flux determinations.




ISBN 0-12-153320-4

# I. OVERVIEW OF THE BASIC FEATURES OF THE RABBIT DESCENDING COLON

## A. Ion Transport

The ion transport properties of the mammalian colon provide numerous issues for investigators using frequency domain techniques. Like other so-called tight epithelia, the rabbit descending colon actively transports $Na^+$ from the lumen to the serosal side of the epithelium. Under normal *in vitro* conditions, a spontaneous transepithelial potential of approximately $-20$ to $-40$ mV (lumen negative) is present and the short-circuit current ($I_{sc}$, i.e., the amount of current required to reduce the transepithelial potential to 0 mV) is largely accounted for by the rate of net $Na^+$ absorption (Frizzell *et al.*, 1976; Schultz *et al.*, 1977). Both $Na^+$ absorption and $I_{sc}$ are inhibited by the diuretic amiloride and are stimulated by the hormone aldosterone (Frizzell and Schultz, 1978).

The colon also actively absorbs $Cl^-$ by a mechanism which is apparently electroneutral. In the presence of adenosine 3′,5′-cyclic monophosphate (cAMP) or elevated intracellular $Ca^{2+}$ levels, this absorption is abolished and electrogenic net $Cl^-$ secretion occurs (Frizzell *et al.*, 1976; Frizzell and Heintze, 1979).

Results concerning potassium transport by this epithelium have been more controversial. While it is generally agreed that the mammalian colon secretes $K^+$ *in vivo* (cf. Powell, 1979), there is disagreement about the mechanism of this transport. Some investigators claimed that potassium transport in the rabbit descending colon is passive and occurs through a potassium-selective paracellular pathway (Frizzell *et al.*, 1976; Frizzell and Schultz, 1978). Recently, Fromm and Schultz (1981) proposed that this pathway becomes less potassium selective when bathing solution potassium concentrations are increased above 15 m*M*. In contrast, several other investigators reported evidence for active secretion of potassium by this epithelium (Yorio and Bentley, 1977; Moreto *et al.*, 1981; Wills and Biagi, 1982; McCabe *et al.*, 1982). Their findings indicate a transcellular route for potassium transport, in addition to possible paracellular movements of this ion. Indeed, two oppositely directed $K^+$ active transport systems may be present. Wills and Biagi (1982) and McCabe *et al.* (1982) found that net $K^+$ secretion was abolished by $10^{-4}$ *M* serosal ouabain and was replaced by active net $K^+$ absorption. A possible model for these transport systems is described later in this chapter.

## B. Structure

One aspect of the colon which complicates studies of its ion transport properties is its ultrastructure. Unlike many other epithelia such as the cornea or urinary

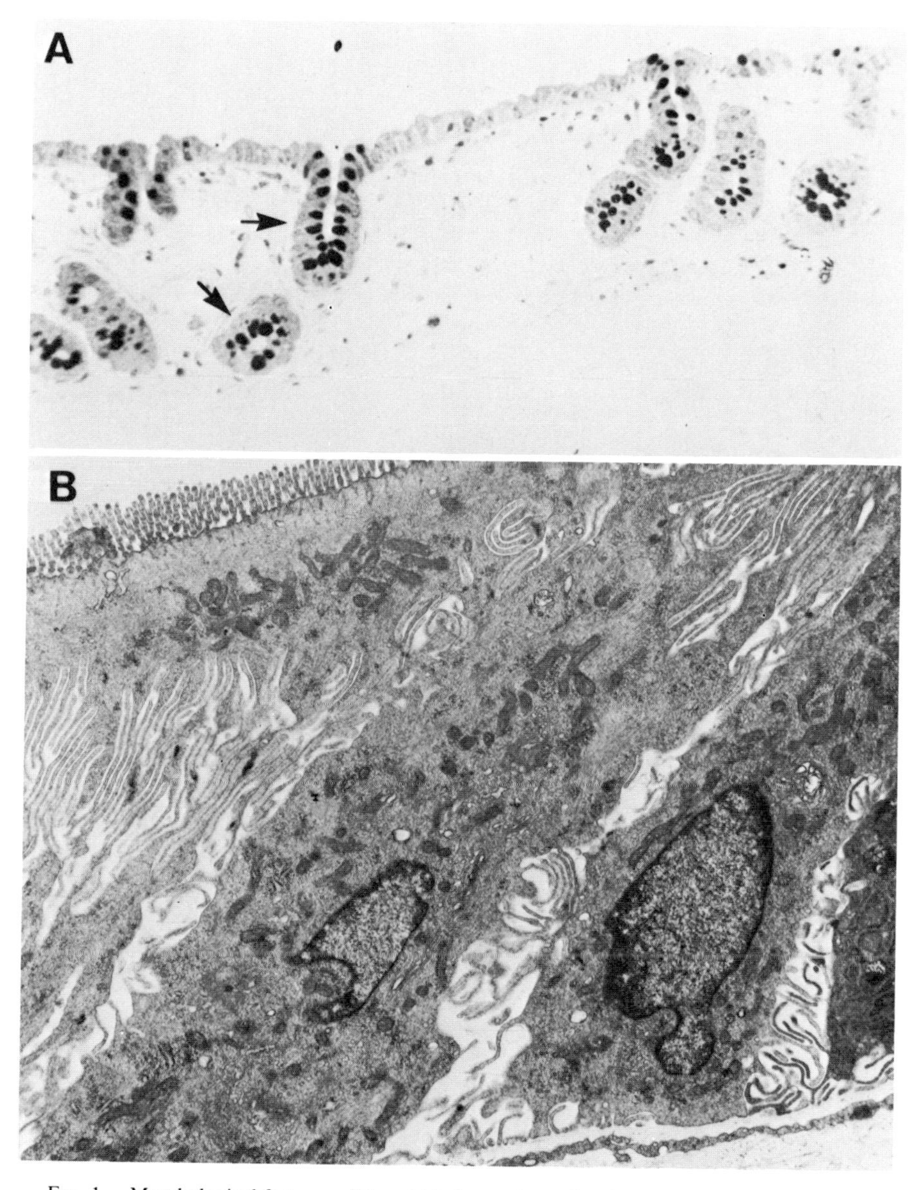


FIG. 1. Morphological features of the rabbit descending colon. The lumen is shown at the top of each section with the serosa below. (A) Light micrograph showing cross section of isolated colon epithelium. Arrows indicate crypts which open to the epithelial surface. ×200. (Courtesy of T. Ardito.) (B) Electron micrograph showing surface epithelial cells. Note amplification of membrane area by microvilli and membrane infolding. ×8600. (Courtesy of D. Biemesderfer.)

bladder, the colon is invaginated with crypts. Figure 1A is a cross-sectional view of the colon as it is studied in the following experiments. The layers of smooth muscle on the serosal side were removed using the method of Frizzell *et al.* (1976), leaving an epithelium which is a single cell layer thick, along with some underlying connective tissue. Several crypts can be seen in this light micrograph as indicated by the arrows. Eventually the crypts open to the lumen, as shown by a few examples in this figure. The presence of crypts complicates studies of the colon because (1) the crypt cells are less easily accessed for microelectrode impalements, (2) the crypt lumen has an unknown resistance and this factor introduces uncertainty into microelectrode measurements of membrane resistance ratios for crypt cells, and (3) a large series resistance in the crypt lumen could conceivably hinder adequate short-circuiting of the epithelium (Tai and Tai, 1981) and thus confound the interpretation of radioisotopic flux measurements.

Because of the technical problems associated with impaling crypt cells, microelectrode studies have generally focused on the surface epithelium. Inspection of micrographs of the colon reveals that the surface of the colon is composed predominantly of one cell type, the so-called absorptive cells. As shown in Fig. 1B, the apical surface of these cells is greatly amplified by microvilli. The lateral membranes are also extensively infolded, similarly increasing the total surface area of this membrane. Because of this infolding, it is important to have information concerning actual membrane areas before the electrophysiological and transport properties of the colon can be compared to other tight epithelia.

## C. Membrane Properties and Intracellular Ion Activities

### 1. Amiloride Blocks Approximately Half of the Apical Membrane Conductance

The apical membrane of surface cells possesses a large $Na^+$ conductance ($G^a_{Na}$) which is blockable by amiloride (Schultz, *et al.*, 1977; Wills *et al.*, 1979b; Thompson *et al.*, 1982a; Welsh *et al.*, 1982). In addition, a nearly equally large amiloride-insensitive or "leak" conductance ($G^a_{leak}$) exists in this membrane (Wills *et al.*, 1979b; Thompson *et al.*, 1982a,b). The properties of the amiloride-sensitive and amiloride-insensitive pathways, determined from equivalent-circuit analysis of the epithelium, are summarized in Table I.

Unfortunately, measurements of membrane potentials alone have not yet identified the ions responsible for the amiloride-insensitive pathway. One difficulty in making such measurements is the electrical coupling that exists between the apical and basolateral membranes by the paracellular conductance. Such coupling attenuates any change in the apical membrane potential produced by mucosal ion replacements.

TABLE I
AVERAGE ESTIMATED APICAL MEMBRANE ELECTRICAL PROPERTIES

| Reference | $R_{Na}$ (kΩ cm²) | $E_{Na}$ (mV) | $R_{leak}$ (kΩ cm²) | $E_{leak}$ (mV) |
|---|---|---|---|---|
| Wills *et al.* (1979a) | ~1.1[a] | 63 | 1.6 | −57 |
| Thompson *et al.* (1982a,b) | ~1.0[a] | 66 | 1.7 | −51 |

[a] Varied over approximately a fourfold range.

A more useful approach to this problem is to examine the electromotive force (emf) for various ions across the apical membrane. As can be seen from Table II, there is a large driving force favoring $Na^+$ entry across the apical membrane under short-circuit conditions. This gradient would be expected to generate a $Na^+$ current across the apical membrane of about 100 μA. In contrast, the driving forces for $Cl^-$ and $K^+$ are small and favor exit of these ions from the cell.

It is important to note that the equilibrium potential for $Cl^-$ is only −38 mV (Wills, 1982). This value is too small to account for the electromotive force of the amiloride-insensitive conductance given in Table I (about −50 to −60 mV). The equilibrium potential for potassium (−69 mV, Wills and Biagi, 1982), however, is large enough to account for this amiloride-insensitive emf in the apical membrane. An estimate of the maximum $Cl^-$ conductance ($g_{Cl}$) of this membrane can be obtained by ignoring the possibility of an amiloride-insensitive $Na^+$ conductance. This procedure is likely to overestimate the $Cl^-$ conductance; however, it provides an upper limit for $g_{Cl}$. Using this approach it is possible to calculate a $Cl^-$ to $K^-$ permeability ratio by using the constant field equations (Goldman, 1943; Hodgkin and Katz, 1949), the estimated emf of the amiloride-insensitive pathway, and the intracellular and extracellular $Cl^-$ and $K^+$ activities. Calculated in this manner, $P_{Cl}/P_K$ is roughly between 0.3 and 0.8. Consequently, $K^+$ may be the major ion responsible for the amiloride-insensitive conductance in the apical membrane. The existence of this conductance is

TABLE II
INTRACELLULAR ION ACTIVITIES AND NET IONIC DRIVING FORCES FOR THE SHORT-CIRCUITED COLON

| | $Na^+$ | $K^+$ | $Cl^-$ |
|---|---|---|---|
| $a_i$, intracellular activity (m*M*) | 12 | 73 | 23 |
| $E_i$, equilibrium potential (mV) | 58 | −69 | −38 |
| $\Delta\mu/F$, net driving force (mV) | −106 | 17[a] | −7[a] |

[a] Exit from cell favored.

important because it provides a possible exit mechanism for transcellular $K^+$ mmmm53

## 2. Basolateral Membrane: $K^+$ Conductance and the Na–K Pump

In contrast to the apical membrane conductance, the basolateral membrane conductance is predominantly due to potassium with small contributions from $Na^+$ and $Cl^-$ (Wills *et al.*, 1979a,b; Thompson *et al.*, 1982b). The potassium conductance of this membrane is decreased by serosal addition of $Ba^{2+}$ (Wills, 1981b) and is blocked by $Cs^+$ in a potential-dependent manner (Wills *et al.*, 1979b).

Intracellular potassium levels are largely maintained by active uptake of this ion across the basolateral membrane by the $Na^+,K^+$-ATPase. Intracellular potassium activity ($a_i^K$) is greatly reduced after serosal addition of ouabain (Wills and Biagi, 1982). Recent evidence suggests that more than one active transport system may contribute to the intracellular potassium activity. As shown in Table III, when $Ba^{2+}$ was added to the serosal solution in ouabain-treated tissues (in order to slow the exit of potassium across the basolateral membrane), $a_i^K$ was increased nearly threefold. Similar results were obtained in the absence of $K^+$ in the serosal bathing solution. Such findings strongly suggest the existence of another active $K^+$ transport mechanism, possibly situated in the apical membrane.

These results complement recent findings of radioisotopic flux measurements in the colon (McCabe *et al.*, 1982; Wills and Biagi, 1982). Thus potassium transport across the colon may be mediated by two oppositely directed active transport systems: (1) a secretory system which includes uptake across the basolateral membrane via the $Na^+,K^+$-ATPase and a passive conductive exit step across the apical membrane and (2) an absorptive system which involves active uptake across the apical membrane and passive exit across the basolateral membrane (see Fig. 2).

The following studies were designed to address several aspects of this model using frequency domain analysis methods. The first set of studies demonstrates how impedance analysis can be used to obtain a better understanding of the conductance properties of the apical membrane. Current fluctuation analysis is

TABLE III

Membrane Potentials and Intracellular $K^+$ Activity after Addition of Ouabain to Serosal Solution

| | $V_a$ (mV) | $V_{bl}$ (mV) | $a_i^K$ (m*M*) | $E_K$ (mV) |
|---|---|---|---|---|
| Control ($10^{-4}$ *M* ouabain serosa) | $-7 \pm 1.5$ | $-12 \pm 1.8$ | $12 \pm 1.9$ | $-21 \pm 4.9$ |
| $BaCl_2$ (5 m*M*, + $10^{-4}$ *M* ouabain serosa) | $-8 \pm 1.8$ | $-6 \pm 1.6$ | $28 \pm 3.1$ | $-44 \pm 2.5$ |

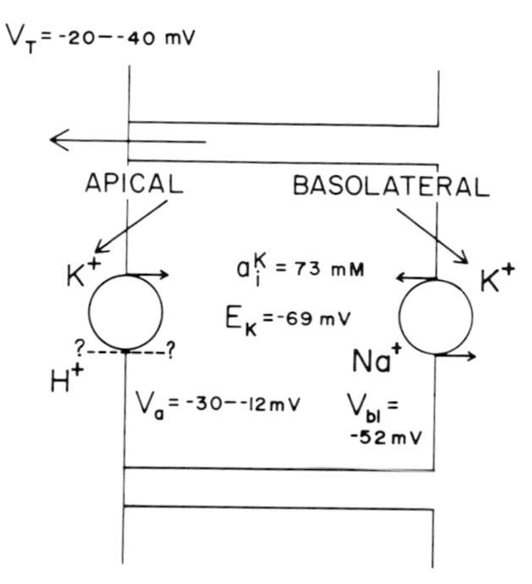


FIG. 2. Model for $K^+$ transport by the mammalian colon under open circuit conditions. Both cell membranes are proposed to have active uptake mechanisms for $K^+$. Uptake across the basolateral membrane is via $Na^+,K^+$-ATPase, whereas uptake across the apical membrane is by a ouabain-insensitive, electrically silent mechanism. Both membranes are conductive to potassium (sodium and chloride conductances are not shown). Under these conditions, the paracellular pathway and apical membrane both contribute to the net $K^+$ secretion. However, under short-circuit conditions, net $K^+$ absorption or secretion can occur, since the direction of net transport is determined by the relative magnitude of $K^+$ exit across the apical and basolateral membranes.

then used to characterize the kinetic properties of the apical membrane amiloride-sensitive $Na^+$ conductance. These methods are also employed to determine whether potassium channels might exist in this membrane, similar to those found in the apical membrane of some other epithelia. Last, a possible method for investigating basolateral membrane channels is proposed.

## II. IMPEDANCE ANALYSIS

Although impedance analysis can be a highly useful technique for assessing the conductance properties of membranes, the true advantage of this method in studies of epithelial transport is that it also provides morphological information such as *in situ* estimates of membrane areas and access series resistances arising from narrow membrane-lined spaces such as the lateral intercellular space (LIS) or crypts. Estimates of these parameters are not readily available from other techniques. Thus a major goal of impedance analysis is the development of an accurate equivalent-circuit model which incorporates the morphological features of the epithelium (see also Clausen and Dixon, Chapter 3, this volume).

As a first step in choosing an appropriate equivalent-circuit model for the colon, two morphological features were taken into consideration: (1) the lateral intercellular space and (2) the crypt lumen. The dimensions of these parameters are important because a narrow or constricted intercellular space or crypt lumen would be expected to have a relatively high electrical resistance. A high resistance in the lateral intercellular spaces can cause the basolateral resistance to be distributed along its lateral borders, as described previously for the rabbit urinary bladder (Clausen *et al.*, 1979) and *Necturus* proximal tubule (Boulpaep and Sackin, 1980). Similarly, in the gastric mucosa the resistance of the crypt lumen has been shown to cause significant distributed effects (Clausen *et al.*, 1983), i.e., if the crypt access resistance is comparable to the apical membrane impedance, then the resistance of the apical membrane will be distributed along the crypt (see Fig. 3A). Consequently, both the LIS and the crypt lumen can conceivably produce measurable distributed effects on membrane resistances. These factors have methodological importance, since membrane areas can be significantly underestimated if they are not taken into account.

Initial attempts to fit the impedance data for the colon by a model which ignores distributed resistance effects (the so-called "lumped model," Fig. 3B) gave reasonable results. However, despite a generally good fit to the data, systematic misfits were obtained over the upper half of the frequency range. A second attempt at fitting the data by a model which included distributed effects of the basolateral membrane along the LIS produced no appreciable improvement in

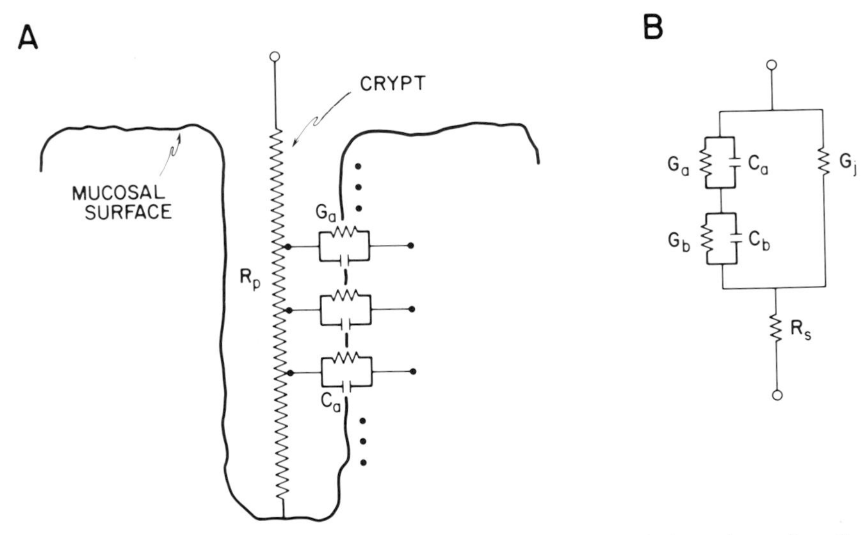


Fig. 3. Equivalent-circuit models (A) Distributed resistance of the apical membrane along the colonic crypts. $R_p$ is the access resistance along the crypts and is dependent on mucosal solution resistivity and the geometry of the crypts. (B) Lumped equivalent-circuit model. Circuit parameters $G_a$ and $C_a$, and $G_b$ and $C_b$ represent the conductances and capacitances of the apical and basolateral membranes, respectively. $G_j$ is the paracellular (or shunt) conductance. A small series resistance $R_s$ is necessary to account for the solution resistance and the unstirred layers on either side of the epithelium.

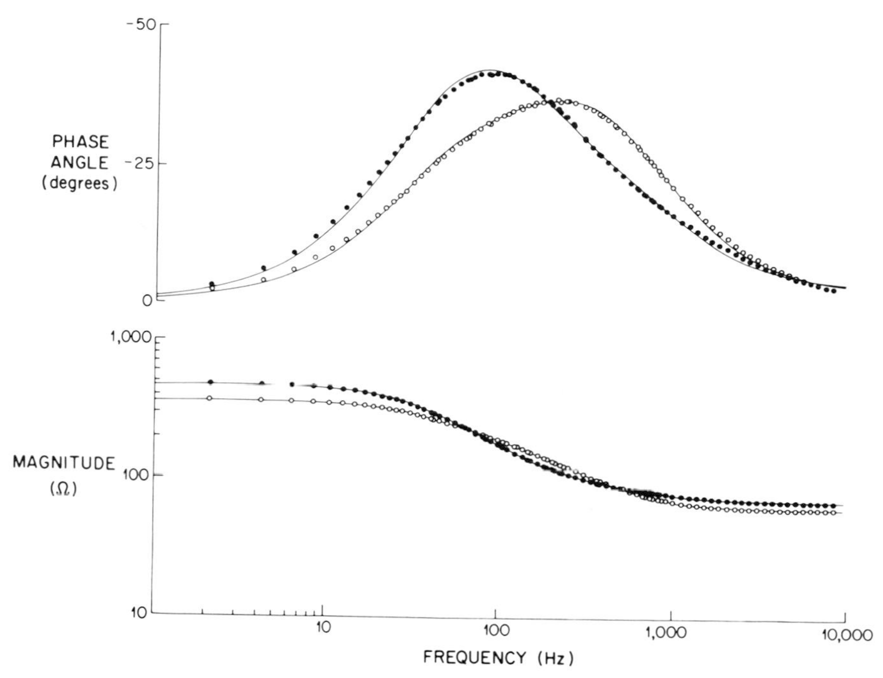


FIG. 4. Impedance measurements from the colon showing the results of fitting the distributed model. Open circles indicate normal NaCl–$HCO_3$ solutions on both sides. Filled circles show results for the same tissue after replacement of the mucosal solution by a potassium sulfate solution containing no NaCl. Reprinted from Clausen and Wills (1981). *In* "Ion Transport by Epithelia; Recent Advances" (S. G. Schultz, ed.), with permission of Raven Press, New York.

the fit. This was not surprising since the lateral spaces for the colon appear relatively wide and thus would be expected to have a lower resistance than the rabbit urinary bladder LIS (Clausen and Wills, 1981). In contrast, use of a model which includes the distributed effects caused by crypt lumen resistance ($R_p$) greatly reduced the misfit of the theory to the data (see Fig. 4). Although this procedure did not significantly alter the estimates of membrane resistances and capacitances, a major advantage of this analysis was that it provided an estimate of crypt lumen resistance ($17 \pm 0.7$ Ω $cm^2$ for six colons). This resistance is in good agreement with values calculated from the dimensions of the crypt lumen estimated from micrographs of fixed, sectioned tissues (Clausen and Wills, 1981).

## A. Is the Colon Adequately Short-Circuited?

The significance of the crypt luminal resistance estimate is that it allows us to calculate whether the epithelium is actually short-circuited during tracer flux experiments. Using the short-circuit currents and transepithelial resistances reported by Wills and Biagi (1982; $I_{sc} = -77.2\ \mu A/cm^2$ and $R_t = 294\ \Omega\ cm^2$), we see that a lack of compensation for crypt resistance ($R_p = 17\ \Omega\ cm^2$) would cause an inadequate short-circuiting of the colon such that a potential of $-1.3$ mV remains across the epithelium (i.e., $V_T = R_p \cdot I_{sc}$). How could this potential affect transcellular potassium fluxes? If we assume that potassium transport is strictly paracellular (Fromm and Schultz, 1981), then the upper limit for passive net $K^+$ flux can be predicted by assuming that all of the parcellular conductance (approximately 770 $\Omega\ cm^2$, Clausen and Wills, 1981) is due to potassium. This is an upper limit because some of the paracellular conductance is due to other ions (Frizzell *et al.*, 1976) and we are assuming that the entire pathway "sees" a $-1.3$-mV driving force, when really this potential should only be present in the crypts. Consequently the actual fluxes may be much less. This calculation indicates that the *maximum* net flux would be only 0.06 $\mu Eq/cm^2 \cdot$ hour or about 2 $\mu A/cm^2$. Thus paracellular $K^+$ movements would be too small to account for the net flux.

One limitation in this calculation is that it ignores the possibility of a passive transcellular $K^+$ movement. Wills *et al.* (1979a) estimated the resistance of the amiloride-insensitive cellular pathway as roughly 1.8 k$\Omega$. If one assumes that potassium can cross the epithelium by *both* crypt cellular and paracellular pathways, the maximum estimated net flux expected from inadequately short-circuiting the epithelium is increased to 0.09 $\mu Eq/cm^2 \cdot$ hour (2.5 $\mu A/cm^2$). The net secretion reported by Wills and Biagi (1982) was over four times this amount, 0.39 $\mu Eq/cm^2 \cdot$ hour. Again these calculations indicate that inadequate short-circuiting of the epithelium (because of crypt access resistance) cannot explain net $K^+$ secretion. Therefore $K^+$ secretion must be mediated by some other, apparently active, transport system.

## B. Estimates of Membrane Parameters Require an Independent Measurement of $R_s$ or $R_a/R_{bl}$

From transepithelial impedance measurements alone, it is impossible to estimate directly membrane resistances and capacitances ($C_m$, where 1 $\mu F \cong 1\ cm^2$ membrane area) without some additional independent measurement such as paracellular resistance ($R_s$) or the resistance ratio ($R_a/R_{bl}$). For this reason, estimates of $R_s$ were obtained using the method of Wills *et al.* (1979b). Briefly, this method consists of using the polyene antibiotic nystatin to selectively increase the apical membrane conductance ($G_a$). The mucosal bathing solution was first replaced by a $K_2SO_4$–Ringer's solution with a $K^+$ activity approximately equal to that measured intracellularly. When nystatin (200 units/ml) was added to the mucosal bath, the transepithelial conductance ($G_t$) and $I_{sc}$ increased linearly as follows:

$$G_t = I_{sc}/E_{bl} + G_s \tag{1}$$

where $E_{bl}$ is the emf of the basolateral membrane and $G_s$ is $1/R_s$. Thus $R_s$ can be conveniently estimated from the inverse intercept of this function (see also Wills, 1981a).

By incorporating this estimate of $R_s$, one can convert the circuit parameters estimated using impedance analysis (and the distributed apical membrane resistance model) into the appropriate apical and basolateral membrane resistances and capacitances. Table IV presents a summary of the results from 10 colons. The apical membrane resistance measured in the presence of normal NaCl-containing mucosal solutions was 460 ± 130 Ω cm² and the basolateral resistance was 110 ± 14 Ω cm². Capacitances for the two membranes were 19 ± 2.4 and 8.0 ± 0.8 μF/cm², respectively.

## C. An Apical Membrane Potassium Conductance

Replacement of NaCl by $K_2SO_4$ in the mucosal bath caused an increase in $R_a$ to 940 ± 250 Ω cm² ($n$ = 5) and a decrease in $R_{bl}$ to 44 ± 3.3 Ω cm² ($n$ = 5). The decrease in $R_{bl}$ appears to be due to a change in membrane area. $C_{bl}$ was increased nearly twofold to 15 ± 1.8 μF/cm², while $C_a$ was not significantly altered. The small increase in $R_a$ indicates that an appreciable conductance remains in this membrane in the absence of mucosal $Na^+$ or $Cl^-$. A similar conclusion was reached from microelectrode measurements of $R_a/R_{bl}$ in surface cells under these conditions (Wills *et al.*, 1979a). Since intracellular chloride activity is less than 2 m*M* when $Cl^-$ is removed from the mucosal bathing solution (Wills, 1982), this conductance is probably due to potassium. The decrease in $R_{bl}$ can be explained by the increase in basolateral membrane area (i.e., capacitance) without postulating any change in specific ionic permeability. The increase in $C_{bl}$ is not fully understood but could indicate cell swelling under these conditions.

Clausen and Wills (1981) used this latter estimate of the apical membrane resistance (for the condition of $K_2SO_4$–mucosal solutions) and the constant field

TABLE IV
MEMBRANE PROPERTIES OF THE COLON ESTIMATED FROM IMPEDENCE ANALYSIS

| | $R_a$ (Ω cm²) | $C_a$ (μF/cm²) | $R_{bl}$ (Ω cm²) | $C_{bl}$ (μF/cm²) |
|---|---|---|---|---|
| NaCl mucosal and serosal solution ($n$=7) | 460 ± 130 | 19 ± 2.4 | 110 ± 14 | 8.0 ± 0.8 |
| $K_2SO_4$ mucosal solution; NaCl serosal solution ($n$=6) | 940 ± 250 | 15 ± 2.6 | 44 ± 3.3 | 15 ± 1.8 |

conductance equation (Goldman, 1943; Hodgkin and Katz, 1949) to estimate the potassium resistance in normal (7 m*M*) $K^+$ solutions. They found the potassium resistance of the apical membrane to be 3.7 kΩ $cm^2$. This value is larger than the value predicted from net transepithelial $K^+$ fluxes and apical membrane net $K^+$ driving force (1.6 kΩ $cm^2$, Wills and Biagi, 1982). It is not known whether the remaining portion of the net potassium secretion is mediated by an electroneutral mechanism. Alternatively, it is possible that the potassium permeability of the apical membrane may have been reduced by some aspect of these experiments and that the "normal" $K^+$ conductance of the membrane may be higher. For example, Wills *et al.* (1979a) observed that the potassium permeability of the basolateral membrane was reduced when $Cl^-$ was replaced by sulfate. Nonetheless, these results indicate that at least part of the observed potassium secretion may involve a passive exit step via a potassium conductance in the apical membrane.

The above finding combined with microelectrode measurements and dc equivalent-circuit analysis of the colon provide converging evidence for an amiloride-sensitive $Na^+$ conductance and an amiloride-insensitive $K^+$ conductance in the apical membranes of the colon. Can these conductances be further characterized by fluctuation analysis?

## III. CURRENT FLUCTUATION ANALYSIS

As discussed elsewhere (see Hoshiko, Chapter 1, Lewis and Alles, Chapter 5, and Palmer, Chapter 6, this volume), considerable evidence has accumulated which validates the use of current fluctuation analysis techniques for studying conductive channels in epithelia. For example, Lindemann and Van Driessche (1977) demonstrated that the $Na^+$ conductance of the apical membrane in frog skin is due to amiloride-blockable $Na^+$ channels. In addition, spontaneously fluctuating $K^+$ channels have been reported for the apical membranes of *Rana temporaria* (Van Driessche and Zeiske, 1980) and toad and *Necturus* gallbladder (Van Driessche and Gögelein, 1978; Gögelein and Van Driessche, 1981a,b). Therefore the following questions arise: (1) Do such channels exist in mammalian epithelia? (2) If so, are they similar to those observed in amphibian tissues?

### A. Amiloride-Induced Current Noise in Rabbit Colon

The diuretic amiloride has been widely used as a specific probe for investigating the properties of $Na^+$ channels in tight epithelia (Cuthbert *et al.*, 1979). Micromolar amounts of this drug have been shown to block $Na^+$ channels in the frog skin (Lindemann and Van Driessche, 1977) and toad bladder (Li *et al.*,

1982; Palmer *et al.*, 1982). As discussed by Hoshiko, Lewis and Alles, and Palmer (Chapters 1, 5, and 6, this volume), the kinetics of amiloride interaction with the channel can be described as a pseudo-first-order reaction. This characteristic allows the calculation of rate constants for amiloride association and disassociation with the $Na^+$ channel, single-channel currents and conductances, and channel density from measurements of the microscopic current fluctuations that are induced by amiloride.

Zeiske *et al.* (1982) examined the kinetics of amiloride action on the rabbit colon and found that the macroscopic short-circuit current (i.e., $I_{sc}$) decreased as a hyperbolic function of amiloride concentration. This result is expected for a reaction having first-order kinetics. Half-maximal inhibition ($K_A$) was obtained at 0.2 μ*M* amiloride. Examination of the short-circuit current with fluctuation analysis techniques revealed a single time constant relaxation noise (or a so-called Lorentzian-type component) in the power spectral density (see Fig. 2). With increasing amiloride concentrations, the plateau value of this component decreased and its corner frequency increased. As can also be seen in Fig. 5, a low-frequency component ($1/f^\alpha$) was also present. The source of this signal has

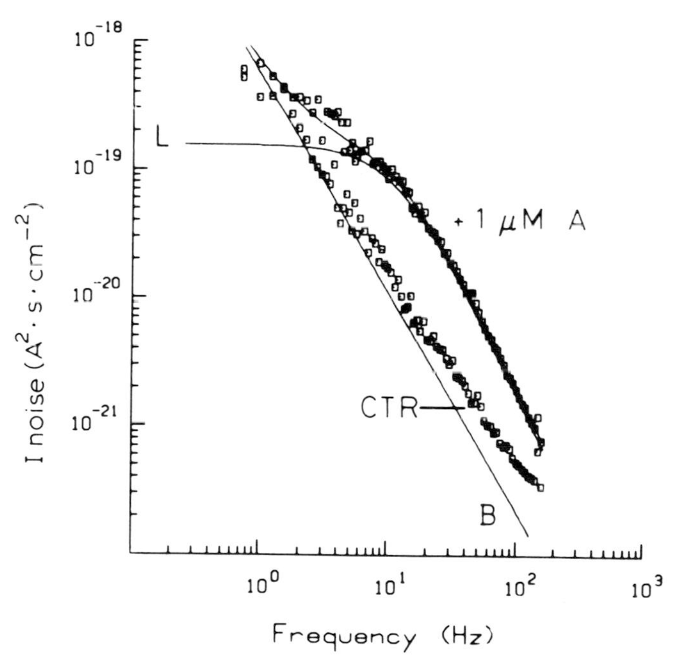


FIG. 5. Power spectral density of the short-circuit current in the absence (CTR) and presence of 1 μ*M* of amiloride (A). The latter spectrum was fitted by a function which was the sum of a linear, low-frequency background component (B) and a Lorentzian function (L). Reprinted from Zeiske *et al.* (1982) with permission from Elsevier Press, Amsterdam.

not yet been identified (see the discussion by Hoshiko, Chapter 1, this volume). In the analyses reported below, the data were simultaneously fitted by an equation which included both components:

$$S(f) = K_b/f^\alpha + S_o/[1 + (f/f_c)^2] \tag{2}$$

where $K_b$ is the spectral density at 1 Hz, $\alpha$ is its slope in a double logarithmic scale, $S_o$ is the low-frequency plateau value, and $f_c$ is the frequency at which $S(f) = S_o/2$.

To our knowledge, these data represented the first measurements of amiloride-induced noise in a mammalian epithelium. Like the action of amiloride on amphibian epithelia, the results in colon were consistent with a two-state mechanism of blocker interaction with the $Na^+$ channel of the form

$$\underset{\text{open}}{A + R} \underset{\alpha_{10} = K_{10}}{\overset{\alpha_{01} = K'_{01}[A]}{\rightleftharpoons}} \underset{\text{closed}}{A \cdot R}$$

Rate constants for the reaction were calculated from six pooled experiments using a linear regression analysis with the equation

$$S_o = \frac{K'_{01}}{\pi^2} \cdot \frac{I_{Na} \cdot [A]}{f_c^2} \cdot i \tag{3}$$

where $K_{10}/K'_{01}$ ($\equiv K'_A$) is the apparent Michaelis constant and $K'_{01}$ [A] + $K_{10}$ = $2\pi f_c$. From the slope of the relationship given in Eq. (3), $K'_{01}$ was estimated as 68.4 ± 5.5 $\mu M^{-1}$ sec$^{-1}$ at 37°C, whereas the dissociation rate constant $K_{10}$ was 7.42 ± 6.08 sec$^{-1}$. The value for $K'_A$ was 0.109, comparable to the value for half-maximal inhibition of the macroscopic current by amiloride.

## A Tentative Estimate of Single-Channel Properties

The single-channel density ($M$) was estimated using the relationship

$$M = \frac{2\pi}{K_{10}} \cdot \frac{I_{Na} \cdot f_c}{i} \tag{4}$$

while the single-channel current ($i$) was calculated from

$$i = \frac{\pi^2 f_c^2 S_o}{K'_{01} \cdot [A] \cdot I_{Na}} \tag{5}$$

where $S_o$ and the amiloride-sensitive macroscropic current ($I_{Na}$) are both normalized to 1 cm$^2$ of tissue area. $M$ and $i$ were 6.1 ± 0.8 $\mu$m$^{-2}$ and 0.4 ± 0.6 pA, respectively (see Table V). In considering these tentative estimates, it is important to note the uncertainty injected into these calculations by the variability in $K_{10}$. While it has not been possible to normalize these values for

TABLE V
ESTIMATED $Na^+$-CHANNEL CHARACTERISTICS FROM CURRENT FLUCTUATION ANALYSIS

| | 37°C | 27°C |
|---|---|---|
| $K'_{01}$ ($sec^{-1}$ $\mu M^{-1}$) | 68.4 | 20.5 |
| $K_{10}$ ($sec^{-1}$) | 7.5 | 19.0 |
| $i$ (pA) | 0.4 | 0.1 |
| $M$ ($\mu m^{-2}$) | 6.1 | 3.2 |
| $K_A$ ($\mu M$) | 0.1 | 0.9 |

membrane area because of the unknown area of the amiloride-sensitive cells, a minimal channel density estimate of 0.3 $\mu m^{-2}$ can be obtained by normalizing to the total apical membrane capacitance (19 $\mu F/cm^2$). Single-channel currents will also be influenced by the intracellular potential and sodium activity levels in the short-circuited state.

The above experiments were conducted at 37°C; however, similar experiments were performed at room temperature (27°C) to permit comparison of these results to those obtained previously from amphibian epithelia. These data are also summarized in Table V. The apparent association rate constant $K'_{01}$ at 27°C was approximately three times lower than at 37°C, while $K_{10}$ was not significantly changed. These rate constants are quite similar to those reported for the frog skin and are an order of magnitude larger than those for the toad urinary bladder. Estimates of $M$ and $i$ also agree within an order of magnitude for the three epithelia. Because of the variability in $K_{10}$ and the lack of information concerning membrane potentials in the colon at 27°C, these parameters require further evaluation. Nonetheless, the microscopic properties of the apical membrane $Na^+$ channels of the colon generally appear to be quite similar to those of amphibian tight epithelia.

## B. Analysis of the Apical Membrane Amiloride-Insensitive Conductance

The finding of an amiloride-insensitive conductance in the apical membrane in parallel to the amiloride-sensitive $Na^+$ channels (Wills *et al.*, 1979b; Clausen and Wills, 1981; Thompson *et al.*, 1982a,b) raised the possibility of an apical membrane potassium conductance. Therefore, Wills *et al.* (1982) determined whether $K^+$ channels could be detected in this membrane and whether these channels resembled those found in the apical membranes of amphibian epithelia. The major findings of these experiments were as follows.

Under short-circuited conditions and in the presence of a maximal dose of amiloride, the power spectral density of the short-circuit current showed no

Lorentzian noise signals. Under these conditions, a "linear" $1/f^{\alpha}$ component was observed in the spectrum. (The slope of this component tended to flatten in the high-frequency range because of amplifier noise, Fishman *et al.*, 1975.) When a potassium gradient was imposed across the colon by replacing the mucosal $Na^+$ with $K^+$ (i.e., 140 m*M* $K^+$), a "shoulder" emerged in the spectrum in the middle frequency range, indicating the presence of a Lorentzian component (i.e., single time constant relaxation noise). The results were analyzed by simultaneously fitting linear ($1/f^{\alpha}$) and nonlinear (i.e., Lorentzian) functions to Eq. (2) as discussed above. An example of such a fit is given in Fig. 6.

The appearance of the Lorentzian component was correlated with the estimated net driving force for $K^+$ across the apical membrane. To illustrate the effects of such $K^+$ transepithelial gradients I have provided a summary of apical membrane $K^+$ driving forces in Fig. 7. These estimates assume as a first approximation that intracellular $K^+$ activity is not significantly altered by increases in

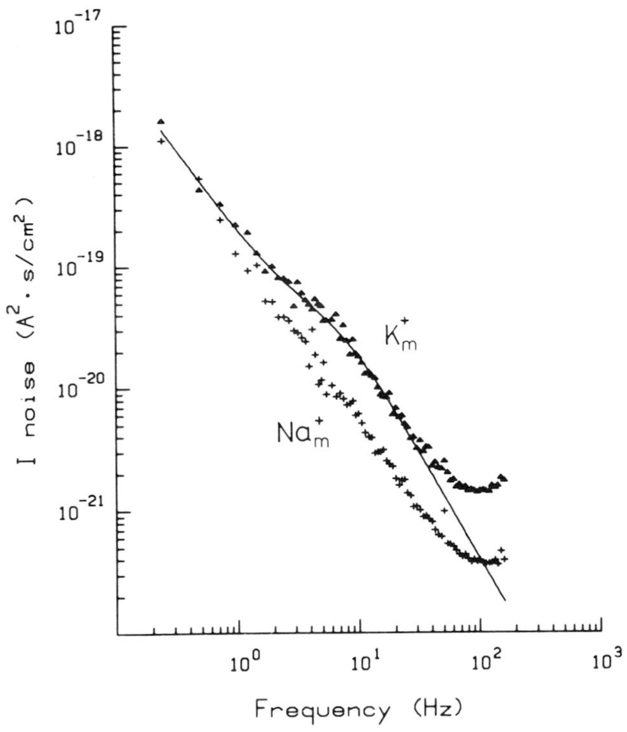


FIG. 6. Power spectral density of the short-circuit current in the absence ($Na_m^+$) and presence ($K_m^+$) of a mucosa-to-serosa potassium gradient across the epithelium. In the upper spectrum, mucosal $Na^+$ was replaced by $K^+$. The data for this spectrum were fitted as in Fig. 5. Reprinted from Wills *et al.* (1982) with permission of Springer-Verlag, New York.

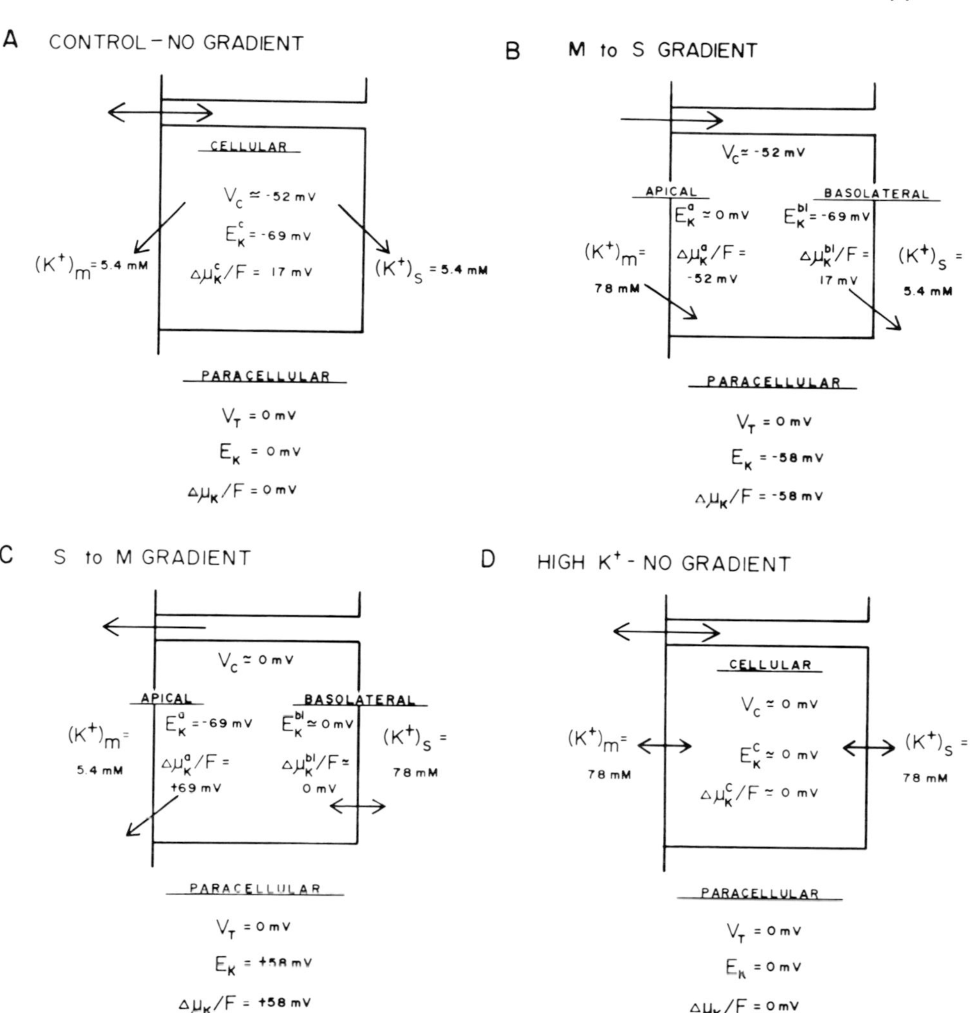


FIG. 7. Estimated net potassium driving forces across the apical and basolateral membranes and the paracellular pathway. The transepithelial potential is voltage clamped to 0 mV under all four conditions; thus no electrical driving force is present across the paracellular pathway. (A) Control conditions (no gradient): normal $Na^+$-containing Ringer's bathing solutions on both sides. (B) Mucosa to serosa (M to S) $K^+$ gradient: replacement of the mucosal $Na^+$ by $K^+$ (140 m*M*). (C) Serosa to mucosa (S to M) $K^+$ gradient: replacement of the serosal $Na^+$ by $K^+$ (140 m*M*). (D) High-potassium solutions on both sides (high $K^+$, no gradient). For further description, see text.

bathing solution $K^+$. The calculations incorporate membrane resistances and intracellular potentials (Schultz *et al.*, 1977) and $K^+$ activity measurements (Wills and Biagi, 1982; Wills, unpublished observations) for the short-circuited condition.

In the absence of transepithelial chemical or electrical driving forces for potassium (Fig. 7A), there is a small net driving force favoring $K^+$ exit from the cell. Replacing the mucosal solution with a high-$K^+$ solution (i.e., M to S gradient, Fig. 7B) would establish a net driving force favoring potassium entry approximately equal to the electrical driving force provided by the apical membrane potential (i.e., the intracellular potential under short-circuit conditions). Replacement of the serosal solution with such a high-$K^+$ solution (i.e., S to M gradient, Fig. 7C) essentially abolishes the basolateral membrane potential (Wills *et al.*, 1979a,b). Thus under these conditions, $K^+$ exit across the apical membrane is favored, in response to the chemical gradient for $K^+$ across this membrane. With high-$K^+$ solutions on both sides of the epithelium (Fig. 7D), no significant gradients are present across the apical membrane.

As shown in Fig. 8, in the absence of a transepithelial $K^+$ gradient a Lorentzian component cannot be observed in the power spectrum. Restoration of a

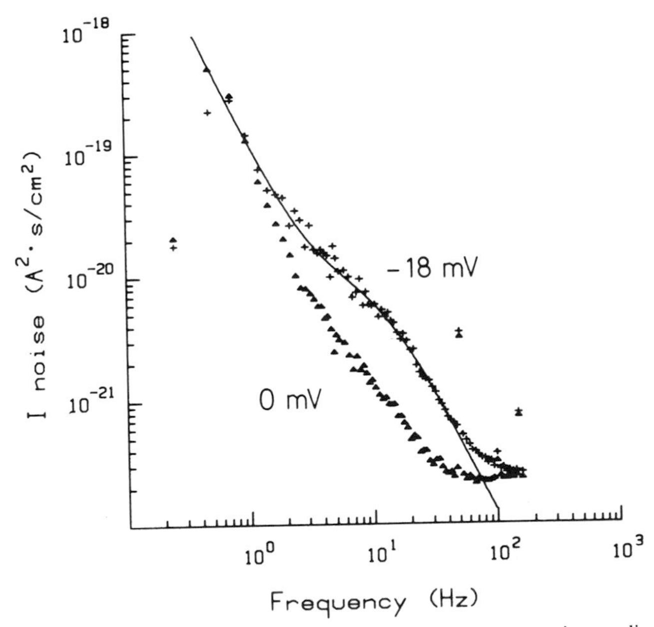


FIG. 8. Power spectra of current noise in the absence of $K^+$ concentration gradients across the colon (see Fig. 7D). At 0-mV clamp potential, only background noise is seen. In contrast, after application of a small electrical potential across the epithelium a Lorentzian component is revealed. Reprinted from Wills *et al.* (1982) with permission of Springer-Verlag, New York.

potassium gradient across the epithelium by replacing the mucosal solution with a normal (7 m*M*) NaCl–Ringer's solution (S to M gradient, Fig. 7B) re-established the Lorentzian component. Also shown in Fig. 8 is the effect of transepithelial potential on the spectrum. The Lorentzian component reemerged when an electrical driving force was applied across the epithelium. Thus the presence of the Lorentzian component is dependent on the existence of a net driving force for either entry or exit of $K^+$ across the apical membrane.

### 1. Location of the Signal Source

One hindrance for studies of potassium transport in the rabbit colon has been the difficulty of resolving cellular and paracellular routes for $K^+$ movement. As can be appreciated from Fig. 7, the net driving forces across the paracellular pathway produced by the imposition of $K^+$ or electrical gradients are similar to those across the apical membrane under short-circuit conditions. For this reason, it was important to localize the source of the Lorentzian noise.

Our first approach to this problem was to examine the effects of serosal $Ba^{2+}$ on the Lorentzian component. In these studies and those which follow, a mucosal

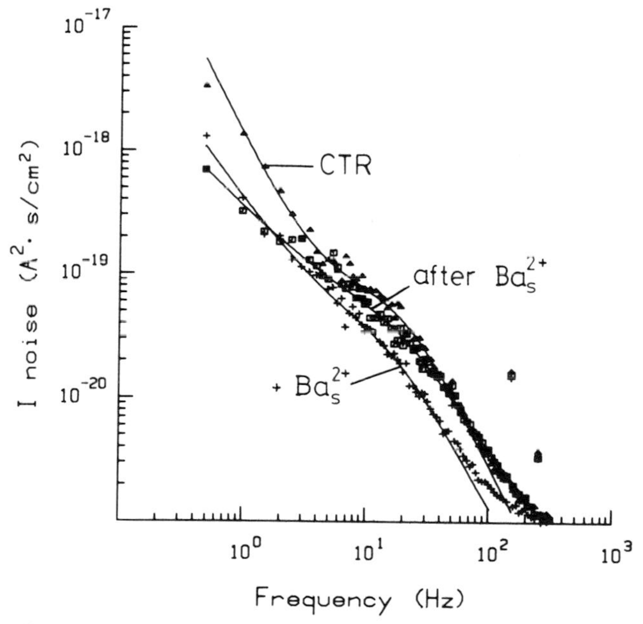


Fig. 9. Application of 5 m*M* serosal $Ba^{2+}$ suppressed the Lorentzian component. The spectra shown are control (CTR, M to S gradient), during $Ba^{2+}$ ($+Ba_s^{2+}$), and postcontrol (after $Ba_s^{2+}$). The effect was partially reversible. For further description, see text. Reprinted from Wills *et al.* (1982) with permission of Springer-Verlag, New York.

to serosal $K^+$ gradient was employed (Fig. 7B). Serosal $Ba^{2+}$ (5 m*M*) is known to decrease the basolateral potential by nearly 75% (Wills, 1981b). Thus under short-circuit conditions, the intracellular potential should be decreased and the net driving force for $K^+$ entry will be reduced. As shown in Fig. 9, serosal $Ba^{2+}$ suppressed the Lorentzian component. This effect is expected for a potassium current transversing the apical membrane. With the reasonable assumption that $Ba^{2+}$ at this concentration does not have a one-sided blocking action on the paracellular pathway for $K^+$, this finding suggests a cellular origin for the Lorentzian noise source.

In order to further localize the source of the Lorentzian noise signal, the effects of nystatin on the Lorentzian component were examined. If the Lorentzian noise signal arises from the basolateral membrane, then it should be possible to enhance this signal by reducing the apical membrane resistance with nystatin. This procedure should reduce the attenuation of the signal caused by the series arrangement of the apical and basolateral membranes. Alternatively, if the signal is produced by fluctuations at the apical membrane, the signal should diminish when nystatin is added to this membrane, since nystatin will reduce both the resistance and the electrical driving force across this membrane. As shown in Fig. 10, addition of 40 units/ml nystatin to the mucosal bathing solution virtually eliminated the Lorentzian component. Serosal addition of the drug had no mea-

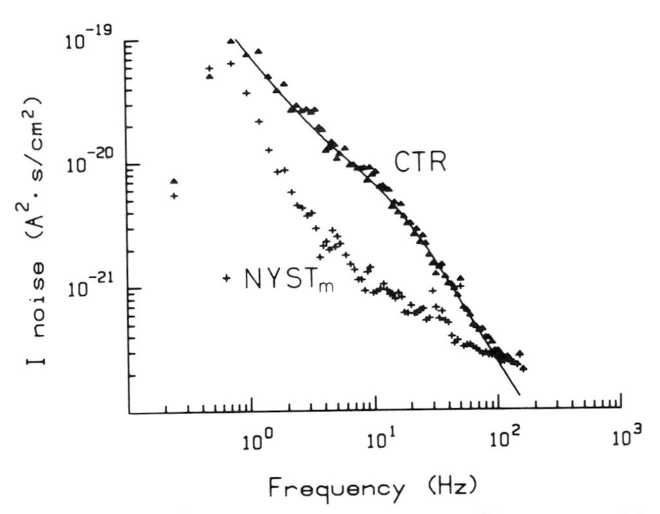


FIG. 10. Current noise spectra before and after nystatin addition to mucosal bath. Mucosal addition of nystatin ($NYST_m$, 40 units/ml) abolished the Lorentzian signal (CTR, M to S gradient). For further description see text. Reprinted from Wills *et al.* (1982) with permission of Springer-Verlag, New York.

surable effect on the signal. Consequently, the source of the Lorentzian noise component is apparently situated in the apical membrane.

2. Effects of $K^+$-Channel Blockers

So far, the characteristics of this apical membrane noise signal are similar to those from spontaneously fluctuating potassium channels in the apical membranes of amphibian epithelia. The signal is suppressed by mucosal addition of $K^+$-channel blockers such as $Cs^+$ and tetraethylammonium (TEA) ions, although neither agent eliminated the signal completely. Unlike $Ba^{2+}$, serosal addition of these agents had no significant effects. Mucosal $Ba^{2+}$ did not have a significant blocking effect, possibly suggesting that apical membrane channels are pharmacologically distinct from the basolateral $K^+$ conductance. In view of the incomplete effects of $Cs^+$ and TEA and the lack of effect of mucosal $Ba^{2+}$, it is possible that more than one population of channels may be identified in the future. These difficulties prohibited calculation of single-channel properties. Nonetheless, current fluctuation analysis methods have permitted us to study apical membrane potassium channels which were relatively inaccessible with other more traditional methods such as transepithelial flux measurements or intracellular microelectrodes.

Are apical membrane $K^+$ channels responsible for the net $K^+$ secretion seen in radioisotopic flux experiments? Preliminary experiments indicate that serosa to mucosa $K^+$ fluxes are decreased by mucosal application of $Cs^+$ and TEA (Wills, unpublished observations; Moreto *et al.*, 1981). Thus, apical membrane potassium channels appear to have an important physiological function in the colon, unlike other epithelia where the function of such channels is unknown. Future work is needed, however, to determine the endogenous factors which regulate these channels.

### C. Do Basolateral $K^+$ Channels Exist?

As discussed in Section I,C,2, the basolateral membrane conductance is largely due to potassium (Wills *et al.*, 1979a,b) and can be decreased by serosal addition of $Ba^{2+}$ (Wills, 1981b). As a continuation of the nystatin experiments reported above, 200 units/ml mucosal nystatin was added to colons bathed on the mucosal side with potassium gluconate–Ringer's solution and on the serosal side with sodium gluconate–Ringer's solution. The dose of nystatin is sufficient to virtually eliminate the apical membrane resistance (Wills *et al.*, 1979a,b, 1981a). Thus under these conditions the epithelium electrically resembles a basolateral membrane in parallel with a paracellular pathway. This so-called "single-membrane" preparation was then voltage clamped at 0 mV and the current noise was recorded. As shown in Fig. 11, in three out of six experiments a novel Lorentzian now appeared in the power spectrum, with a higher $f_c$ and a lower $S_o$ value than seen for the previous apical membrane channels. The Lorentzian component was reversibly blocked by addition of 4 m*M* $Ba^{2+}$ (Fig. 12), and serosal $Ba^{2+}$ caused a low-frequency component similar to the $Ba^{2+}$-induced noise seen for frog skin $K^+$ channels. Preliminary calculations using the method of Van Driessche and Zeiske (1980) indicate that the single-channel current is roughly 0.01 pA and that the channel density is on the order of 180 $\mu m^{-2}$ normalized to chamber area or about 12 $\mu m^{-2}$ normalized to membrane area (15 $\mu F/cm^2$). This calculation assumes as a first approximation that the probability of the channel opening is equal to that of the channel closing. Consequently, these estimates may be significantly changed when estimates for these probabilities become available.

At this point this approach has also been successfully applied to the basolateral membrane of the frog and tadpole skins (Van Driessche *et al.*, 1982). Therefore

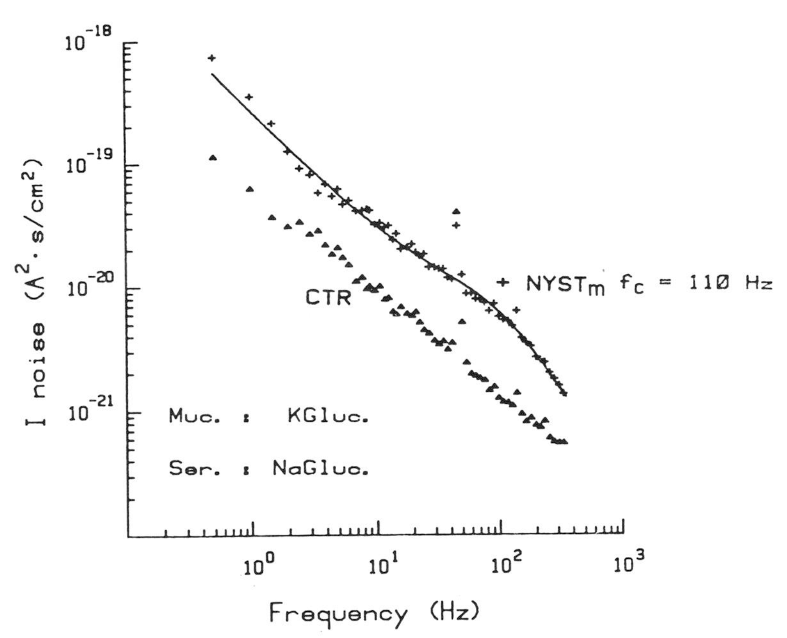


FIG. 11. Effects of different levels of nystatin on current noise. The lower spectrum (CTR) was taken under the same conditions as Fig. 10 (i.e., M to S gradient with 40 units/ml nystatin in the mucosal solution). In the upper spectrum ($NYST_m$), the nystatin level was increased to 200 units/ml in order to essentially abolish the apical membrane resistance. In this "single membrane" preparation, a novel Lorentzian component appeared with a corner frequency of 110 Hz and a plateau value of $7.178 \times 10^{-21}$ $A^2 \cdot sec/cm^2$. Wills, Van Driessche, and Zeiske (unpublished observations).

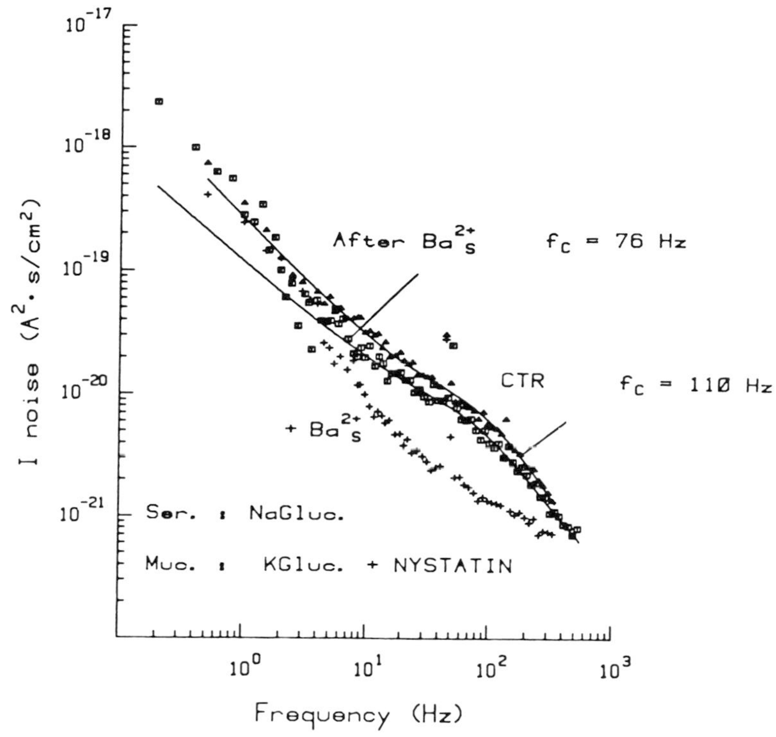


FIG. 12. Effects of serosal $Ba^{2+}$ on the "single membrane" preparation (as described in Fig. 11). Control condition (CTR, M to S gradient) with nystatin (200 units/ml) is indicated by triangles, serosal $Ba^{2+}$ (4 m*M*) is indicated by plus marks, and the post-$Ba^{2+}$ control is indicated by squares. $Ba^{2+}$ abolished the Lorentzian component, similar to its effects on apical membrane channels in amphibian epithelia. Wills, Van Driessche, and Zeiske (unpublished observations).

this method promises to be a fruitful technique for investigating the properties of basolateral membrane channels in intact epithelia.

## IV. SUMMARY

Frequency domain analysis techniques have been used to characterize the mechanisms of $Na^+$ and $K^+$ transport in the mammalian colon. Both of these transport systems involve conductive channels in the apical membrane. $Na^+$ channels appear to be generally similar to the amiloride-sensitive channels in amphibian tight epithelia. Apical membrane $K^+$ channels are partially blockable by mucosal $Cs^+$ and TEA. Future research is needed to evaluate the selectivities of these channels and their possible regulation by hormones or other metabolic factors.

## ACKNOWLEDGMENTS

I gratefully acknowledge the collaboration of Drs. C. Clausen, W. Van Driessche, and W. Zeiske. I particularly wish to thank Dr. Van Driessche for the use of his laboratory facilities and his support. I am indebted also to Drs. B. Biagi, D. Eaton, J. Hanrahan, M. Ifshin, and S. Lewis for their contributions. Superb technical assistance was provided by Lieve Handschutter-Janssens and Bill Alles. This work was supported by NIH Grant AM 29962 to N. K. Wills.

## REFERENCES

Boulpaep, E. L., and Sackin, H. (1980). Electrical analysis of intraepithelial barriers. *Curr. Top. Membr. Trans.* **13,** 169–198.

Clausen, C., and Wills, N. K. (1981). Impedance analysis in epithelia. *In* "Ion Transport by Epithelia: Recent Advances (S. G. Schultz, ed.), pp. 79–92. Raven, New York.

Clausen, C., Lewis, S. A., and Diamond, J. M. (1979). Impedance analysis of a tight epithelium using a distributed resistance model. *Biophys. J.* **26,** 291–317.

Clausen, C., Machen, T., and Diamond, J. M. (1983). Use of A.C. impedance analysis to study membrane changes related to acid secretion in amphibian gastric mucosa. *Biophys. J.* **41,** 167–179.

Cuthbert, A. W., Fanelli, G. M., and Scriabine, A. (1979). "Amiloride and Epithelial $Na^+$ Transport." Urban & Schwarzenberg, Munich.

Fishman, H., Moore, L. E., and Poussart, D. J. M. (1975). Potassium-ion conductance noise in squid axon membrane. *J. Membr. Biol.* **24,** 305–328.

Frizzell, R. A., and Heintze, K. (1979). Electrogenic chloride secretion by mammalian colon. *Kroc Found. Ser.* **12,** 101–110.

Frizzell, R. A., and Schultz, S. G. (1978). Effect of aldosterone on ion transport by rabbit colon in vitro. *J. Membr. Biol.* **39,** 1–26.

Frizzell, R. A., Koch, M. J., and Schultz, S. G. (1976). Ion transport by rabbit colon: I. Active and passive components. *J. Membr. Biol.* **27,** 297–316.

Fromm, M., and Schultz, S. G. (1981). Potassium transport across rabbit descending colon *in vitro*: Evidence for single file diffusion through a paracellular pathway. *J. Membr. Biol.* **63,** 93–98.

Gögelein, H., and Van Driessche, W. (1981a). Noise analysis of the $K^+$ current through the apical membrane of *Necturus* gall bladder. *J. Membr. Biol.* **60,** 187–198.

Gögelein, H., and Van Driessche, W. (1981b). The effects of electrical gradients on current fluctuations and impedance recorded from *Necturus* gall bladder. *J. Membr. Biol.* **60,** 199–209.

Goldman, D. E. (1943). Potential, impedance, and rectification in membranes. *J. Gen. Physiol.* **27,** 37–60.

Hodgkin, A., and Katz, B. (1949). The effect of sodium ions on the electrical activity of the giant axon of the squid. *J. Physiol. (London)* **108,** 37–77.

Li, J. H., Palmer, L. G., Edelman, I. S., and Lindemann, B. (1982). The role of sodium-channel density in the natriferic response of the toad urinary bladder to an antidiuretic hormone. *J. Membr. Biol.* **64,** 77–89.

Lindemann, B., and Van Driessche, W. (1977). Sodium-specific channels of frog skin are pores: Current fluctuations reveal high turnover. *Science* **195,** 292–294.

McCabe, R., Cooke, H., and Sullivan, L. (1982). Potassium transport by rabbit descending colon. *Am. J. Physiol.* **11,** 81–86.

Moreto, M., Planas, J. M., and Naftalin, R. J. (1981). Effects of secretagogues on the $K^+$ permeability of muscosal and serosal borders of rabbit colon. *Biochim. Biophys. Acta* **648,** 215–224.

Palmer, L. G., Li, J. H., Lindemann, B., and Edelman, I. S. (1982). Aldosterone control of the density of sodium channels in the toad urinary bladder. *J. Membr. Biol.* **64,** 91–102.

Powell, D. W. (1979). Transport in large intestine. *Membr. Transp. Biol.* **4B,** 781–809.

Schultz, S. G., Frizzell, R., and Nellans, N. H. (1977). Active sodium transport and the electrophysiology of rabbit colon. *J. Membr. Biol.* **33,** 351–384.

Tai, Y. H., and Tai, C. Y. (1981). The conventional short-circuiting technique under-short-circuits most epithelia. *J. Membr. Biol.* **59,** 173–177.

Thompson, S., Suzuki, Y., and Schultz, S. G. (1982a). The electrophysiology of the rabbit descending colon: I. Instantaneous transepithelial current–voltage relations and the current–voltage relations of the $Na^+$-entry mechanism. *J. Membr. Biol.* **66,** 41–54.

Thompson, S., Suzuki, Y., and Schultz, S. G. (1982b). The electrophysiology of the rabbit descending colon: II. Current–voltage relations of the apical membrane, the basolateral membrane and the parallel pathways. *J. Membr. Biol.* **66,** 55–61.

Van Driessche, W., and Gögelein, H. (1978). Potassium channels in the apical membrane of the toad gallbladder *Nature (London)* **275,** 665–667.

Van Driessche, W., and Zeiske, W. (1980). Spontaneous fluctuations of potassium channels in the apical membrane of frog skin. *J. Physiol. (London)* **299,** 101–116.

Van Driessche, W., Wills, N. K., Hillyard, S. D., and Zeiske, W. (1982). $K^+$ channels in an epithelial single-membrane preparation. *Arch. Int. Physiol. Biochim.* **90**(2), P12–P14.

Welsh, M. J., Smith, P. L., Fromm, M., and Frizzell, R. A. (1982). Crypts are the site of intestinal fluid and electrolyte secretion. *Science* **218,** 1219–1221.

Wills, N. K. (1981a). Antibiotics as tools for studying the properties of tight epithelia. *Fed. Proc. Fed. Am. Soc. Exp. Biol.* **40,** 2202–2205.

Wills, N. K. (1981b). Mechanisms for potassium transport across rabbit descending colon. *Proc. Int. Biophys. Congr., 7th,* p. 183.

Wills, N. K. (1982). Intracellular $Cl^-$ activity in rabbit descending colon. *Physiologist* **25,** 334.

Wills, N. K., and Biagi, B. (1982). Active potassium transport by rabbit descending colon epithelium. *J. Membr. Biol.* **64,** 195–203.

Wills, N. K., Eaton, D. C., Lewis, S. A., and Ifshin, M. (1979a). I–V relationship of the basolateral membrane of a tight epithelium. *Biochim. Biophys. Acta* **555,** 519–523.

Wills, N. K., Lewis, S. A., and Eaton, D. C. (1979b). Active and passive properties of rabbit descending colon: A microelectrode and nystatin study. *J. Membr. Biol.* **45,** 81–108.

Wills, N. K., Zeiske, W., and Van Driessche, W. (1982). Noise analysis reveals $K^+$ channel conductance fluctuations in the apical membrane of rabbit colon. *J. Membr. Biol.* **69,** 187–197.

Yorio, T., and Bentley, P. S. (1977). The permeability of the rabbit colon, *in vitro. Am. J. Physiol* **232,** F5–F9.

Zeiske, W., Wills, N., and Van Driessche, W. (1982). $Na^+$ channels and amiloride-induced noise in the mammalian colon. *Biochim. Biophys. Acta* **688,** 201–210.

# Chapter 5

# Analysis of Ion Transport Using Frequency Domain Measurements

*SIMON A. LEWIS AND WILLIAM P. ALLES*

*Department of Physiology*
*Yale University School of Medicine*
*New Haven, Connecticut*



## I. INTRODUCTION

Classically the methods that have been employed in the study of ion transport by epithelia have been limited to the time domain. Thus, the addition of a pharmacological agent to either side of an epithelium and monitoring of the time response yield information concerning the macroscopic response of the epithelium as a whole but in many cases indicate nothing about the microscopic mechanism(s) which elicits the total response. One approach to investigating this microscopic mechanism(s) is the use of frequency domain analysis. The emphasis of this chapter then is to outline and illustrate from recently acquired data on




ISBN 0-12-153320-4

the rabbit urinary bladder what additional information is gained from studying ion transport processes in the frequency domain as opposed to steady state or time domain analysis.

The two frequency analysis systems we concentrate on are (1) the measurement of membrane area using impedance analysis to determine membrane capacitance (in this instance, 1 $cm^2$ of true membrane area is approximately equal to 1 $\mu F$ of capacitance), and (2) the calculation of single $Na^+$-channel density, currents, and conductances by measurement of the microscopic current fluctuations which represent the sum of either spontaneously fluctuating (opening and closing) channels or channels induced to open and close by pharmacological agents.

Before describing the advantages of and information available from these methods, we first describe the basic transport properties of this epithelium.

## II. BLACK BOX ANALYSIS OF EPITHELIAL TRANSPORT

Of the two broad categories of epithelia, i.e., tight and leaky, we restrict our comments to the former. As outlined by Frömter and Diamond (1972), tight epithelia possess high electrical resistances, develop spontaneous transepithelial potentials up to 120 mV, maintain concentration gradients of up to six orders of magnitude, and can support large osmotic gradients (at least 10:1). A subset of tight epithelia are the so-called "$Na^+$ transport" epithelia, in which transport can be inhibited by the diuretic amiloride. As first studied by Ussing and his collaborators and formalized by Koefoed-Johnsen and Ussing (1958), the model for $Na^+$ transport across frog skin was visualized as a two-step process. The first step involved the entry of $Na^+$ across the apical membrane through a highly $Na^+$-selective pathway down its net electrochemical gradient. Once inside the cell, the $Na^+$ ion is transported across the basolateral membrane via an ATP-utilizing $Na^+$–$K^+$ pump, in which one $Na^+$ ion is exchanged for one $K^+$ ion. The $K^+$ that is pumped into the cell then diffuses out into the plasma down its net electrochemical gradient. This process results in a net movement of $Na^+$ from the mucosal solution or "pond" into the plasma. Since one cannot continually transport a cation without a counterion, it was speculated that either $K^+$ would move into the pond or $Cl^-$ would diffuse into the plasma. The experimental measurements that allowed the development of this model were performed not by assessing the role of each of the membranes using microelectrode techniques, but by making simple but astute measurements of the transepithelial potential response to ion replacements in either mucosal or serosal solutions.

Remarkably similar transport properties are found for the rabbit urinary bladder (Lewis and Diamond, 1976) when compared to the frog skin and the toad urinary bladder. Electrically, the rabbit urinary bladder has a spontaneous transepithelial potential ranging from −40 to −120 mV (referenced to the serosal solution), a transepithelial resistance that varies (among preparations) from 3000 to 75,000 Ω $cm^2$ and a $Na^+$ transport rate that spans a range of 5–155 pEq/sec · $cm^2$. This $Na^+$ transport (measured using isotopes) is proportional to the amount of transepithelial current required to reduce the spontaneous transepithelial potential to zero (the short-circuit current or $I_{sc}$). This short-circuit current can be rapidly inhibited by the addition of amiloride to the mucosal solution, more slowly reduced by serosal ouabain addition, and stimulated after a 45-minute lag period by aldosterone. In all cases, inhibition of $Na^+$ transport causes an increase in transepithelial resistance, while a stimulation of $I_{sc}$ results in a decrease in transepithelial resistance. Although one is tempted to speculate that these changes in resistance are associated with alteration of the apical membrane $Na^+$ permeability, in the absence of a measurement of the individual membrane resistances as a function of $Na^+$ transport, it is not clear whether just the apical or basolateral membrane permeability is changing, or if both membrane resistances change in concert.

## III. DISSOCIATION OF MEMBRANE PROPERTIES

Because of the uncertainty in being able to ascribe a resistance change to an individual membrane or membranes from transepithelial measurements, microelectrodes must be used in an attempt to calculate the values for not only the membrane resistances but also the parallel tight junction resistance. In addition to determining the membrane resistances, one also measures the spontaneous potentials across each membrane, which allows (in conjunction with the membrane resistances) a calculation of the equivalent electromotive force (emf) of each membrane.

To determine the values for the three resistors of the epithelial equivalent circuit, we measure the transepithelial resistance and resistance ratio ($R_a/R_{bl}$) before and during amiloride action on the apical membrane (assuming that amiloride alters *only* the apical resistance). From these measurements we also calculate the emf of the *two* cell membranes. The conductances and emf are summarized in Table I as a function of transport (Lewis and Wills, 1981). Of interest is that only the apical conductance responds to either an increase or a decrease in $Na^+$ transport rate, and that even at zero transport rate there is still a finite conductance in the apical membrane (12 μS/$cm^2$). Using ion-specific microelectrodes we calculate the intracellular ion activities for $Na^+$, $K^+$, and $Cl^-$, and from these activities we can determine the apical membrane ionic permeability using the constant field equation, emf, and conductance. In agreement with our previous speculations, an increase in apical $Na^+$ transport is

TABLE I
MEMBRANE CONDUCTANCE AND emf OF RABBIT URINARY BLADDER AT TWO DIFFERENT RATES OF $Na^+$ TRANSPORT

| Transport rate ($\mu A$) | $E_a$ (mV) | $G_a$ ($\mu S/cm^2$) | $E_{bl}$ (mV) | $G_{bl}$ ($\mu S/cm^2$) |
|---|---|---|---|---|
| ~2 | −7.5 | 49.1 | 53.3 | ~670 |
| ~4 | +9.1 | 55.3 | 52.5 | ~670 |

directly correlated with an increase in apical $Na^+$ permeability. The leak or amiloride-insensitive pathway, on the other hand, remains reasonably constant and is not strongly correlated with the magnitude of the $Na^+$ transport rate.

The potential (−52 mV) and resistance (1500 $\Omega$ $cm^2$) across the basolateral membrane, unlike the apical membrane, are not strongly influenced by the net rate of $Na^+$ transport. By performing ion replacement studies, the relative permeabilities to $Na^+$, $K^+$, and $Cl^-$ were determined. The $P_{Na}/P_K$ was 0.04 and $P_{Cl}/P_K$ was 1.2. Although $Cl^-$ has a greater permeability than $K^+$, it does not contribute to the resting basolateral membrane potential since it is passively distributed (as determined from $Cl^-$ ion-specific microelectrode measurements; see Lewis *et al.*, 1978). Thus, the basolateral membrane potential of this epithelium can be described by the combined diffusion of $Na^+$ and $K^+$.

Since neither $Na^+$ nor $K^+$ is in electrochemical equilibrium across the basolateral membrane (there is 120 mV driving $Na^+$ into the cell and 20 mV favoring $K^+$ exit), energy must be continually expended to maintain the ion gradients for $Na^+$ and $K^+$. This energy, in the form of ATP, is used by the $Na^+$,$K^+$-ATPase, which extrudes $Na^+$ and accumulates $K^+$ in the cells. To determine the stoichiometry of this $Na^+$–$K^+$ pump, Wills (1981) and Lewis and Wills (1981) performed a series of electrical measurements employing conventional and ion-specific microelectrode transepithelial electrical measurements in conjunction with the antibiotics nystatin and gramicidin D. In brief, they demonstrated that the pump was not electroneutral as originally suggested by Koefoed-Johnsen and Ussing (1958), but rather could generate a current. The activation of the pump is best described by a model of highly cooperative binding with binding coefficients ($n$) of 2.8 and 1.8 ($Na^+$ and $K^+$, respectively) and $K_m$ (concentration for half-maximal stimulation) of 14 and 2.3 m$M$ ($Na^+$ and $K^+$, respectively). The binding coefficient indicates how many ions must bind to activate the pump, not the number transported. However, further analysis indicates that the stoichiometry is indeed three $Na^+$ ions to two $K^+$ ions (Wills, 1981). From purely energetic considerations it is unlikely that more than three $Na^+$ ions can be transferred per pump cycle.

The data we have outlined above agree well with the original model and indeed have verified and extended the model. Perhaps the major alterations are as follows: (1) A leak pathway exists in the apical membrane, with a finite $Na^+$ and $Cl^-$ permeability in the basolateral membrane. (2) The pump is not electroneutral but rather electrogenic with a stoichiometry of three $Na^+$ to two $K^+$. (3) An increase in $Na^+$ transport is related to an increase in apical $Na^+$ permeability and not alterations in the net driving force for $Na^+$ entry. (4) Inhibition of transport from either the mucosal solution by amiloride or the serosal solution by ouabain results in an increase in apical resistance and thus a decrease in apical permeability. The latter effect of pump inhibition might be mediated by an increase in cytoplasmic $Na^+$ activity which then inhibits either directly or indirectly the apical $Na^+$ pathways. Indirect inhibition might be a result of increased cytoplasmic $Ca^{2+}$ activity (Taylor and Windhager, 1979).

Although this is certainly not the end of the information available from transepithelial and microelectrode recordings during transport perturbations, it represents, at least, the major findings for the rabbit urinary bladder.

There is a certain subset of questions concerning epithelial transport that cannot be addressed using the methods outlined above but might be answered by investigating epithelial transport in the frequency domain.

## IV. QUESTIONS FOR ANALYSIS IN THE FREQUENCY DOMAIN

1. What are the true membrane areas of the apical and basolateral membranes, and do they change as a function of transport rate?
2. Does conductive $Na^+$ transport (as outlined above) occur by a shuttle-like carrier mechanism similar to the antibiotic valinomycin or, alternatively, is there a protein pore that spans the lipid bilayer of the apical membrane?
3. During hormone stimulation, does the number of conductive pathways change or do the kinetic properties of a single pathway change so as to transfer more charge or ions per unit time?
4. Can we learn something, by pharmacological means, about the structure and regulation of these transport proteins?

## V. METHODS FOR FREQUENCY ANALYSIS

As illustrated in other chapters of this volume, two types of frequency data are available. First, one can apply small current or voltage signals composed of either single or multiple sinusoidal waves. By measuring the transepithelial voltage or current response to the input signal, insight is gained into the resistances and, more importantly (for the determination of area), the capacitance values of the apical and basolateral membranes. Since capacitance is directly correlated with membrane area (1 $cm^2$ is approximately equal to 1 μF) this analysis offers a noninvasive approach to measuring surface area.

Second, one can look for intrinsic microscopic deviations from either macroscopic spontaneous voltages or short-circuiting currents, i.e., one can measure spontaneous "fluctuations" of the protein pathways responsible for net ion transport. By a statistical analysis of these spontaneous or induced microscopic deviations from the macroscopic mean, and the development of an appropriate (and perhaps simplistic) model, one can attempt to determine (1) whether the ion enters or exits the cell by a carrier or pore mechanism, and (2) the density and single-pathway currents of each transport protein.

In the remainder of this chapter we describe experiments using both of these techniques, with an emphasis on the power of these methods to provide answers not readily available using other techniques.

## VI. IMPEDANCE ANALYSIS

### A. Determination of Membrane Resistances

In Section III we indicated that the three resistors of an epithelial equivalent circuit could be easily calculated by perturbing only one of the three resistors (usually the apical or basolateral membrane) while simultaneously monitoring the transepithelial resistance and the ratio of apical to basolateral membrane resistances. There are two built-in assumptions: first, that the method for altering a membrane resistance does indeed affect only that membrane, and second, that the simple equivalent circuit used is morphologically accurate. Both of these assumptions can be assessed using impedance analysis.

For the mammalian urinary bladder the above assumptions have already been tested (see Clausen *et al.*, 1979). In brief, these investigators found that the simple three-resistor model was *not* adequate to describe the impedance of the epithelium. The deviation of the model from the data occurred at high frequencies. The inadequacy of the model was a result of the finite resistance of the lateral intercellular space (LIS) compared to the bordering lateral membrane. This additional resistance then demanded that the lateral membrane and the LIS be a distributed network. At very low frequencies, the LIS resistance is negligible compared to that of the lateral membrane. As the frequency of the wave form increases, the lateral membrane impedance decreases (because the parallel capacitors allow a greater current flow at high frequencies) and becomes comparable to that of the LIS. However, for microelectrode experiments (in mammalian urinary bladder), the resistance ratio, measured as a dc value, will represent the actual ratio of apical to basolateral membranes. Thus, for dc measurements, the simple model reasonably describes the resistance properties of this epithelium.

Using microelectrodes in conjunction with impedance analysis yields adequate information to determine the three resistors and two capacitors of the equivalent circuit without having to perturb a membrane circuit element. This allows one to determine whether the basolateral membrane resistance changes as a function of alteration of the apical membrane resistance. For open circuit conditions, the basolateral membrane resistance remained independent of the net $Na^+$ transport, while the apical resistance was inversely proportional to the net rate of $Na^+$ transport. Both apical and basolateral membrane areas (i.e., capacitances) did not change as a function of $Na^+$ transport and yielded values of 1.8 and 8.6 $\mu F/cm^2$ for the apical and basolateral membranes, respectively. The ratio of apical to basolateral capacitance (1:5) is in agreement with the morphology of this preparation in which the cells can be considered as cubes. To summarize, impedance analysis has verified the assumptions required for analysis of microelectrode experiments and in addition yielded information concerning the relative and absolute areas of apical and basolateral membranes.

## B. Determination and Control of Membrane Areas

The apical membrane capacitance in the studies by Clausen *et al.* (1979) varied between 0.99 and 2.4 $\mu F/cm^2$, a range similar to that previously reported by Lewis and Diamond (1976). These latter investigators attempted to correlate tissue dry weight with the apical membrane capacitance (Fig. 1). They found that at high tissue dry weight there was also a large apical capacitance. As the tissue was stretched both capacitance and dry weight decreased linearly until the capacitance reached a value near 1 $\mu F/cm^2$. At this point, further stretch did *not* result in a decrease in capacitance but did cause a continuing decrease in dry weight (by ~245%). Such a large decrease in weight without a decrease in apical area can be most easily explained by the addition of new membrane to the apical surface from some cytoplasmic store. This interpretation has a morphological basis. Using morphometric analysis, Minsky and Chlapowski (1978) demonstrated that there is a higher density of cytoplasmic fusiform vesicles in contracted mammalian urinary bladders than in distended bladders. They speculated that during stretch these vesicles are transported into the apical membrane and upon micturition (and subsequent bladder collapse) these vesicles are rapidly removed. Close inspection of the freeze fracture images of the apical membrane and of the cytoplasmic vesicles demonstrates remarkable similarities. The apical membrane of the mammalian urinary bladder has an unusually thickened membrane, ~12 nm. Staehelin *et al.* (1972) demonstrated that this thickened membrane occupied 73% of the apical surface and was composed of polygonal plaques (~1 μm in diameter) with a unit structure of hexagonally arranged particles. The remaining 27% of the apical membrane had a thickness of about 7 nm, comparable to the thickness of a normal lipid bilayer. Freeze fracture of the cytoplasm (Staehelin *et*

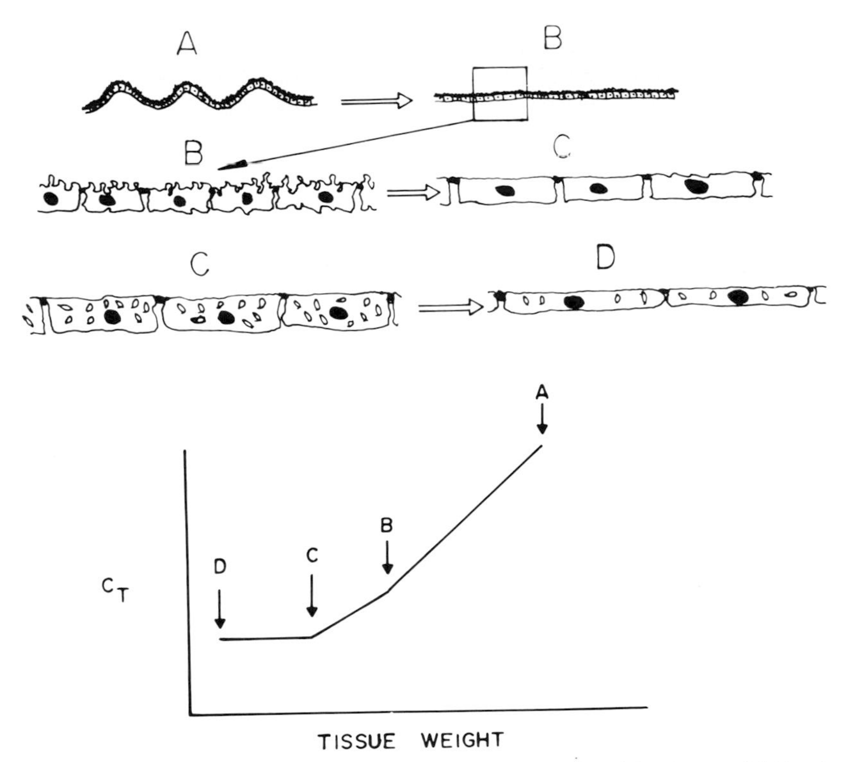


FIG. 1. Predicted correlation between capacitance $C_T$ ($\mu F/cm^2$) and dry tissue weight ($mg/cm^2$) for preparations mounted at various degrees of stretch. The area ($cm^2$) is the nominal area of the chamber opening. For stretch (between A and B), the capacitance decreases approximately linearly with dry weight, as macroscopic folding of the tissue is eliminated. As the stretch is increased (from B to C), the capacitance decreases much more slowly than does the dry weight, with the microscopic folds disappearing and the epithelial cells becoming flatter. Further stretching (C to D) requires addition of cytoplasmic vesicles into the apical membrane. Dry weight decreases because the number of cells per chamber area decreases.

*al.*, 1972) revealed that the fusiform vesicles were composed of two closely opposed plaques containing the same hexagonal particles. Most fascinating is that the cytoplasmic vesicles, as well as the apical membrane plaques, are all tethered by a dense network of filaments 7 nm in diameter (Minsky and Chlapowski, 1978), which coalesce in the region of desmosomes in both the lateral and basal membranes. Minsky and Chlapowski (1978) speculated that these filaments might provide the directed motive force for both vesicle incorporation into the removal from the apical membrane during an expansion/ contraction cycle of the urinary bladder.

Two questions come to mind at this point. First, can the measurement of membrane capacitance demonstrate that these vesicles are, indeed, translocated into the apical membrane as a function of stretch? Second, what is the control or propulsion system which allows an orderly insertion/withdrawal of these lipid–protein vesicles?

There are a number of methods which one can use to "convince" cytoplasmic vesicles to migrate toward and fuse with the apical membrane of the urinary bladder. First and most obvious is to simply stretch the bladder once it is mounted in the modified Ussing chambers (Lewis and Diamond, 1976). Any increase in capacitance from baseline indicates an addition of membrane. This experiment has been performed and demonstrated an 18% increase in capacitance by simply bowing the epithelium into the serosal chamber by an excess hydrostatic pressure head. After removal of this pressure head the capacitance returned to within 6% of control within 5 minutes. The value of 18% is not overly large, so an alternative approach to investigating the stimulation of vesicle movement was decided upon. In this approach the cell volume was perturbed by exposing both mucosal and serosal solutions to a Ringer's solution of one-half osmotic strength (Lewis and de Moura, 1982). This procedure caused an increase in capacitance of ~78% after 60 minutes (Fig. 2). Returning the epithelium to isosmotic solutions revealed a very rapid exponential-like return of the capacitance to near control values (with a time constant of 11 minutes). This very reproducible and large increase and decrease in capacitance offers a convenient

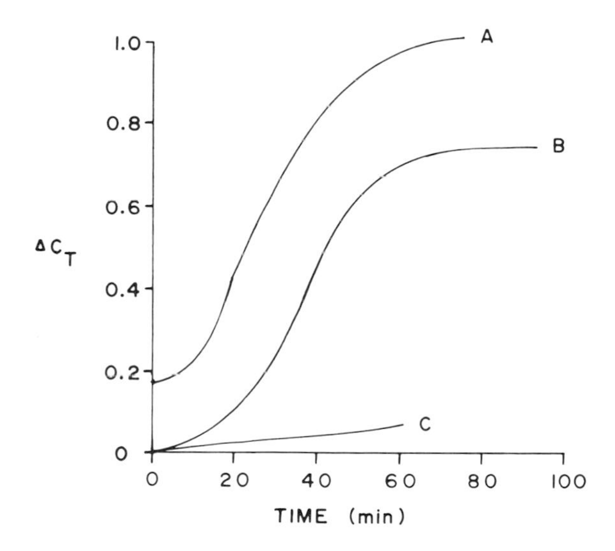


FIG. 2. Increase in the normalized capacitance as a function of time following the replacement of both serosal and mucosal bathing solutions with Ringer's solution of one-half normal osmolarity. (A) Pretreatment with 0.1 m*M* colchicine for 3 hours, (B) control, and (C) pretreatment with 35 μ*M* cytochalasin B for 90 minutes.

method for testing which cytoplasmic elements might control vesicle mobilization resulting in fusion and withdrawal. The two predominant cytoplasmic elements for exocytosis or endocytosis are microtubules and microfilaments. Microtubules can be blocked by the pharmacological agent colchicine, and similarly microfilaments can be disrupted by cytochalasin B. Two possible experimental series are obvious. The first series involves determining the mechanism of vesicle insertion into the apical membrane. After pretreating the epithelium with either colchicine (0.1 m*M* for 3 hours) or cytochalasin B (35 μ*M* for 1 hour), one can monitor the capacitance changes during an osmotic challenge and determine from the absence or occurrence of an apical membrane area change whether microtubules or microfilaments are involved in vesicle translocation. These experiments have been performed (Lewis and de Moura, 1981), and demonstrated the following results (Fig. 2). First, colchicine, in the absence of an osmotic gradient, caused an increase in the apical capacitance by 18%. Upon osmotic challenge, the capacitance increased at a more rapid rate than the control situation. In contrast, preincubation with cytochalasin B and ultimate osmotic challenge resulted in a complete inhibition of capacitance increase. It was concluded that microfilaments are responsible for vesicle transfer into the apical membrane and that microtubules are essential for either cell volume regulation or the harnessing and control of microfilaments and, ultimately, vesicle translocation.

The second series of experiments is directed at resolving the steps involved in vesicle retrieval into the cytoplasm. This was determined in the following manner. After an hour of osmotic challenge the epithelium was treated for an additional hour with cytochalasin B, the osmotic challenge removed, and the rate of capacitance change recorded. It was found that the rate of apical vesicle return to its cytoplasmic compartment required a much longer period of time than the control period (at least 22 minutes). Since these return experiments were conducted in the absence of cytochalasin B, the return might reflect the reaggregation of microfilaments to reestablish a normal membrane area–cell volume regime rather than some other secondary mechanism establishing the control condition.

## VII. FLUCTUATION ANALYSIS

### A. $Na^+$ Pathways: Channels or Carriers?

To this point we have not considered the particular transport protein which allows a ready diffusion of $Na^+$ from the mucosal solution to the cell interior nor have we considered how the incorporation of cytoplasmic vesicles might facilitate this movement of ions from urine to cytoplasm.

It is just such a question which leads us to the second part of this chapter, i.e., does $Na^+$ enter the cell by a carrier or a pore protein and do the cytoplasmic vesicles contain either of these transport proteins? As is obvious from other chapters, many of the cation ($Na^+$ or $K^+$) transport pathways seem to be pores rather than carriers. The evidence which discriminates between these mechanisms is based on a statistical analysis of the microscopic variations in short-circuit current. These fluctuations are recorded and then decomposed into a sum of sine and cosine waves of different frequencies (Fourier analysis). From these data one calculates the so-called power spectral density ($A^2 \cdot$ second vs frequency). It is the shape of this curve that allows a discrimination between carriers and pores (or channels). As an example, a channel that spontaneously and randomly opens and closes has a single relaxation time constant and demonstrates the (by now) well-known Lorentzian shape which has a plateau at low frequencies and a slope of $-2$ (log–log plot) at high frequencies. The transition between the plateau and $-2$ slope has a point called the corner frequency, which is proportional to the sum of the open and close rate constants and is easily determined, as it has a power equal to one-half of the plateau value. Carriers, on the other hand, have a very different spectral shape. Apparently, at low frequencies the power is small and as the frequency increases, so does the power (Kolb and Läuger, 1978).

It is possible, then, to differentiate between a carrier and a pore solely on the basis of the spectral shape. In addition, using an appropriate model, one can determine the kinetic properties of the opening and closing of a channel (either spontaneously or by a pharmacological intervention) and the current carried in a unitary event. The number of ions transferred by a carrier will be quite small (1–10), whereas the number of ions that can move through a channel while it is open is expected to be in the thousands.

Because of this sort of differentiation between channels and carriers, one can simply measure the spontaneous current fluctuations and observe the power spectral density (PSD). Recently, Loo *et al.* (1982) have measured the PSD of the rabbit urinary bladder. In the presence of normal NaCl–$NaHCO_3$ Ringer's solution, the PSD was a straight line with a slope of $-1.2$. Thus they could not measure any channel or carrier noise over the frequency range 0.4–100 Hz (Fig. 3). As first demonstrated by Lindemann and Van Driessche (1977), the addition of the specific channel blocker amiloride to the apical bathing solution causes a marked change in the shape of the PSD. In addition to the linear component there is a superimposed Lorentzian curve. They concluded that this Lorentzian component was induced by the random and total blockade of individual amiloride-sensitive $Na^+$ channels which followed the simple scheme

$$A + R \underset{K_{10}}{\overset{K_{01}}{\rightleftarrows}} AR$$

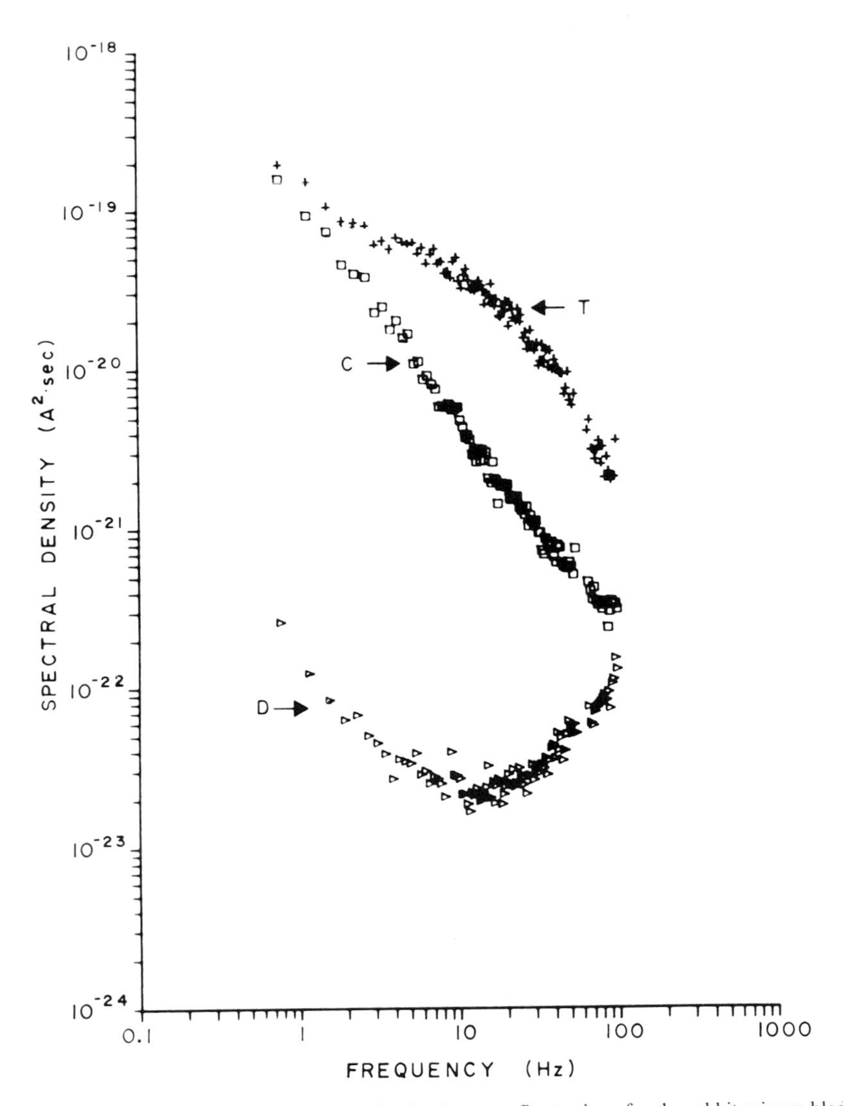


FIG. 3. Power spectrum of the short-circuited current fluctuations for the rabbit urinary bladder in the absence (C) and presence of 20 $\mu M$ triamterene (T) in the mucosal bathing solution. Also shown is the power spectrum of the current fluctuations for the voltage clamp used with a 5000-$\Omega$ resistor and 2.47-$\mu$F capacitor (D).

The relationship between this equilibrium scheme and the PSD results in the Lorentzian curve which is described by the following equations:

$$S(f) = S_o/[1 + (f/f_c)^2]$$

where the plateau value ($S_o$) is

$$S_o = 4Mi^2a\, K_{01}K_{10}\, [\text{Amil}]/(K_{01}\, [\text{Amil}] + K_{10})^3$$

and the corner frequency ($f_c$) is

$$2\, \pi\, f_c = K_{10} + K_{01}\, [\text{Amil}]$$

where $K_{01}$ is the open to close rate constant (in units of $\sec^{-1} M^{-1}$), $K_{10}$ is the close to open rate constant (in units of $\sec^{-1}$), [Amil] is the blocker concentration (in molar), $M$ is the total channel density, $a$ is membrane area, and $i$ is the single-channel current.

The object is to determine the four parameters $K_{01}$, $K_{10}$, $i$, and $M$. First we can solve for $K_{01}$ and $K_{10}$ by measuring the corner frequency as a function of amiloride concentration. A plot of $2\pi f_c$ vs [Amil] will have an intercept of $K_{10}$ and a slope of $K_{01}$. This still does not allow us to calculate $i$ or $M$ (two unknowns but only one equation); we require another equation which relates macroscopic short-circuit current ($I_o$ remaining after a suboptimal dose of amiloride) to single-channel current and channel density. The equation is then

$$I_o = i \cdot M \cdot P_o$$

where $P_o$ (probability of the channel being open) is given by

$$P_o = K_{10}/(K_{01}\, [\text{Amil}] + K_{10})$$

Thus the single-channel currents and channel density can be determined at each amiloride concentration.

In the absence of blockers there is no obvious Lorentzian component in the PSD. When a blocker is added, a Lorentzian component will appear if (1) the model is correct and (2) the rate constants at the blocker concentration used will yield a corner frequency within the measured frequency range (0.1–100Hz). Using triamterene (a known blocker, Christensen and Bindslev, 1982) at a final concentration of 20 $\mu M$ induces a definite Lorentzian component (Fig. 3), with an additive linear component identical to that before blocker addition.

It is reassuring that a Lorentzian component is induced by a specific blocker. This however does not prove that the model is correct. Perhaps the only acceptable proof would be recording from a single channel and correlating the open–close rate constants and single-channel conductances. A less rigorous proof for acceptance of the model is to determine whether, over a range of amiloride concentrations, there is a linear relationship between $2\pi f_c$ and amiloride. Additionally, the plateau value must start at zero, peak at one-half of the $K_D$ ($K_D = K_{10}/K_{01}$) and then decrease toward zero as the blocker concentration approaches infinity. Using amiloride as a blocker, we have verified both of these predictions for the rabbit urinary bladder. Table II summarizes the values for amiloride binding kinetics and single-channel currents from control bladders.

TABLE II

EFFECT OF PUNCHING ON SINGLE-CHANNEL CURRENT, AMILORIDE BINDING, AND ESTIMATED CHANNEL DENSITY[a]

| | $i$ (pA) | $K_{01}$ ($sec^{-1}$ $\mu M^{-1}$) | $K_{10}$ ($sec^{-1}$) | $M$ ($\times 10^6$) |
|---|---|---|---|---|
| Control bladders | 0.64 | 52.1 | 11.6 | 1.2 |
| Bladders after punching | 0.74 | 43.9 | 13.2 | 9.1 |

[a] $i$, Single-channel current; $K_{01}$, amiloride association rate constant; $K_{10}$, amiloride dissociation rate constant; $M$, number of channels per square centimeter.

## B. Do the Vesicles Contain $Na^+$ Channels?

Having measured the $Na^+$ channels in the apical membrane, we now ask whether the vesicles contain $Na^+$ channels. If so, are they "identical" in terms of binding kinetics and single-channel currents to those in the apical membrane, and is the channel density the same as in the apical membrane?

Although swelling the cells is a reasonable method for the insertion of vesicles into the apical membrane, close inspection of electron micrographs indicates that the mitochondria are very swollen. Consequently, we looked for an alternative method for the stimulation of vesicle translocation and fusion. One obvious method is to simply distend and collapse a piece of bladder after mounting it in the *in vitro* chambers. An alternate method is to apply hydrostatic pressure pulses to the epithelium (Lewis and de Moura, 1982). Such pressure pulses might momentarily squash the cells, causing the vesicles and apical membrane to fuse. Removal of the pressure pulse will cause a rebound and perhaps removal of some of the apical membrane vesicles.

Measurements of the amiloride-sensitive current before and after the pressure pulses (termed "punching") demonstrated a 10-fold increase in the current after punching without a measurable change in the apical capacitance (Fig. 4). From these experiments it was concluded that punching caused both fusion and removal of vesicles. The increase in transport also indicates that the vesicles contain a higher density of $Na^+$ channels than the vesicles which make up the apical membrane. A different conclusion might be that the channel density is the same for the vesicles and apical membrane but that the vesicles contain channels with higher permeability. To resolve this question, the single-channel currents and channel density before and after punching the preparation were determined using fluctuation analysis. The data (Table II) support the hypothesis that there is

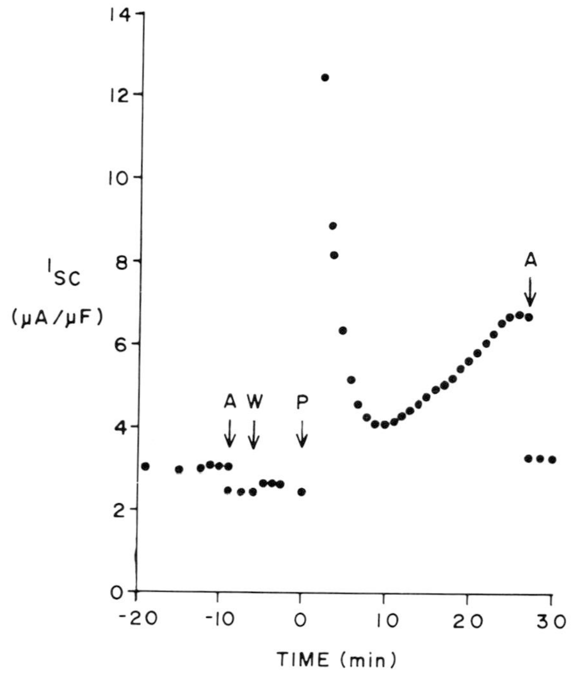


FIG. 4. Effect of "punching" on the short-circuit current. Addition of 10 μ*M* amiloride (A) to the mucosal bathing solution produced a small (0.60 μA/μF) decrease in $I_{sc}$. Within 30 minutes after punching the tissue (P), $I_{sc}$ went through a rapid increase and decrease, followed by a slower increase until reaching a steady state of approximately three times the prepunch value. Amiloride (10 μ*M*) was again added (A), producing a decrease in $I_{sc}$ of 3.5 μA/μF, approximately a sixfold increase over the prepunch value. W, wash of mucosal bathing solution with normal Ringer's solution to remove amiloride.

a greater *density* of channels in the vesicles than in the apical membrane since single-channel currents and amiloride binding kinetics were not significantly different pre- and postpunching.

## C. Do Other Pathways Exist in the Apical Membrane?

Lewis and Wills (1980) reported that in parallel with the amiloride-sensitive pathway there is an amiloride-insensitive pathway. Using microelectrodes these investigators found that this pathway did not discriminate between $Na^+$ and $K^+$. Can fluctuation analysis tell us anything about this transport system?

In the absence of a blocker there is a pronounced low-frequency noise component. Addition of a saturating level of a blocker (e.g., amiloride) does *not* alter this low-frequency noise component. We found, however, a direct linear relationship between the amiloride-insensitive macroscopic current and the power of the low-frequency noise at 1 Hz. At the present time, we have found two methods for decreasing the macroscopic current (and also the low-frequency noise). First, if one simply replaces the mucosal solution with fresh Ringer's solution, there is a spontaneous decrease in both the macroscopic current and power at 1 Hz. Our tentative conclusion is that these pathways are "washed out" of the apical membrane. Second, serotonin (which also blocks the amiloride-sensitive $Na^+$ channels, Legris *et al.*, 1982) causes a decrease in both the power at 1 Hz and the macroscopic leak current. It is tempting to speculate that this leak pathway is nothing more than a degraded or sick amiloride-sensitive channel. The following evidence points toward such a speculation.

1. A transport protein should be stable within the bilayer. This is not the case, as this pathway can be washed out, while the amiloride channel cannot.
2. The density of amiloride channels is greater in the vesicles than in the apical membrane.
3. The $Na^+$-to-$K^+$ selectivity of the amiloride channels is lower than the selectivity of the vesicle channels, suggesting a progressive alteration in channel properties during exposure to urine (Lewis and Wills, 1981).
4. The proteolytic enzyme trypsin acts as an irreversible amiloride, suggesting the susceptibility of these channels to degradation by tryptic-like enzymes known to occur in urine.
5. Serotonin reversibly inhibits both amiloride-sensitive and -insensitive pathways.
6. In a few experiments, the spontaneous decrease in amiloride current was accompanied by an increase in leak current.

Further experiments will be required before a definite conclusion can be made about the source of this leak pathway.

## VIII. SUMMARY

This chapter has attempted to demonstrate that frequency analysis is a powerful tool in assessing ion transport processes. By measuring membrane capacitance, we find that the bladder accommodates stretch by the insertion of cytoplasmic vesicles into the apical membrane, the movement of these vesicles being dependent upon an intact microfilament system. Fluctuation analysis allows a measurement of single-channel properties, a comparison of apical channels to those in the vesicles, and lastly, some insight into the life of a channel. Further application of these techniques will certainly expand our understanding of the regulation of ion transport.

## ACKNOWLEDGMENTS

We would like to express our thanks to Drs. Clausen, Diamond, Ifshin, Loo, de Moura, and Wills for their collaborations over the past years. This work was supported by NIH Grant AM20851.

## REFERENCES

Christensen, O., and Bindslev, N. (1982). *J. Membr. Biol.* **65,** 19–30.
Clausen, C., Lewis, S. A., and Diamond, J. M. (1979). *Biophys. J.* **26,** 291–318.
Frömter, E., and Diamond, J. M. (1972). *Nature (London), New Biol.* **235,** 9–13.
Koefoed-Johnsen, V., and Ussing, H. H. (1958). *Acta Physiol. Scand.* **42,** 298–308.
Kolb, H. A., and Läuger, P. (1978). *J. Membr. Biol.* **41,** 167–187.
Legris, G. J., Will, P. C., and Hopfer, U. (1982). *Proc. Natl. Acad. Sci. U.S.A.* **79,** 2046 2050.
Lewis, S. A., and de Moura, J. L. C. (1982). *Nature (London)* **297,** 685–688.
Lewis, S. A., and Diamond, J. M. (1976). *J. Membr. Biol.* **28,** 1–40.
Lewis, S. A., and Wills, N. K. (1980). *Biophys. J.* **31,** 127–138.
Lewis, S. A., and Wills, N. K. (1981). *Ann. N.Y. Acad. Sci.* **372,** 56–63.
Lewis, S. A., Wills, N. K., and Eaton, D. C. (1978). *J. Membr. Biol.* **41,** 117–148.
Lindemann, B., and Van Driessche, W. (1977). *Science* **195,** 292–294.
Loo, D. D. F., Lewis, S. A., and Diamond, J. M. (1982). *Biophys. J.* **37,** 267a.
Minsky, B. D., and Chlapowski, F. J. (1978). *J. Cell Biol.* **77,** 685–697.
Staehelin, L. A., Chlapowski, F. J., and Bonneville, M. A. (1972). *J. Cell Biol.* **53,** 73–91.
Taylor, A., and Windhager, E. E. (1979). *Am. J. Physiol.* **236,** F505–F512.
Wills, N. K. (1981). *Fed. Proc. Fed. Am. Soc. Exp. Biol.* **40,** 2202–2205.

# Chapter 6

# Use of Potassium Depolarization to Study Apical Transport Properties in Epithelia

*LAWRENCE G. PALMER*

*Department of Physiology*
*Cornell University Medical College*
*New York, New York*



## I. INTRODUCTION

The ability of epithelia to transport solutes and water vectorially depends on the differences in the transport properties of the inward-facing (serosal or basolateral) and outward-facing (mucosal or apical) cell membranes. Understanding the nature of epithelial transport therefore involves the study of each membrane as well as the interactions between them. Several techniques have been used in electrophysiological studies of epithelial transport to derive infor-




ISBN 0-12-153320-4

mation on the individual membranes. They can be classified into three general categories, each having its own advantages and disadvantages.

Microelectrodes have been used to record from the interiors of epithelial cells. This provides direct measurements of the electrical potential difference (PD) across each of the two membranes, and of the ratio of the two membrane resistances. One disadvantage of this technique is the possibility of membrane damage that can result from the impalement itself, thereby altering the properties of the cell being studied (Lindemann, 1975; Higgins *et al.*, 1977).

Alternating current impedance analysis using extracellular electrodes is also capable of distinguishing the electrical properties of the two membranes, when their electrical time constants are different (see Clausen and Dixon, Chapter 3, this volume). A great advantage of this technique is that the measurements are made noninvasively. A disadvantage is that the individual membranes cannot be voltage clamped. Another possible limitation is the speed with which measurements can be made, which depends on the membrane time constants. In general, several seconds will be required.

An alternative method is to functionally eliminate one membrane in order to study the properties of the other. This involves manipulating the composition of the bathing media to minimize the resistance and potential across the membrane to be eliminated, and/or using ionophores to shunt the native transport properties of that membrane. This approach has the advantage of being simple to use, once the appropriate conditions are worked out. In addition, it allows the membrane of interest to be voltage clamped. The major disadvantage is that the elimination of the series membrane will, in general, be incomplete. A second problem is that the experimental manipulations involved may alter the properties of the membrane being studied. Finally, information about the membrane that is being eliminated is lost.

In this chapter I discuss the use of high concentrations of $K^+$ in the serosal solution to depolarize the basolateral membranes of tight epithelia and to electrically isolate the apical membrane. I first review some of the information that has been obtained using this approach, and then discuss the extent of the elimination of basolateral PD and resistance, and the problem with this procedure of the preservation of the properties of the apical membrane.

## II. USES OF $K^+$ DEPOLARIZATION

### A. Current–Voltage Relationship

The current–voltage ($I$–$V$) relationship of the epithelial $Na^+$ channels was first measured in the $K^+$-depolarized frog skin (Fuchs *et al.*, 1977). This method permitted simultaneous estimates of the $Na^+$ permeability ($P_{Na}$) and the intracellular $Na^+$ activity. Subsequently, the technique was adapted to the toad urinary bladder (Palmer *et al.*, 1980) and to the rabbit descending colon (Thompson *et al.*, 1982) and has been useful in studying the regulation of apical $Na^+$ permeability. Using *I–V* measurements in conjunction with fast-flow techniques, Fuchs *et al.* (1977) demonstrated a down-regulation of $Na^+$ permeability with increasing mucosal $Na^+$, which was independent of intracellular $Na^+$. In the toad bladder, stimulation of $Na^+$ permeability by oxytocin and aldosterone was associated with small increases in intracellular $Na^+$ activity (Li *et al.*, 1982; Palmer *et al.*, 1982). This addressed the suggestion that hormones might increase $P_{Na}$ indirectly by stimulating $Na^+$ extrusion and lowering cell $Na^+$ (Frizzell and Schultz, 1978).

## B. Single-Channel Properties

The area density and turnover rate of apical $Na^+$ channels were determined in the depolarized frog skin by fluctuation analysis (Lindemann and Van Driessche, 1977). Subsequently, the modulation of channel densities by external $Na^+$ ions and, in the toad urinary bladder, by oxytocin and aldosterone was also assessed using depolarized preparations (Van Driessche and Lindemann, 1979; Li *et al.*, 1982; Palmer *et al.*, 1982).

In these studies, depolarization of the epithelia being studied simplified the interpretation of the results, because changes in the electrical driving forces across the apical membrane were minimized. In addition, interpretation of noise spectra across two membranes in series is rather complex, except in the limiting case where the impedance of the membrane that is the noise source is much greater than the impedance of the series membrane. The presence of a series resistance can attenuate the noise signal, giving rise to underestimates in single-channel currents and conductances. If the series membrane impedance is complex, as is expected for the basolateral membrane, distortions in the shape of the noise spectra will result (Van Driessche and Gögelein, 1980; Lindemann and DeFelice, 1981).

## C. $Na^+$-Channel Characteristics

Morel and LeBlanc (1975) found that in frog skin, with high $K^+$ on the serosal side, the amiloride-sensitive short-circuit current could be reversed, provided that the cells were first allowed to accumulate $Na^+$ and that the mucosal $Na^+$ concentration was rapidly lowered. Presumably the high $K^+$ concentration facilitated the net inward movement of $K^+$, completing the electrical circuit and replacing the net loss of $Na^+$ from the cell. This demonstrated that the apical $Na^+$ channels could support $Na^+$ transport in either direction, and that the direction of net ion movement was down an electrochemical activity gradient. In the toad bladder, Palmer (1982a) used a similar approach to establish electrochemical equilibrium for $Na^+$ across the apical membrane and to measure the flux ratio exponent for $Na^+$ transport. This provided direct evidence that apical $Na^+$ transport satisfied Ussing's flux ratio criterion for independent ion movements. Similarly, when a $K^+$ gradient was established across the apical membrane, an outward amiloride-sensitive $K^+$ current could be measured, and the permeability ratio for $Na^+$ and $K^+$ of the apical $Na^+$ channel could be estimated (Palmer, 1982b).

### D. Apical Membrane Capacitance

To investigate the role of membrane fusion events in the action of antidiuretic hormone (ADH), Palmer and Lorenzen (1983) measured the electrical time constant of the $K^+$-depolarized toad bladder to obtain estimates of apical membrane capacitance. The response of the depolarized epithelium to a constant-current pulse could be described by a single exponential process. As shown by Lewis and Diamond (1976), when the apical resistance is much larger than that of the basolateral membrane and the parallel tight junctional resistance is high, the epithelium should behave very nearly as a single RC network, and the apparent capacitance will be approximately that of the apical membrane.

The ADH-induced change in apical capacitance measured in this way (mean increase of 28%) was similar to previously reported values obtained with ac impedance analysis (23%, Warncke and Lindemann, 1981) and with measurements of total effective transepithelial capacitance (25–36%, Stetson *et al.*, 1982). In this case, ac impedance analysis is the method of choice, as more complete information is obtained and fewer assumptions about the tissue are made. Transient analysis of depolarized epithelia can, however, provide a simple, convenient alternative.

In summary, the use of $K^+$-depolarized epithelia in the study of apical membrane function has proven to be a useful, convenient alternative to the more difficult techniques of intracellular recording and ac impedance analysis. However, as mentioned above, the use of such a preparation involves assumptions about both the extent of serosal membrane elimination and the physiological state of the depolarized tissue. In the studies described above, both the resistance and PD of the basolateral membrane are assumed to be zero. These assumptions are considered in detail below.

## III. BASIS FOR THE $K^+$-DEPOLARIZATION TECHNIQUE

According to the model of Koefoed-Johnsen and Ussing (1958), the basolateral membrane of the tight epithelial cell is selectively permeable to $K^+$. This idea was based on the dependence of the transepithelial electrical potential on serosal $K^+$, a phenomenon also observed in the toad bladder (Gatzy and Clarkson, 1965; Leb *et al.*, 1965). Thus, application of high concentrations of $K^+$ in the serosal solution should decrease the $K^+$ diffusion potential and, hence, the electrical PD and resistance across the basolateral membrane.

Other investigators, however, suggested that an electrogenic $Na^+$–$K^+$ exchange pump may account for at least some of the electrical properties of this membrane (Bricker *et al.*, 1963; Frazier and Leaf, 1963; Finn, 1974). Recent investigations of a variety of tight epithelia have provided evidence for both a $K^+$-diffusive pathway, blockable by $Ba^{2+}$, and an electrogenic pump, blockable by ouabain, as well as a $Cl^-$ permeability in the basolateral membrane (Lewis *et al.*, 1978; Helman *et al.*, 1979; Nielson, 1979; Kirk *et al.*, 1980; Nagel, 1979, 1980; Nagel *et al.*, 1980). Since the quantitative contributions of these parallel elements to the PD and conductance of the membrane are difficult to assess, there is no clear theoretical basis for predicting the extent of the voltage and resistance changes that would result from the use of high-$K^+$ solutions. These must be determined experimentally.

Bricker *et al.* (1963), using the isolated frog skin, showed that replacement of serosal $Na^+$ with $K^+$ resulted in a reduction of the transepithelial PD and resistance. Using strophanthidin and $CN^-$, they provided evidence that the short-circuit current under these conditions represented, in part, the active transport of $Na^+$, even though the net flux of $Na^+$ was down an electrochemical activity gradient. The short-circuit current fell rapidly after exposure to high $K^+$ and then recovered. Similar findings were reported in the toad bladder by Robinson and Macknight (1976). Replacement of serosal $Na^+$ with $K^+$ resulted in a transient decrease in short-circuit current followed by recovery. In the steady state, transepithelial PD and resistance were decreased, while short-circuit current returned to control levels.

The electrophysiological observations of the depolarized toad urinary bladder discussed in Section II, and assessed below, were made with a high concentration of KCl (85 m*M*) in the serosal medium (Palmer *et al.*, 1980). $Cl^-$ was chosen as the anion because the bladders maintained larger and more stable rates of $Na^+$ transport than when an "impermeant" anion, such as sulfate or isethionate, was used. In addition, the presence of $Cl^-$ might further decrease basolateral membrane resistance, as was reported for frog skin (Rawlins *et al.*, 1970), if the membrane has a significant conductance for that ion. To prevent swelling of the epithelial cells, 50 m*M* sucrose was added to the serosal medium. This solution is isotonic to the normal NaCl–Ringer's solution used and has a $K^+$ ion activity similar to that reported for toad bladder epithelial cells (DeLong and Civan, 1978, 1980).

In the next section, I discuss the completeness of the functional elimination of the basolateral membrane in this preparation.

## IV. EVALUATION OF DEPOLARIZATION

### A. Basolateral Membrane Voltage

Palmer *et al.* (1980) made use of a voltage-dependent conductance in the apical membrane to estimate the shift in the series basolateral PD after application of the KCl medium. The transepithelial voltage at which the apical conductance changed shifted by about 55 mV, indicating the change of this magnitude in the basolateral PD. Unfortunately, the starting membrane potential, in normal Ringer's solution, was not known, and there is no general agreement on the magnitude of this potential in the toad bladder. Under short-circuited conditions, most studies with intracellular microelectrodes have indicated intracellular potentials of 0–10 mV, cell negative (Frazier, 1962; DeLong and Civan, 1978). With improved microelectrode techniques, DeLong and Civan (1980) reported cell potentials of −10 to −20 mV, and Sudou and Hoshi (1977) found basolateral potentials of −17 mV in the presence of mucosal amiloride.

The basolateral PD is apparently much larger in closely related epithelia such as *Necturus* urinary bladder (−80 to −90 mV, Higgins *et al.*, 1977) and frog skin (around −100 mV, Nagel, 1976; Helman and Fisher, 1977). These differences may be due in part to artifacts arising from impalement damage of the toad bladder cells (Lindemann, 1975; Higgins *et al.*, 1977; DeLong and Civan, 1980). Indeed, Leader and Macknight (1982) used a noninvasive technique—accumulation of a lipophilic cation—to measure membrane potential in the toad bladder, and reported a mean cell potential of −62 mV under short-circuited conditions. On the other hand, Narvarte and Finn (1980) impaled the toad bladder cells from the serosal side, a technique believed to minimize impalement damage (Higgins *et al.*, 1977), and reported results similar to those obtained with impalements from the mucosal side. Clearly the nature and magnitude of the basolateral PD in the toad bladder are yet to be completely elucidated.

Direct measurements of cell potentials in toad bladders exposed to high serosal $K^+$ were made by Frazier and Leaf (1963). They reported no significant effect of high $K^+$ on the basolateral PD, which was found to be 10–30 mV, cell negative. They found, in addition, that a large fraction of the total transcellular resistance (around 80%) was between the microelectrode and the serosal solution under these conditions. In light of the discussion above, and the resistance results presented below, it is questionable whether these measurements were made from the cytoplasm of undamaged cells.

Direct measurements of basolateral PD in $K^+$-depolarized epithelia other than the toad bladder have been reported. In the frog skin, Nagel (1977) found cell potentials of −10 to −35 mV after depolarization with either KCl or $K_2SO_4$. Lewis *et al.* (1978) measured basolateral potentials of 10–15 mV in the rabbit urinary bladder with high serosal $K^+$. Thompson *et al.* (1982) found that the cell potential in the rabbit descending colon was virtually abolished by high serosal $K^+$.

In summary, the change in the basolateral PD estimated by Palmer *et al.* (1980) is comparable in magnitude to the largest potentials that have been reported using either microelectrodes or lipophilic cation distribution. Thus the assumption of a basolateral PD of zero in the depolarized state is not unreasonable. The uncertainty involved in this assumption is large, however—perhaps as much as 20 mV.

## B. Resistance

A second assumption that is made in using depolarized epithelia is that the basolateral membrane resistance becomes negligible relative to that of the apical membrane. In nondepolarized toad bladders, resistance ratios (apical:basolateral) of 1:1 to 2:1 are generally obtained, using intracellular microelectrodes (Reuss and Finn, 1974; DeLong and Civan, 1978; Narvarte and Finn, 1980). These estimates are also subject to some uncertainties from possible impalement damage, and ratios of 3:1 to 10:1 have been reported in frog skin and *Necturus* urinary bladder (Nagel, 1977; Helman and Fisher, 1977; Higgins *et al.*, 1977). Several lines of evidence support the notion, however, that the basolateral resistance is substantial in nondepolarized toad bladders and that this resistance decreases markedly with depolarization.

First, application of the KCl medium substantially increased the conductance of the active transport pathway, defined as the conductance sensitive to mucosal amiloride or removal of mucosal $Na^+$ in the short-circuited state (Palmer *et al.*, 1980). The rate of transport, monitored as the short-circuit current, was reduced by about 25%, whereas the active conductance increased by about 100%. The amiloride-insensitive conductance, thought to be largely paracellular, was unchanged.

The conductances of both membranes were probably increased. Since the apical membrane $Na^+$ conductance can be described by the constant field equation (Fig. 1), the expected change in conductance and permeability for an estimated 55-mV change in membrane potential and a 25% decrease in net flux can be calculated. The intracellular $Na^+$ activity is assumed to remain low (one-tenth that of the mucosal solution) under both conditions. The calculated apical conductance increase is 34%. This leaves a substantial conductance change that can be attributed to the basolateral membrane. Modeling the active pathway with two resistors in series and assuming an initial resistance ratio of 1:1, the basolateral conductance must have increased by about fourfold. For an initial resistance ratio of 2:1, the basolateral resistance must disappear completely to account for the data, at least with this simple model.

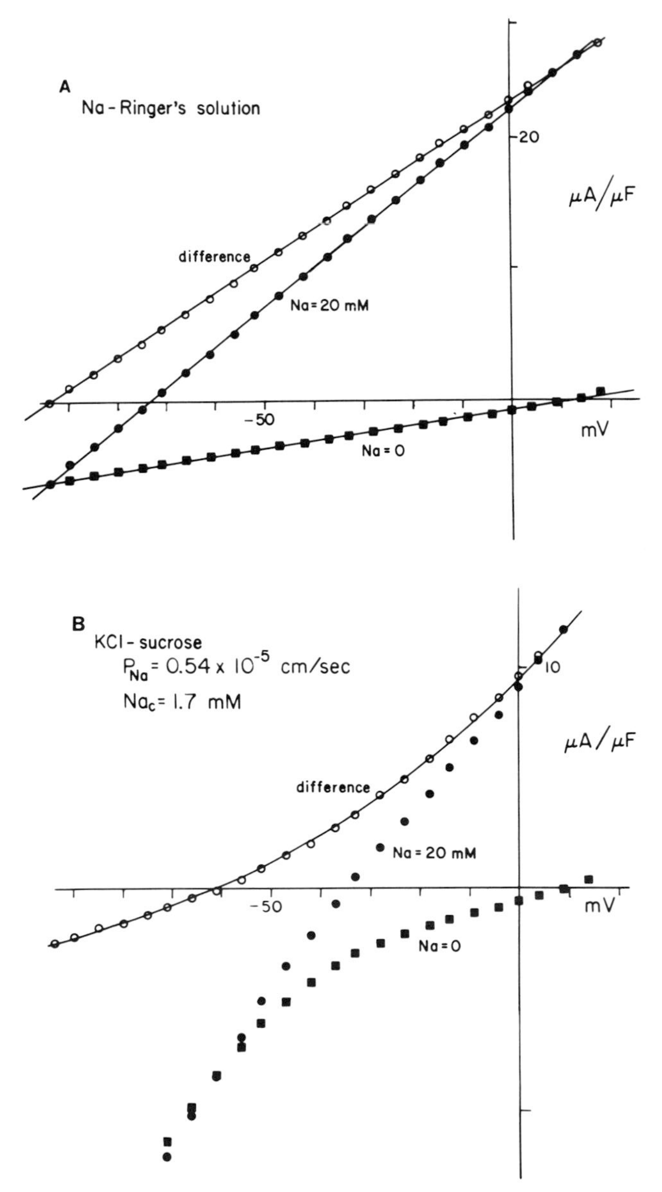


FIG. 1. "Instantaneous" transepithelial current–voltage relationships of toad urinary bladder measured with $Na^+$-free (filled squares) and with 20 m*M* $Na^+$ (filled circles) mucosal solutions. The difference curve (open circles) was obtained by subtracting the two currents at each voltage. (A) The serosal medium was NaCl–Ringer's solution. The solid lines were drawn by eye through the data points. (B) Data from the same bladder 60 minutes after the serosal solution was replaced with KCl–sucrose medium. The solid line was obtained by fitting the data with the constant field equation for a single permeant ion. $Na^+$ permeability was $0.54 \times 10^{-5}$ cm/second and intracellular $Na^+$ activity was 1.7 m*M*. From Palmer *et al.* (1980). Reprinted with permission from *Journal of Membrane Biology*.

Further evidence for a large increase in the resistance ratio is indicated by the current–voltage characteristics of the epithelium (Fig. 1). Under control (nondepolarized) conditions, the *I–V* plots of the transepithelial active transport pathway are roughly linear over a substantial voltage range. After depolarization, the relationship becomes curvilinear and can be described by the constant field equation. Presumably a decrease in the basolateral membrane resistance unmasks the true apical *I–V* relationship. An alternative explanation—that the apical *I–V* characteristic is itself altered by the high-serosal-$K^+$ solution—is unlikely, in light of the findings of Thompson *et al.* (1982) that very similar *I–V* plots are obtained with extracellular electrodes after depolarization and with an intracellular electrode using nondepolarized tissues, at least in the rabbit colon.

An independent approach to this problem involves the use of ac impedance analysis to estimate individual membrane resistances. Using this technique, Warncke and Lindemann (1981) observed two impedance loci under control conditions in the toad bladder. The higher frequency impedance was identified with that of the apical membrane, which had a resistance comparable to, but a capacitance smaller than, that of the basolateral membrane. After depolarization with high serosal $K^+$, the low-frequency component disappeared, implying that under these conditions the transepithelial resistance is dominated by that of the apical membrane.

A similar experiment, in which the response to a transepithelial constant-current pulse was analyzed, is shown in Fig. 2. In the nondepolarized condition, the voltage transient can be described by the sum of two exponentials, as expected for the charging of two capacitors in series, corresponding to the apical and basolateral membranes. In the same tissue after $K^+$ depolarization, a single exponential describes the voltage transient rather well, although at short times, corresponding to high frequencies, deviations are often observed. Once again, this indicates that either the basolateral resistance has become negligibly small relative to that of the apical membrane, or that the time constants of the two membranes have become too similar to be distinguished. In the latter case, a very large fall in basolateral resistance is still required, assuming that the membrane capacitance remains constant.

It is not clear why the basolateral conductance increases so dramatically. If the basolateral $K^+$ permeability can be modeled by the constant field equation, as in the rabbit urinary bladder (Lewis *et al.*, 1978), a three- to fourfold increase in the $K^+$ conductance of this membrane is predicted from an 85 m*M* increase in $K^+$ concentration and a 55-mV depolarization. This probably does not entirely account for the observed conductance changes. Another possibility is that a $Cl^-$ conductance in the basolateral membrane is activated at depolarizing voltages. This idea was suggested by Ussing (1982) for frog skin to explain certain features of volume regulation in that tissue.

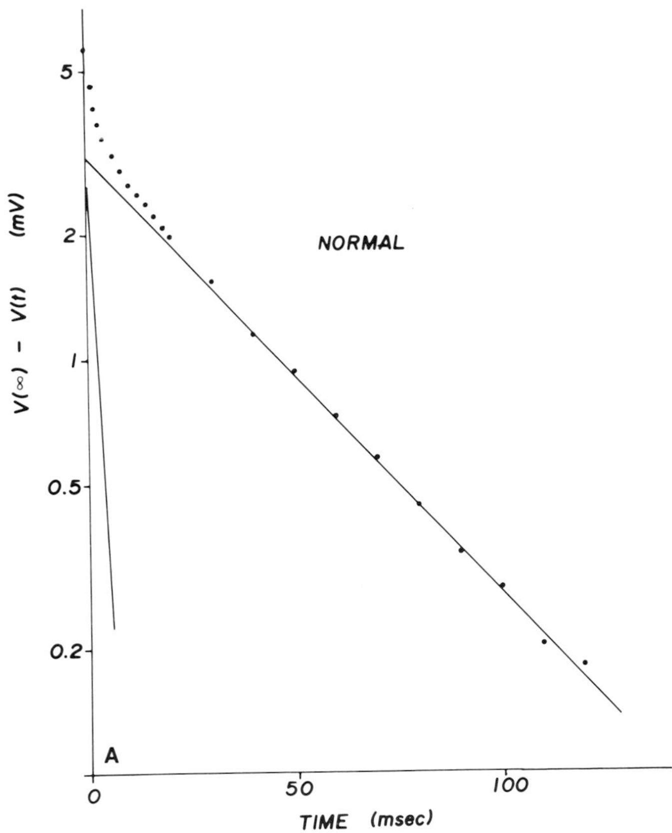


FIG. 2. Impedance analysis of toad urinary bladder. Bladders were open circuited, except for brief (250 msec) pulses of constant current sufficient to change the transepithelial voltage by 5–6 mV. Details of the experimental setup are given by Palmer and Lorenzen (1983). The difference between the steady state voltage $V_\infty$ and the pre-steady state voltage $V(t)$ is plotted semilogarithmically as a function of time. The data are fitted with functions of the form $A \exp(-t/\tau_m)$. (A) Mucosal solution contained 115 m*M* NaCl; the serosal solution was NaCl–Ringer's solution. The data could be described by two exponentials with $A = 3.1$ mV and $\tau_m = 42$ msec, and $A = 2.65$ and $\tau_m = 2.26$ msec. The open-circuit PD was 60 mV, and the transepithelial resistance was 4.35 kΩ

## C. Analysis of Errors

Incomplete elimination of the basolateral membrane potential and resistance will result in errors in the estimation of apical membrane permeability, single-channel conductances, and the intracellular $Na^+$ activity ($[Na^+]$) obtained from reversal potentials. Of these parameters, the conductance and permeability will probably be the least affected. The true apical membrane $Na^+$ conductance $g_a$ will be related to the measured conductance $g'_a$ by the relationship

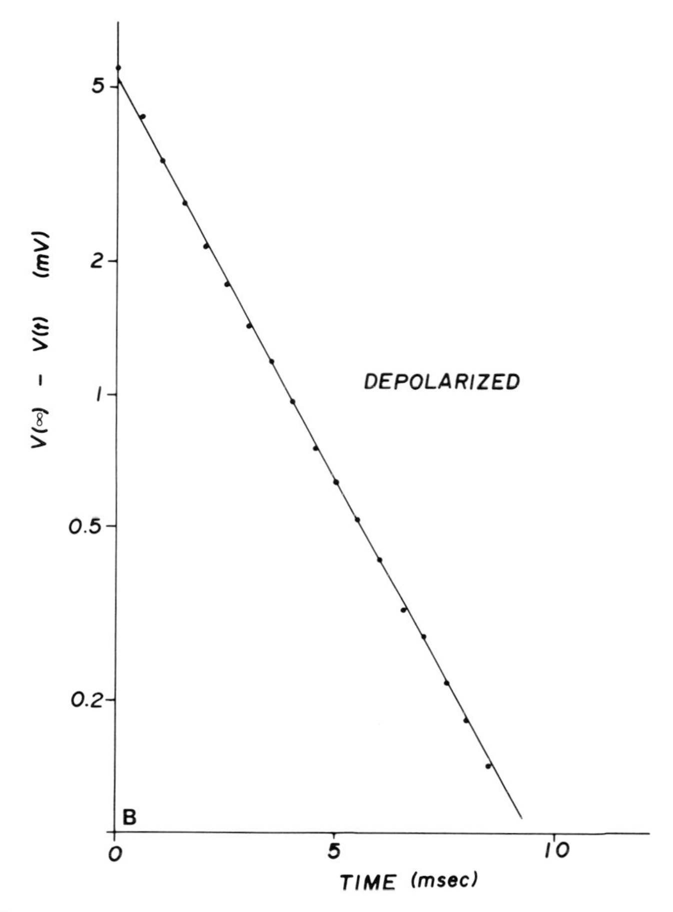


cm². (B) The same bladder as in (A), 90 minutes after replacing the serosal solution with KCl–sucrose medium. The mucosal solution was the same as in (A). The data could be described by a single exponential with $A = 5.2$ mV and $\tau_m = 2.43$ msec. (Small deviations from this curve were seen at times less than 1 msec.) The open circuit PD was 48 mV, and the transepithelial resistance was 2.47 kΩ cm².

$$g_a/g'_a = 1 + r$$

where $r$ is the ratio of basolateral to apical resistance. A similar equation holds for the single-channel conductance or current. From the dc and ac impedance measurements discussed above, $r$ must be at most 1:4 and is probably less than 1:10. For many types of experiments this degree of systematic error is tolerable.

The measurement of $[Na^+]$ is more susceptible to error. A series potential difference $V_s$ would shift the measured $I$–$V$ curves along the voltage axis with respect to the true apical $I$–$V$ relationship, so that the reversal potential would be in error by the same amount. The relationship between the true $[Na^+]$ and measured $[Na^+]'$ will be given by

$$[Na^+]/[Na^+]' = \exp(FV_s/RT)$$

where $R$, $T$, and $F$ have their usual meanings. If $V_s$ is 10 mV, cell negative with respect to serosa, $[Na^+]$ will be underestimated by 33%. It should be stressed, however, that changes in $[Na^+]$ would still be measured at least qualitatively correctly.

## D. Preservation of Apical Membrane Properties

The $K^+$-depolarized epithelium maintains its ability to utilize its active transport system to promote the net movement of $Na^+$ ions from mucosa to serosa. This is inferred from the sensitivity of the short-circuit current to drugs such as amiloride, which blocks $Na^+$ entry, ouabain, which blocks $Na^+$ extrusion from the cell, and metabolic inhibitors (Bricker *et al.*, 1963; Fuchs *et al.*, 1977; Palmer *et al.*, 1980). In the toad bladder, transport also remains sensitive to hormonal modulation by both aldosterone and antidiuretic hormones (Li *et al.*, 1982; Palmer *et al.*, 1982; Palmer and Lorenzen, 1983).

On the other hand, at least one regulatory mechanism governing $Na^+$ transport may be lost after depolarization. Weinstein *et al.* (1980), in a study of current transients resulting from changes in the voltage-clamped PD across the toad bladder, suggested that these transients reflect at least in part a negative feedback system whereby $Na^+$ entry across the apical membrane inhibited $Na^+$ permeability. They found that a number of experimental manipulations, including raising the serosal $K^+$ concentration, affected the magnitude of the transients. An example of this phenomenon is illustrated in Fig. 3, which shows the current response as the transepithelial clamp voltage is changed from 0 to −10 mV and back. With normal NaCl–Ringer's solution on the serosal side, the clamp current showed a modest overshoot before relaxing to a steady state level. After depolarization with KCl–sucrose solution on the serosal side, this overshoot completely vanished. In our studies, we have observed the transient overshoot currents in most, although not all, bladders before depolarization. We have never seen this type of overshoot in a depolarized bladder. If this behavior does reflect a regulatory system for $Na^+$ transport, this system apparently does not operate in depolarized bladders.

One way in which increased $Na^+$ entry might down-regulate $Na^+$ permeability is by $Na^+/Ca^{2+}$ exchange at the basolateral membrane (Taylor and Windhager, 1979; Chase and Al-Awqati, 1981). The subsequent increase in cytoplasmic $Ca^{2+}$ would reduce $Na^+$ permeability. The absence of such a feedback loop in the depolarized epithelium would imply either that the apical

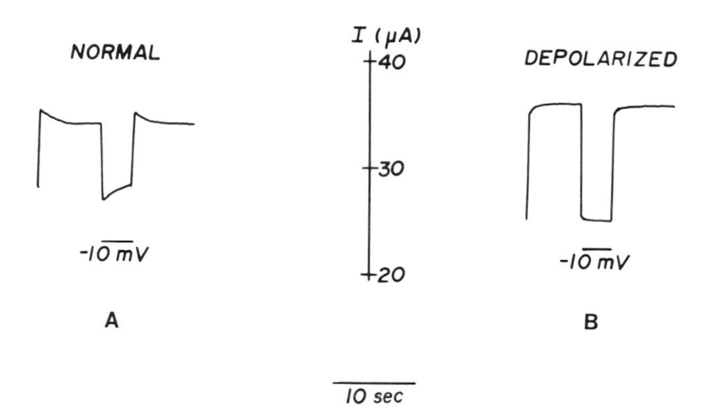


FIG. 3. Slow current transients in toad urinary bladder. Bladders were voltage clamped at 0 mV except for 5-second periods denoted by horizontal bars, when the potential was −10 mV (mucosa negative). In trace A, the mucosal solution contained 115 m*M* NaCl. The serosal solution was NaCl–Ringer's solution. Trace B was of the same bladder, 60 minutes after replacing the serosal solution with KCl–sucrose medium. The mucosal solution was the same as on the left. Note that the short-circuit currents were similar in the two cases, but the transepithelial conductance, given by the steady state deflection in current after changing the voltage, was larger with KCl–sucrose.

membrane becomes insensitive to $Ca^{2+}$ or, more likely, that processes other than $Na^+/Ca^{2+}$ exchange can maintain low cytoplasmic $Ca^{2+}$ activities under these conditions. For example, a $Na^+$-independent ATP-driven $Ca^{2+}$ pump, such as those described in the plasma membranes of red cells (Schatzmann, 1966) and squid axons (DiPolo, 1978), might be operative. This mechanism, if it is electrogenic, might be stimulated by depolarization of the membrane. This idea would be consistent with the observations that $Na^+$ transport is not inhibited when serosal NaCl is replaced by KCl (Robinson and Macknight, 1976; Palmer *et al.*, 1980), whereas replacement with choline chloride reduces $Na^+$ transport, presumably via inhibition of $Na^+/Ca^{2+}$ exchange (Taylor and Windhager, 1979; Chase and Al-Awqati, 1981).

Other recent reports have suggested that the permeabilities and transport properties of the apical and basolateral membranes may be intimately related by mechanisms which are not yet understood (Schultz, 1981; Davis and Finn, 1981). Such interactions may be suppressed, altered, or missing altogether when the basolateral membrane is depolarized.

For example, in many tight epithelia the rate of transport of $Na^+$ across the apical membrane saturates as mucosal $Na^+$ concentration is increased, implying a decrease in the apparent apical membrane permeability to $Na^+$. From this simple observation, it is not clear whether the mucosal $Na^+$ itself or an increase in cytoplasmic $Na^+$ is responsible for the decreased permeability. Fuchs *et al.* (1977), using fast-flow techniques in conjunction with current–voltage analysis on the depolarized frog skin, observed a decrease in $Na^+$ permeability upon increasing mucosal $Na^+$, before significant changes in intracellular Na occurred. In the toad bladder, Li *et al.* (1982) reported similar decreases in permeability in the steady state, associated with small increases in cell $Na^+$. Since much larger changes in cell $Na^+$ in ouabain-treated bladders had little effect on $Na^+$ permeability (Palmer *et al.*, 1980), it was concluded that extracellular $Na^+$ itself could down-regulate $Na^+$ permeability. While it seems reasonable to suppose that the same process operates in nondepolarized tissues, the possibility of an additional regulatory system which involves increases in cell $Na^+$ and is suppressed in the depolarized state should not be ruled out.

Similar problems arise in the study of hormonal regulation of transport. Palmer *et al.* (1982) found that the stimulation of $Na^+$ transport by aldosterone in the depolarized toad bladder could be accounted for by an increase in $Na^+$ permeability of the apical membrane arising from the recruitment of additional $Na^+$ channels. Increased extrusion of $Na^+$ from the cell could be explained by an increase in cell $Na^+$, which in turn would stimulate the turnover rate of the basolateral $Na^+$ pumps. In normal preparations the action of the hormone may not be so simple. Spooner and Edelman (1975) reported that during a late phase of hormone action, several hours after administration of aldosterone, $Na^+$ transport increased without a significant change in transepithelial resistance. In cultured epithelial cells from toad kidney, Handler *et al.* (1981) found that the number of $Na^+$ pump units in the basolateral membrane was increased in response to steroids, although this increase was apparently secondary to an increase in apical $Na^+$ permeability. Similar findings have been reported in the mammalian cortical collecting duct (Petty *et al.*, 1981). These results again point to significant interactions between apical and basolateral membranes which may well be lost with $K^+$ depolarization.

## V. CONCLUSIONS

In the toad bladder, depolarization with serosal KCl appears to be an effective method for reducing both the PD and resistance of the basolateral membrane, making possible studies of the electrical properties of the apical membrane using extracellular electrodes. This approach can be viewed as a stopgap measure for experiments in which other techniques such as intracellular recording and ac impedance analysis are impractical. The results should be viewed with regard to the more physiological state of the tissue, in which more complex regulatory processes may be present. Finally, the results obtained from this technique should be confirmed in nondepolarized preparations using more sophisticated methods or more suitable model epithelia.

## ACKNOWLEDGMENTS

I am grateful for support from U.S. Public Health Service Grant AM 27847, and from a grant from the Whitaker Foundation.

## REFERENCES

Bricker, N. S., Biber, T., and Ussing, H. H. (1963). Exposure of the isolated frog skin to high potassium concentrations at the internal surface. I. Bioelectric phenomena and sodium transport. *J. Clin. Invest.* **42,** 88–99.

Chase, H. S., Jr., and Al-Awqati, Q. (1981). Regulation of the sodium permeability of the luminal border of toad bladder by intracellular sodium and calcium. *J. Gen. Physiol.* **77,** 693–712.

Davis, C. W., and Finn, A. L. (1981). Sodium transport inhibition by amiloride reduces baso-lateral membrane potassium conductance in tight epithelia. *Science* **216,** 525–527.

DeLong, J., and Civan, M. M. (1978). Dissociation of cellular $K^+$ accumulation from net $Na^+$ transport by toad urinary bladder. *J. Membr. Biol.* **42,** 19–43.

DeLong, J., and Civan, M. M. (1980). Intracellular chemical activity of potassium in toad urinary bladder. *Curr. Top. Membr. Transp.* **13,** 93–105.

DiPolo, R. (1978). Ca pump driven by ATP in squid axons. *Nature (London)* **274,** 390–392.

Finn, A. L. (1974). Transepithelial potential difference in the toad urinary bladder is not due to ionic diffusion. *Nature (London)* **250,** 495–496.

Frazier, H. S. (1962). The electrical potential profile of the isolated toad bladder. *J. Gen. Physiol.* **45,** 515–528.

Frazier, H. S., and Leaf, A. (1963). The electrical characteristics of active sodium transport in the toad bladder. *J. Gen. Physiol.* **46,** 491–503.

Frizzell, R. A., and Schultz, S. G. (1978). Effect of aldosterone on ion transport by rabbit colon *in vitro. J. Membr. Biol.* **39,** 1–26.

Fuchs, W., Hviid Larsen, E., and Lindemann, B. (1977). Current–voltage curve of sodium channels and concentration dependence of sodium permeability in frog skin. *J. Physiol. (London)* **267,** 137–166.

Gatzy, J. T., and Clarkson, T. W. (1965). The effect of mucosal and serosal solution cations on bioelectric properties of isolated toad bladder. *J. Gen. Physiol.* **48,** 647–671.

Handler, J. S., Preston, A. S., Perkins, F. M., and Matsumura, M. (1981). *Ann. N.Y. Acad. Sci.* **372,** 442–454.

Helman, S. I., and Fisher, R. S. (1977). Microelectrode studies of the active Na transport pathway of frog skin. *J. Gen. Physiol.* **69,** 571–604.

Helman, S. I., Nagel, W., and Fisher, R. S. (1979). Ouabain on active transepithelial sodium transport in frog skin. Studies with microelectrodes. *J. Gen. Physiol.* **74,** 105–127.

Higgins, J. T., Jr., Gebler, B., and Frömter, E. (1977). Electrical properties of amphibian urinary bladder epithelia II. The cell potential profile in *Necturus maculosus. Pfluegers Arch.* **371,** 87–97.

Kirk, K. L., Halm, D. R., and Dawson, D. C. (1980). Active Na transport by turtle colon via an electrogenic Na–K exchange pump. *Nature (London)* **287,** 237–239.

Koefoed-Johnsen, V., and Ussing, H. H. (1958). The nature of the frog skin potential. *Acta Physiol. Scand.* **42,** 298–308.

Leader, J. P., and Macknight, A. D. C. (1982). Alternative methods for measurement of membrane potentials in epithelia. *Fed. Proc. Fed. Am. Soc. Exp. Biol.* **41,** 54–59.

Leb, D. E., Hoshiko, T., and Lindley, B. D. (1965). Effects of alkali metal cations on the potential across toad and bullfrog urinary bladder. *J. Gen. Physiol.* **48,** 527–540.

Lewis, S. A., and Diamond, J. M. (1976). $Na^+$ transport by rabbit urinary bladder, a tight epithelium. *J. Membr. Biol.* **28,** 1–40.

Lewis, S. A., Wills, N. K., and Eaton, D. C. (1978). Basolateral membrane potential of a tight epithelium: Ionic diffusion and electrogenic pumps. *J. Membr. Biol.* **41,** 117–148.

Li, J. H.-Y., Palmer, L. G., Edelman, I. S., and Lindemann, B. (1982). The role of sodium-channel density in the natriferic response of the toad urinary bladder to an antidiuretic hormone. *J. Membr. Biol.* **64,** 77–89.

Lindemann, B. (1975). Impalement artifacts in microelectrode recordings of epithelial membrane potentials. *Biophys. J.* **15,** 1161–1164.

Lindemann, B., and DeFelice, L. J. (1981). On the use of general network functions in the evaluation of noise spectra obtained from epithelia. *In* "Ion Transport by Epithelia" (S. G. Schultz, ed.), pp. 1–13. Raven, New York.

Lindemann, B., and Van Driessche, W. (1977). Sodium specific membrane channels of frog skin are pores: Current fluctuations reveal high turnover. *Science* **195,** 292–294.

Morel, F., and LeBlanc, G. (1975). Transient current changes and Na compartmentalization in frog skin epithelium. *Pfluegers Arch.* **358,** 135–157.

Muller, J., Kachadorian, W. A., and DiScala, V. A. (1980). Evidence that ADH-stimulated intramembrane particle aggregates are transferred from cytoplasmic to luminal membranes in toad bladder epithelial cells. *J. Cell Biol.* **85,** 83–95.

Nagel, W. (1976). The intracellular electrical potential profile of the frog skin epithelium. *Pfluegers Arch.* **365,** 135–143.

Nagel, W. (1977). Effect of high K upon the frog skin intracellular potential. *Pfluegers Arch.* **368,** R22.

Nagel, W. (1979). Inhibition of potassium conductance by barium in frog skin epithelium. *Biochim. Biophys. Acta* **552,** 346–357.

Nagel, W. (1980). Rheogenic sodium transport in a tight epithelium, the amphibian skin. *J. Physiol. (London)* **302,** 281–295.

Nagel, W., Pope, M. B., Peterson, K., and Civan, M. M. (1980). Electrophysiological changes associated with potassium depletion of frog skin. *J. Membr. Biol.* **57,** 235–241.

Narvarte, J., and Finn, A. L. (1980). Microelectrode studies in toad urinary bladder. Effects of Na concentration changes in the mucosal solution on equivalent electromotive forces. *J. Gen. Physiol.* **75,** 323–344.

Nielson, R. (1979). A 3 to 2 coupling of the Na–K pump in frog skin disclosed by the effect of Ba. *Acta Physiol. Scand.* **107,** 189–191.

Palmer, L. G. (1982a). $Na^+$ transport and flux ratio through apical $Na^+$ channels in toad bladder. *Nature (London)* **297,** 688–690.

Palmer, L. G. (1982b). Ion selectivity of the apical membrane Na channel in the toad urinary bladder. *J. Membr. Biol.* **67,** 91–98.

Palmer, L. G., and Lorenzen, M. (1983). Antidiuretic hormone-dependent membrane capacitance and water permeability in the toad urinary bladder. *Am. J. Physiol.* **244,** F195–F204.

Palmer, L. G., Edelman, I. S., and Lindemann, B. (1980). Current-voltage analysis of apical Na transport in the toad urinary bladder: Effects of inhibitors of transport and metabolism. *J. Membr. Biol.* **57,** 59–71.

Palmer, L. G., Li, J. H.-Y., Lindemann, B., and Edelman, I. S. (1982). Aldosterone control of the density of sodium channels in the toad urinary bladder. *J. Membr. Biol.* **64,** 91–102.

Petty, K. J., Kokko, J. P., and Marver, D. (1981). Secondary effect of aldosterone on Na-K ATPase activity in the rabbit cortical collecting tubule. *J. Clin. Invest.* **68,** 1514–1521.

Rawlins, F., Maten, L., Fragachan, F., and Whittemburg, G. (1970). Isolated toad skin epithelium: Transport characteristics. *Pfluegers Arch.* **316,** 64–80.

Reuss, L., and Finn, A. L. (1974). Passive electrical properties of toad urinary bladder epithelium:

Intracellular electrical coupling and transepithelial cellular and shunt conductances. *J. Gen. Physiol.* **64,** 1–25.

Robinson, B. A., and Macknight, A. D. C. (1976). Relationships between serosal medium K concentration and Na transport in toad urinary bladder. *J. Membr. Biol.* **26,** 217–238.

Schatzmann, H. J. (1966). ATP-dependent $Ca^{++}$ extrusion from human red cells. *Experientia* **22,** 364–368.

Schultz, S. G. (1981). Homocellular regulatory mechanisms in sodium-transporting epithelia: Avoidance of extinction by "flush-through." *Am. J. Physiol.* **241,** F579–F590.

Spooner, P. M., and Edelman, I. S. (1975). Further studies on the effect of aldosterone on electrical resistance of toad bladder. *Biochim. Biophys. Acta* **406,** 304–314.

Sudou, K., and Hoshi, T. (1977). Mode of action of amiloride in toad urinary bladder: An electrophysiological study of the drug action on sodium permeability of the mucosal border. *J. Membr. Biol.* **32,** 115–132.

Stetson, D. L., Lewis, S. A., Alles, W., and Wade, J. B. (1982). Evaluation by capacitance measurements of antidiuretic hormone induced membrane area changes in toad bladder. *Biochim. Biophys. Acta* **689,** 267–274.

Taylor, A., and Windhager, E. E. (1979). Possible role of cytosolic calcium and Na-Ca exchange in regulation of transepithelial sodium transport. *Am. J. Physiol.* **236,** F505–F512.

Thompson, S. M., Suzuki, Y., and Schultz, S. G. (1982). The electrophysiology of rabbit descending colon. I. Instantaneous transepithelial current-voltage relations and the current-voltage relations of the Na-entry mechanism. *J. Membr. Biol.* **66,** 41–54.

Ussing, H. H. (1982). "Pathways for Transport in Epithelia in Functional Regulation at the Cellular and Molecular Level" (R. A. Corradino, ed.), pp. 285–297. North-Holland Publ., Amsterdam.

Van Driessche, W., and Gögelein, H. (1980). Attenuation of current and voltage noise signals recorded from epithelia. *J. Theor. Biol.* **86,** 629–648.

Van Driessche, W., and Lindemann, B. (1979). Concentration dependence of currents through single sodium-selective pores in frog skin. *Nature (London)* **282,** 519–520.

Warncke, J., and Lindemann, B. (1981). Effect of ADH on the capacitance of apical epithelial membranes. *Adv. Physiol. Sci. Proc. Int. Congr., 28th, 1980* **3,** 128–133.

Weinstein, F. C., Rosowski, J. J., Peterson, K., Delalic, Z., and Civan, M. M. (1980). Relationship of transient electrical properties to active Na transport by toad urinary bladder. *J. Membr. Biol.* **52,** 25–35.

# Part II

# Use of Antibodies to Epithelial Membrane Proteins

# Chapter 7

# Biosynthesis of $Na^+,K^+$-ATPase in Amphibian Epithelial Cells

*B. C. ROSSIER*

*Institut de Pharmacologie*
*de l'Université de Lausanne*
*Lausanne, Switzerland*



## I. INTRODUCTION AND GENERAL BACKGROUND

The sodium- and potassium-dependent adenosinetriphosphatase $Na^+,K^+$-ATPase (ATP phosphohydrolase, EC 3.6.1.3) is a plasma membrane protein which has been identified as the enzymatic expression of the sodium pump (Robinson and Flashner, 1979; Wallick *et al.*, 1979; Jørgensen, 1980; Levitt, 1980). $Na^+,K^+$-ATPase serves two functions: the extrusion of sodium out of the intracellular compartment and the accumulation of potassium into the same compartment. The cell devotes a large fraction (2–40%) of its ATP production for this single function (Smith, 1979). In the most extreme cases (epithelial cells), up to 70% of the oxydative metabolism can be coupled to ouabain-sensitive sodium and potassium transport (Le Bouffant *et al.*, 1982). It is, therefore, not surprising that this process plays a critical role in the homeostasis of sodium and potassium not only in the intracellular but also in the extracellular




ISBN 0-12-153320-4

milieu. In all cells, $Na^+,K^+$-ATPase maintains a high intracellular $K^+$ concentration (~140 m*M*) required for the macromolecule biosynthesis machinery. In epithelial cells, $Na^+,K^+$-ATPase is one of the major factors allowing *transcellular* salt transport, thereby controlling the osmolarity of the extracellular compartment. Epithelia are, therefore, able to create and maintain a large hydroosmotic gradient between the extracellular space and the external environment.

At the cellular level, one can explain such basic asymmetry by the differentiation of the epithelial plasma membrane into three major domains: the basolateral membrane, the junctional apparatus, and the apical membrane. These three domains, together with specific elements of the underlying cytoskeleton, determine the so-called "polarity" of epithelial cells. Physiological and pharmacological evidence suggests that $Na^+,K^+$-ATPase is selectively distributed in the basolateral membrane while specific sodium channels are concentrated in the apical membrane. Such cellular organization is compatible with the physiological model of transepithelial sodium transport first developed by Ussing in the early 1950s (see review in Ussing *et al.*, 1974).

Transepithelial sodium transport and the sodium pump have been extensively studied over the last 30 years and are rather well understood in physiological and pharmacological terms. For instance, in the erythrocyte model (Sen and Post, 1964) and the giant axon of the squid (Caldwell, 1969), the stoichiometry of the coupled $Na^+$–$K^+$ transport has been fairly well established:

$$MgATP + 2K^+_{ext} + 3Na^+_{int} \rightleftharpoons 2K^+_{int} + 3Na^+_{ext} + MgADP + P_i$$

The same stoichiometry has also been observed in reconstitution experiments using highly purified $Na^+,K^+$-ATPase (Hilden and Hokin, 1975; Sweadner, 1979; Anner *et al.*, 1977; Hokin and Dixon, 1979).

The affinity constant ($K_m$ = 0.1–0.6 m*M*) for the overall hydrolytic activity of ATP for the enzyme is remarkably similar in all species from the different classes of vertebrates (Geering and Rossier, 1979; Girardet, 1982). The activation constant ($K_{1/2}$) for $Na^+$ is also very similar (12–20 m*M*) as is the $K_{1/2}$ for $K^+$. By contrast, the $K_i$ (inhibition constant) of ouabain, a noncompetitive inhibitor for the $K^+$-dependent reaction, varies greatly from species to species or even from tissue to tissue within the same animal. Contrary to the highly conserved sites for ATP, $Na^+$, and $K^+$, the ouabain binding site is highly variable.

Much less is known about the molecular structure and organization of $Na^+,K^+$-ATPase. After the discovery of the enzyme in microsomes from peripheral nerves (Skou, 1957), it took more than 10 years before $Na^+,K^+$-ATPase could be obtained in a reasonably pure form (Kyte, 1971; Uesugi *et al.*, 1971; Nakao *et al.*, 1973; Hokin *et al.*, 1973; Dixon and Hokin, 1974; Lane *et al.*, 1973; Jørgensen, 1974a,b; Geering and Rossier, 1979; Winter and Moss, 1979; Esmann *et al.*, 1979; Koepsell, 1978; Peterson and Hokin, 1980) from different organs of various species from fish to mammals. The enzyme has been amazingly conserved throughout the evolution.

On SDS–PAGE, the enzyme is resolved in two subunits: the α subunit ($M_r$ = 96,000), which carries the catalytic determinant and the ouabain binding sites, and the β subunit ($M_r$ = 45,000–65,000), a glycoprotein. Both subunits are transmembranous polypeptides (Karlish *et al.*, 1977; Farley *et al.*, 1980; Girardet *et al.*, 1981, 1983). The physiological role of the β subunit remains uncertain, but its presence seems required in order to reconstitute enzymatic function into liposomes. The amino acid composition of both polypeptides has been determined in a number of preparations (Peterson and Hokin, 1980; Lane *et al.*, 1979; Hopkins *et al.*, 1976; Kyte, 1972; Perrone *et al.*, 1975), but the sequence is not known. With highly purified enzyme preparations it should have been possible to obtain specific immunological probes required for the study of the biosynthesis of the enzyme. Indeed, a number of such probes have been prepared against the holoenzyme and/or the subunits (Kyte, 1974, 1976a,b; Jørgensen *et al.*, 1973; Smith and Wagner, 1975; Askari, 1974; Michael *et al.*, 1977; Koepsell, 1978, 1979; Rhee and Hokin, 1975, 1979; Jean and Albers, 1976; Jean *et al.*, 1975). But for reasons discussed in the next sections, none of these early probes could be used to immunoprecipitate the enzyme complex or each subunit selectively. Thus, very little information on the biosynthesis of the enzyme has become available (Knox and Sen, 1974; Peterson *et al.*, 1978; Churchill and Hokin, 1979; Pollack *et al.*, 1981). In a few studies carried out in the rat (Lo and Edelman, 1976; Lo *et al.*, 1976; Lo and Lo, 1980), the kidney enzyme was labeled with [$^{35}$S]methionine *in vivo*, and the incorporation of the precursor into a partially purified enzyme was then analyzed. In this system the authors were able to demonstrate that thyroid hormone increased in parallel with the relative rate of synthesis of the α and the β subunits without affecting their rate of degradation. *In vitro*, Churchill and Hokin (1979) studied the biosynthesis of $Na^+,K^+$-ATPase by incorporation of labeled valine into highly purified eel electroplax membranes and estimated that the time of insertion into the plasma membrane was roughly 2 hours. The major drawback of the labeling approach is that the incorporation can be studied only in purified enzyme preparations. No information can be obtained on the translational and early posttranslational events since these precursors cannot be distinguished from the general pool of labeled proteins. To study the biosynthesis of a membrane-bound enzyme localized in the basolateral membrane of an epithelial cell, the following basic questions should be answered: How are the two subunits (1) translated; (2) inserted into membranes; (3) processed co- and/or posttranslationally (glycosylation, phosphorylation, etc.); (4) assembled in a mature and potentially active form, directed to the right location, i.e., the basolateral membrane, and finally expressed as a physiologically active unit (sodium pump); (5) degraded, and how does their turnover function? (6) Finally, how is each of these previous steps controlled and regulated?

In order to tackle these questions, specific probes for the α and the β subunits of the enzyme and a suitable experimental model are needed. Four years ago, my colleagues and I decided to study the biosynthesis of $Na^+,K^+$-ATPase in epithelial cells and its possible hormonal control. The general strategy was first to purify $Na^+,K^+$-ATPase and its two subunits, second, to produce monospecific antibodies against each subunit, and third, with the use of these specific probes, to investigate the biosynthesis of our $Na^+,K^+$-ATPase in a suitable epithelial system.

We selected the toad (*Bufo marinus*) urinary bladder for the following reasons. First, it is an *in vitro* physiological model which is very well defined in terms of transepithelial sodium transport (Leaf, 1958). Moreover, it responds *in vitro* to aldosterone and vasopressin, the major hormones controlling sodium and water permeability. Second, highly differentiated epithelial cell lines have been derived from the toad bladder (Handler *et al.*, 1979). This gives the opportunity to study the biosynthesis and its hormonal control in a cell culture system which can be controlled for its responsiveness to steroid hormones. Third, we postulated that the large phylogenetic distance between the toad (our source of antigen) and the rabbit (our source of antibody) would favor production of immunological probes otherwise difficult to obtain between closely related species (intramammalian), since $Na^+,K^+$-ATPase appears to be so highly conserved throughout evolution.

In this chapter, we therefore describe (1) the preparation of toad $Na^+,K^+$-ATPase and its subunits, (2) the preparation of the immunological probes, (3) the biosynthesis of $Na^+,K^+$-ATPase in intact cells (toad urinary bladder in organ culture conditions), and (4) the effects of aldosterone and of thyroid hormone on the biosynthesis of $Na^+,K^+$-ATPase. Finally, we discuss a model describing the major site of action of hormone control of the synthesis and the expression of the enzyme in a polarized epithelial cell.

## II. PURIFICATION OF $Na^+,K^+$-ATPase

The purification of $Na^+,K^+$-ATPase has proven to be a difficult task. Two strategies have been used. The first is termed *positive*, and the aim is to solubilize the enzyme from microsomal membranes with tensioactive reagents such as nonionic detergent (for instance, Lubrol-WX). The solubilized enzyme is separated from the membrane fraction by ultracentrifugation. The detergent can then be removed either by ultracentrifugation (Perrone *et al.*, 1975) or by ion-exchange column chromatography (Nakao *et al.*, 1973). The purified enzyme can then be precipitated by glycerol or ammonium sulfate (Uesugi *et al.*, 1971). The $Na^+,K^+$-ATPase of the rectal gland of the dogfish (Hokin *et al.*, 1973) and that of the electroplax of the eel (Dixon and Hokin, 1974) have been purified according to this strategy. The second procedure is termed *negative*. The aim is to peel off the membrane from contaminants by the use of various detergents at well-defined concentrations that leave the $Na^+,K^+$-ATPase inserted in the membrane. The $Na^+,K^+$-ATPase of the dog kidney has been purified by this technique (Kyte, 1971; Lane *et al.*, 1973). Jørgensen (1974a) published a method using sodium dodecyl sulfate (SDS) as detergent, which allows the purification of the enzyme by a single-step procedure through a sucrose gradient. This technique has been used successfully for the purification of the enzyme from various sources (Hopkins *et al.*, 1976; Geering and Rossier, 1979; McDonough *et al.*, 1982; Winter and Moss, 1979).

The conditions of protein concentration and detergent concentration vary greatly depending on the extracted tissue, and should be adapted in each case. In the case of amphibian $Na^+,K^+$-ATPase (Geering and Rossier, 1979; Girardet *et al.*, 1981), it is possible to partially purify an enzyme having an activity of up to 1200 $\mu M$ $P_i$/mg protein · hour. Three major polypeptide bands were resolved by SDS–PAGE with $M_r$s of 96,000, 60,000, and 32,000, respectively (Girardet *et al.*, 1981). The 96,000-d polypeptide was identified as the α subunit because of its phosphorylation by [$\gamma$-$^{32}$P]ATP in the presence of sodium but not in the presence of potassium. Recently the toad kidney α subunit was shown to exactly comigrate with the rabbit kidney α subunit purified by the same technique (Geering *et al.*, unpublished observation). The 60,000-d polypeptide was identified as a glycoprotein by staining with periodic acid–Schiff and is termed the β subunit. The 32,000-d polypeptide is a proteolipid of unknown function but does not seem to be related (at least immunochemically) to either of the two $Na^+,K^+$-ATPase subunits. Low-molecular-weight contaminants ($M_r$ < 12,000) were frequently observed. Anti-holoenzyme sera were obtained, but the specificity of such sera is always questionable in view of these contaminants. In order to obtain monospecific antibodies, it was mandatory to purify each subunit. This was achieved by preparative SDS–PAGE followed by electrodialysis. The results of such a purification are shown in Fig. 1.

These highly purified subunits can now be used as a source of antigens. They can also be used to check the specificity of the antisera (immunocompetition). Finally, when labeled with $^{125}$I, they are used for the development of a radioimmunoassay.

## III. PREPARATION AND CHARACTERIZATION OF IMMUNOLOGICAL PROBES

Characteristics of antibodies to various $Na^+,K^+$-ATPases (holoenzymes) have been reported since 1972 (Askari and Rao, 1972; Kyte, 1976a,b; Jørgensen *et al.*, 1973; Askari, 1974). These anti-holoenzyme antibodies have different effects on the sodium pump activity, inhibitory in some cases, partially inhibitory or totally inactive in others. Antibodies against the α subunit (Kyte, 1974;

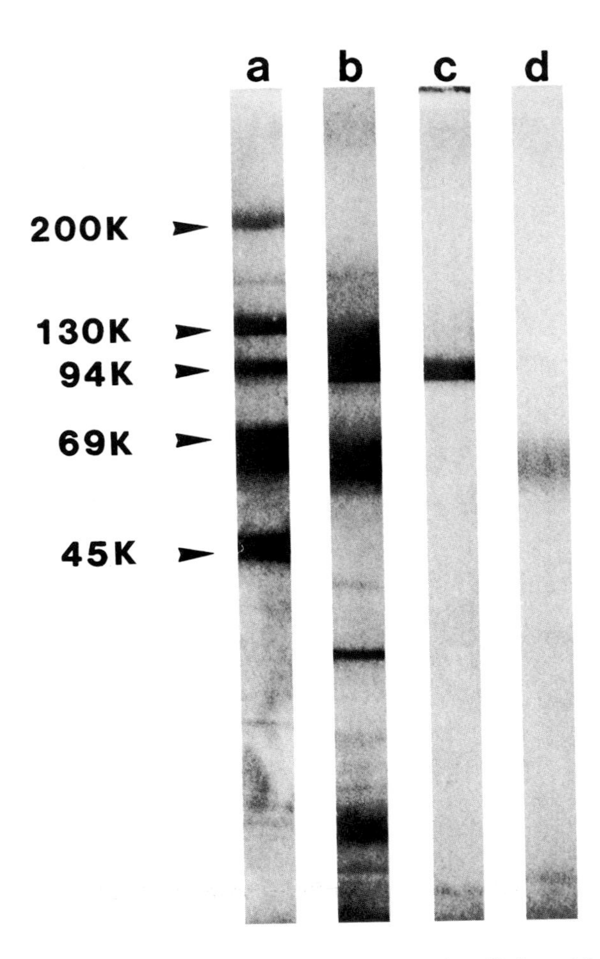


FIG. 1. Partially purified $Na^+,K^+$-ATPase holoenzyme and purified α and β subunits from toad kidney analyzed by 5–13% SDS–PAGE and stained with Coomassie brilliant blue (a–d). (a) Molecular weight markers: myosin (200,000), galactosidase (130,000), phosphorylase *b* (94,000), bovine serum albumin (69,000), and ovalbumin (45,000); (b) partially purified $Na^+,K^+$-ATPase from toad kidney (50 μg SDS detergent-purified enzyme); (c) purified α subunit; (d) purified β subunit. Modified from Girardet *et al.* (1983).

Rhee and Hokin, 1979; Jean and Albers, 1976) and against the β subunit (Jean *et al.*, 1975; Rhee and Hokin, 1979) have also been developed. They have been used to study their effect on the enzymatic activity with various, often contradictory, results. This situation is mostly due to the lack of precise immunochemical characterization when using current techniques such as indirect immunoprecipitation, immunoreplication, or bidimensional immunoelectrophoresis.

The preparation of monospecific antibodies against the α and the β subunits of the toad enzyme has been described in detail (Girardet *et al.*, 1981). The anti-α serum reacted exclusively with a 96,000-d protein, as assessed by two-dimensional immunoelectrophoresis and by indirect immunoprecipitation of detergent-solubilized $Na^+,K^+$-ATPase subunits from radioiodinated or biosynthetically labeled kidney holoenzyme microsomes or postnuclear supernatants. The anti-β serum was rendered specific to this subunit by absorption with purified α subunit. This study also provided evidence that both antisera reacted with antigenic determinants exposed at the cell surface and with cytoplasmic and/or intramembranous antigenic sites. These antisera can thus be used to selectively label the surface-exposed site of each subunit of the living cell and also to immunoprecipitate radioiodinated or biosynthetically labeled enzymes in the intact cell. In order to further assess the specificity of our antisera we studied the stoichiometric relationship between the radioactivity ([$^{35}$S]methionine) specifically incorporated into the immunoprecipitate and the total cellular enzyme ac-

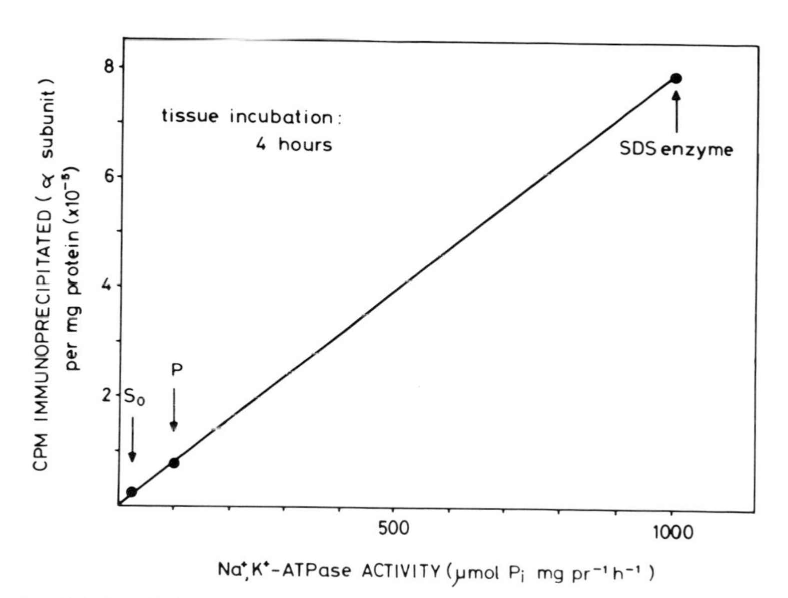


FIG. 2. Relationship between enzyme activity and immunoprecipitated radioactivity in the α subunit of $Na^+,K^+$-ATPase in three cell fractions of toad kidney cells incubated *in vitro* in the presence of [$^3$H]leucine (120 μCi/ml) for 4 hours. Preparation of cell fractions and immunoprecipitation with antiserum of $Na^+,K^+$-ATPase as described in Geering *et al.* (1982a). $S_0$, Postnuclear fraction; P, microsomal fraction; SDS enzyme, detergent-purified enzyme. Each fraction was divided into two aliquots for separate determination of enzyme activity (maximally stimulated by SDS) and immunoprecipitation of the α subunit. Abscissa, enzyme activity in each fraction maximally stimulated by SDS. Ordinate, radioactivity incorporated into the α subunit. Reprinted with permission from Geering *et al.* (1982a). Copyright 1982 Alan R. Liss, Inc.

tivity (Geering *et al.*, 1982a). There was a linear relationship between enzyme activity and immunoprecipitated radioactivity in the α subunit in three cell fractions of biosynthetically labeled toad kidney slices. The three fractions (labeled at steady state) used for this study were a postnuclear fraction ($S_0$), a microsomal fraction (P), and purified SDS enzyme. Enzyme activity increased from 12 to 1000 μ*M* $P_i$/mg protein · hour (Fig. 2).

In addition, the rate of incorporation of [$^{35}$S]methionine into the α and the β subunits was linear for up to 60 minutes of incubation in the toad bladder system (Geering *et al.*, 1982b). Throughout the labeling periods, the anti-β serum immunoprecipitated four times fewer counts than the anti-α serum. This ratio 1:4 was constant and indicates an identical rate of synthesis of the two proteins. If our antisera were not specific for two proteins only, one should expect a change in this ratio with time—unless the contaminating peptides had exactly the same molecular weight and identical rate of synthesis as the two subunits. This is possible but quite unlikely. Finally, in immunocompetition studies, biosynthetically labeled α, pre-, and β subunits (5- to 30-minute pulse with [$^{35}$S]methionine) can always be completely displaced by an excess of unlabeled mature purified α and β subunits. As a conclusion to these biosynthetic studies, we feel confident that our antisera can be considered monospecific. However, as we are dealing with polyclonal antibodies, it is never possible to rule out the

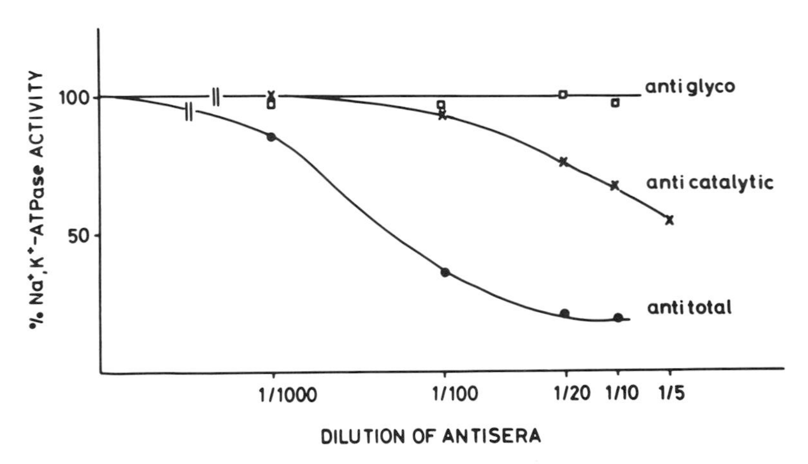


FIG. 3. Effects of various antisera on the enzyme activity of $Na^+,K^+$-ATPase. Partially purified holenzyme (2 μg) was incubated in the presence of increasing concentration of anti-holoenzyme serum (●), anti-α serum (×), and anti-β serum (□). Sera were dialyzed 8 hours with buffer containing 30 m*M* histidine, 5 m*M* EDTA, 70 m*M* Tris, and 200 m*M* sucrose (pH 7.4). Controls were incubated in the presence of the corresponding preimmune sera. As preimmune sera consistently increased the enzyme activity (up to 60%) compared to a standard assay (without serum), the inhibition curve was always calculated by taking this effect into account. The specific activity of the enzyme measured in standard conditions was 750 μ*M* $P_i$/mg protein · hour.

presence of antibodies against low-level contaminants which could be detected by different experimental approaches.

Finally, it was of interest to test whether our antisera could affect the hydrolytic function of $Na^+,K^+$-ATPase. As shown in Fig. 3, the anti-β serum was totally inactive and did not inhibit the enzyme activity at all, even at the lowest dilution (1:5). The anti-α clearly inhibited enzyme activity (up to 50%) but was much less potent than an anti-holoenzyme used as a positive control. One of the major problems in interpreting these data is the fact that inhibition of enzyme activity does not imply per se that an antibody hits an active site. Steric hindrance or allosteric effect of such large ligands could explain the antagonistic effect of polyclonal antibodies. In addition, it is frequently observed that preimmune sera have variable effects on basal enzyme activity. In the present case, an enhancement of the activity (up to 60%) was observed and had to be taken into account. Clearly, polyclonal antibodies are not well suited for mapping specific domains of the two subunits, and monoclonal antibodies will be of great help in this respect.

## IV. BIOSYNTHESIS OF $Na^+,K^+$-ATPase IN AMPHIBIAN EPITHELIAL CELLS

Biosynthesis of the α and β subunits was studied in toad urinary bladder incubated *in vitro* during a 5- or 10-minute pulse in the presence of [$^{35}$S]methionine. This pulse was followed by a 2-hour chase period in the presence of a vast excess of unlabeled methionine and cycloheximide (Geering *et al.*, 1982c). Epithelial cells were scraped off the underlying tissue and fractionated into a crude microsomal fraction, a 100,000 *g* soluble fraction, and an SDS-treated plasma membrane fraction. The enzyme activity of $Na^+,K^+$-ATPase was 6.5 $\mu M$ $P_i$/mg protein · hour in the homogenate, 28 $\mu M$ $P_i$/mg protein · hour in the crude microsomal fraction, and 124 $\mu M$ $P_i$/mg protein · hour in the plasma membrane-enriched fraction. No detectable activity was found in the soluble fraction. After a 10-minute pulse the labeled α subunit was immunoprecipitated from the soluble fraction (in small amount) and from the crude microsomal fraction, but not from the plasma membrane-enriched fraction. After a 120-minute chase, the α subunit was clearly incorporated into this fraction. These results suggest that the α subunit may be synthesized on free polysomes. After a 10-minute pulse, the β subunit was recovered as a 32,000-d unglycosylated precursor in the presence of tunicamycin from the crude microsomal fraction but not from the plasma membrane-enriched fraction. Upon release of the tunicamycin blockage, the 32,000-d precursor was processed into a 42,000-d core-glycosylated precursor, as evidenced by its sensitivity to endoH [endoribonuclease H (calf thymus)].

After a 120-minute chase, this 42,000-d precursor was terminally glycosylated into a 60,000-d β subunit or its 120,000-d dimer recovered from the crude microsomal fraction and the plasma membrane-enriched fraction. These results suggest that the β subunit is synthesized, as expected, on bound polysomes and processed like other glycoproteins (Sabatini *et al.*, 1982).

The striking observation that the two subunits might be synthesized in two distinct pools of polysomes is consistent with some findings obtained recently in cell-free translation systems (McDonough *et al.*, 1982; Sabatini *et al.*, 1982).

Clearly much more information is required before this new pathway can be definitely proven, but it already raises the interesting question of where and how the two subunits are assembled into a functional enzyme unit.

## V. HORMONAL CONTROL OF $Na^+,K^+$-ATPase SYNTHESIS IN AMPHIBIAN EPITHELIAL CELLS

Until now, my colleagues and I have mainly studied the effects of aldosterone and thyroid hormone on the rate of synthesis of $Na^+,K^+$-ATPase in toad bladder epithelial cells. Aldosterone increases $Na^+$ reabsorption across these cells (for recent reviews, see Ludens and Fanestil, 1976; Crabbé, 1977; Marver, 1980). After a lag period of about 60 minutes, the transepithelial sodium transport increases two- to threefold until a maximum is reached 6–8 hours after the addition of aldosterone. The mineralocorticoid response includes at least two distinct phases which differ electrophysiologically (Spooner and Edelman, 1975). First there is a rapid increase in transepithelial sodium transport (early response) with a concomitant decrease in transepithelial electrical resistance. Second, there is a further increase in sodium transport (late response) with little change in electrical resistance. The physiological response is thought to be mediated by a classic protein induction pathway (Edelman, 1975) which includes (1) binding of aldosterone to cytoplasmic receptors and translocation of the hormone receptor complex into the nucleus where it binds to chromatin acceptor sites and (2) induction of a few specific mRNAs and proteins which in turn are believed to mediate directly or indirectly the changes in sodium transport.

Consistent with this working hypothesis are the findings that spirolactone (a competitive antagonist for the mineralocorticoid receptor), actinomycin D (a transcriptional inhibitor), and cycloheximide (a translational inhibitor) are all able to completely antagonize the aldosterone response. Interestingly, we have observed that thyroid hormone can antagonize the late response (Rossier *et al.*, 1979a,b; Rossier *et al.*, 1982) in a rather selective manner. With the methodology described in the previous section, it was therefore tempting to study the effects of these two hormones on the biosynthesis of the enzyme, since previous studies on the influence of aldosterone on enzyme activities have yielded conflicting results. In some studies (Schmidt *et al.*, 1975; Petty *et al.*, 1981) aldosterone increased the enzyme activity as early as 1–3 hours after hormone treatment, while in other studies no modification of $Na^+,K^+$-ATPase activity was observed during this period (Jørgensen, 1972; Chignell and Titus, 1966; Hill *et al.*, 1973). Long-term effects of corticosteroids, however, have often been observed (Jørgensen, 1972; Chignell and Titus, 1966; Charney *et al.*, 1974; Garg *et al.*, 1981). On the other hand, thyroid hormones increased the rate of $Na^+,K^+$-ATPase synthesis in the kidney of hypothyroid rats (Lo and Edelman, 1976), an effect which appears to be related to the acquisition of homeothermy. One study showed that triiodothyronine ($T_3$) did not increase the enzyme activity in tissues of the toad, a poikilothermic animal (Rossier *et al.*, 1979a). The possibility remained that in this species $T_3$ was not simply inactive but rather a repressor.

We measured the relative rates of synthesis of both subunits in tissues treated by aldosterone, $T_3$, or both (Geering *et al.*, 1982b). After 18 hours of exposure to the hormones, aldosterone significantly increased the biosynthesis rate of α and β subunits (2.8- and 2.4-fold, respectively). The hormonal effect was insignificant after 3 hours of exposure to aldosterone and became highly significant at 6 hours. The effect was dose dependent with an apparent $K_{1/2}$ of 3 n$M$. It was completely abolished by spirolactone (500-fold excess), suggesting that it was a receptor mediated process. Recent studies showed that the induction is closely related to Type I (high affinity, low capacity) aldosterone binding sites rather than Type II (low affinity, high capacity) binding sites (Geering *et al.*, 1983). Finally, as shown in Fig. 4, the effect of aldosterone was completely abolished by actinomycin D at a dose which fully antagonized the mineralocorticoid response and 75% of the incorporation of [$^3$H]uridine into poly(A)$^+$ RNA (Rossier *et al.*, 1978). The effect of aldosterone was not dependent on the increased entry of $Na^+$ at the apical membrane, since amiloride did not prevent the induction of the α and β subunits by aldosterone. Finally, $T_3$ had no effect on either basal or aldosterone-stimulated synthesis rates of both subunits.

Interestingly, my colleagues and I have recently demonstrated that sodium butyrate, a rather selective inhibitor of steroid action at the transcriptional level (by a mechanism clearly different from that of actinomycin D), is able to selectively and reversibly inhibit the effect of aldosterone on the late response but leaves the early mineralocorticoid response intact (Truscello *et al.*, 1983). As shown in Fig. 5, sodium butyrate was also able to block the induction of the α subunits (also β subunits; data not shown), whereas overall protein synthesis was not at all inhibited.

At present, our data indicate that aldosterone acts mainly by controlling the expression of the $Na^+,K^+$-ATPase gene, probably at the transcriptional level. Of course this does not rule out the possibility that the hormone may also be

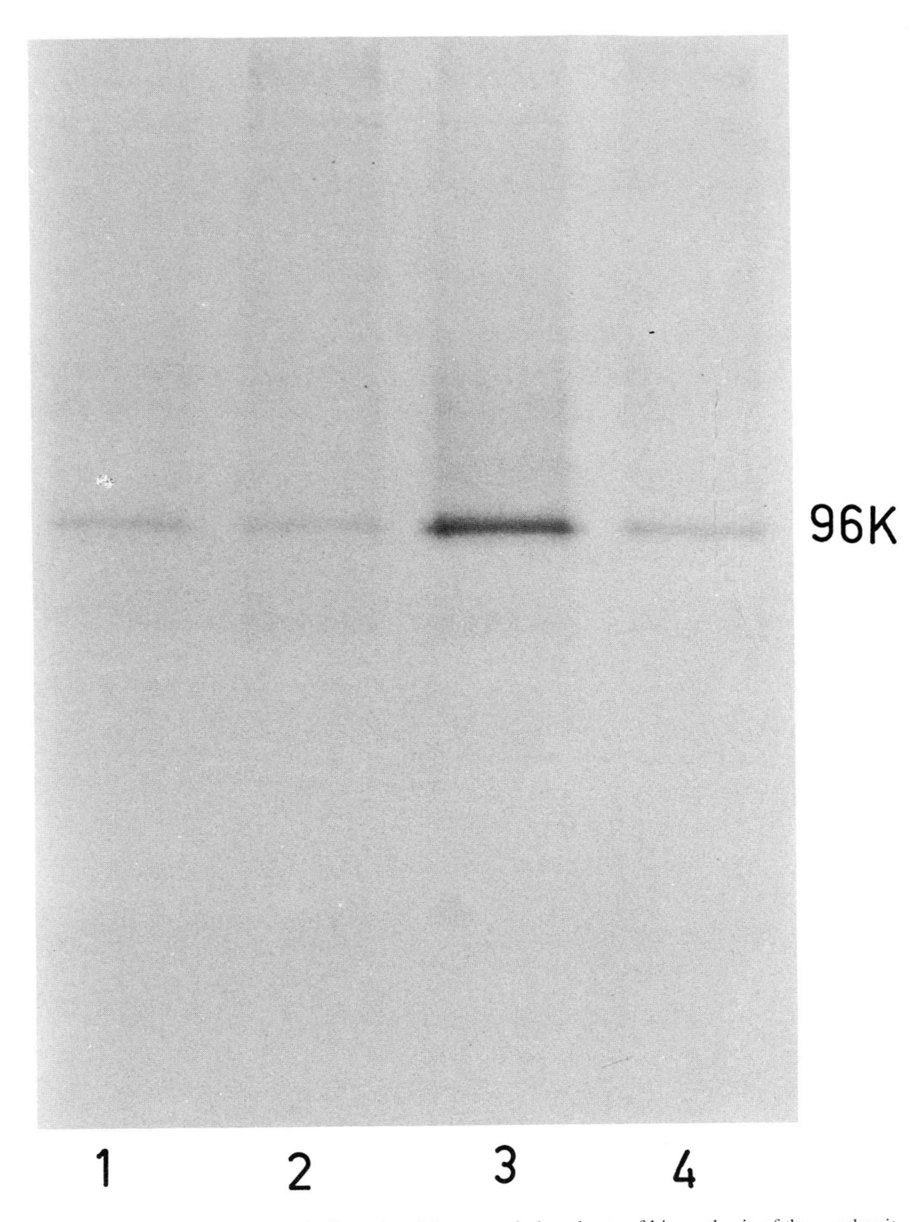


FIG. 4. Effect of actinomycin D on the aldosterone-induced rate of biosynthesis of the α subunit of $Na^+$, $K^+$-ATPase. Bladders were incubated for 6 hours either without hormone (1), with 2 μg/ml actinomycin D (2), with 80 n*M* aldosterone (3), or with aldosterone + actinomycin D (4). Tissue labeling, preparation of cell extracts, and immunoprecipitation were as described in Geering *et al.* (1982b). Autoradiography patterns of immunoprecipitated subunits resolved on 5–13% SDS–PAGE are shown. Immunoprecipitation was performed on total cell extracts containing $1.025 \times 10^6$ cpm. The specific activities of the total cell extracts were control, $5.05 \times 10^6$ cpm/mg protein; actinomycin D, $5.34 \times 10^6$ cpm/mg protein; aldosterone, $6.26 \times 10^6$ cpm/mg protein; and aldosterone + actinomycin D, $4.01 \times 10^6$ cpm/mg protein.

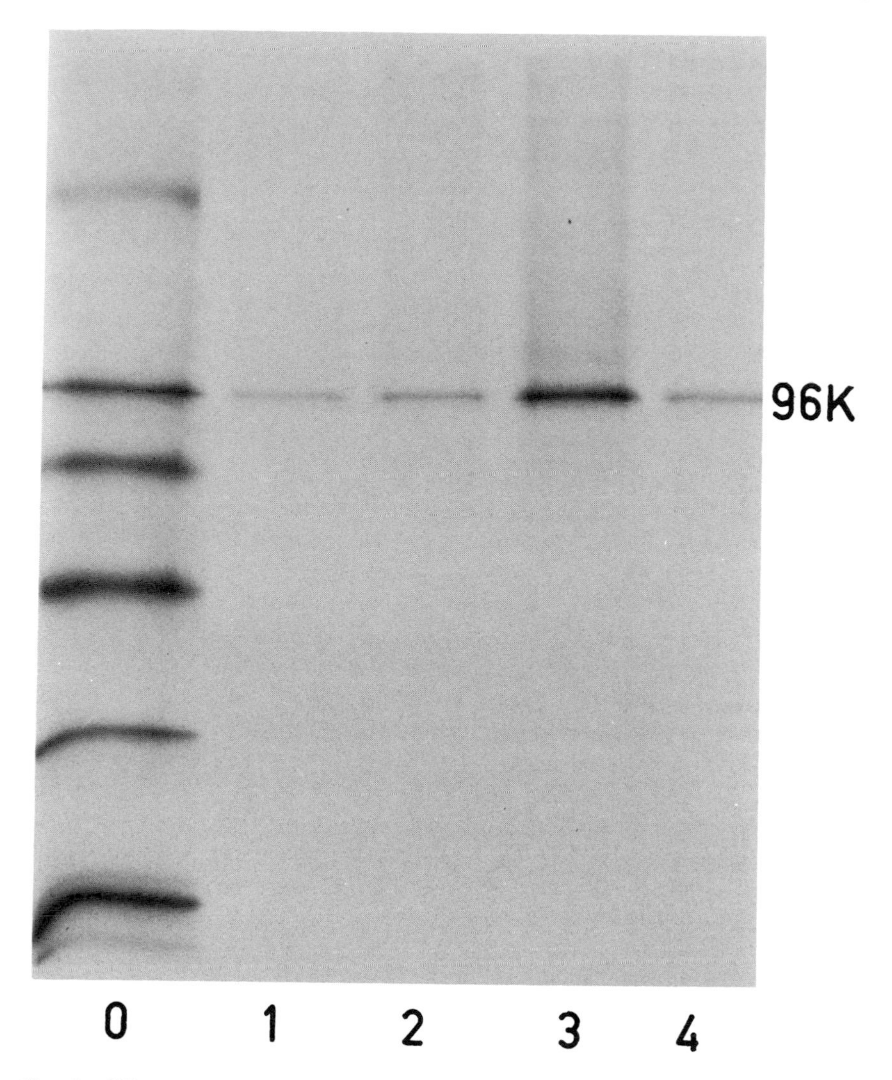


FIG. 5. Effect of sodium butyrate on the aldosterone-dependent rate of biosynthesis of the α subunit of $Na^+,K^+$-ATPase. Toad bladder tissues were incubated either for 18 hours without drug or hormone (1), for 22 hours with 3 m*M* sodium butyrate alone (2), for 18 hours with 80 n*M* aldosterone (3), or for 22 hours with 3 m*M* sodium butyrate plus 80 n*M* aldosterone for the last 18 hours of incubation (4). Tissues were then pulse labeled for 30 minutes with L-[$^{35}S$]methionine at 25°C. Preparation of cell extract and immunoprecipitation were as described in Truscello *et al.* (1983). Pattern of autoradiography in immunoprecipitated preparation resolved on SDS–PAGE is shown. (0), $^{14}C$-Labeled protein marker. Reprinted with permission from Truscello *et al.* (1983).

active (either simultaneously or not) at other levels, namely, translational or posttranslational. This aspect is presented as a model in the next section.

## VI. POSSIBLE SITES OF ACTION OF HORMONES ON THE CONTROL OF SYNTHESIS AND EXPRESSION OF $Na^+,K^+$-ATPase: A MODEL

In the model shown in Fig. 6, three distinct sites of action of aldosterone are depicted. In this model, the hormone could control the synthesis and/or the expression of the enzyme at the cell surface, thereby modifying the overall capacity of the epithelial cell to reabsorb sodium.

1. The first site of action is at the *transcriptional* level, for which we have the best—though indirect—evidence. Upon interaction between the aldosterone receptor complex and specific chromatin acceptor sites situated on the $Na^+,K^+$-ATPase gene, the level of $Na^+,K^+$-ATPase mRNA is increased. This, in turn, increases the synthesis of the enzyme, a process which is not sensitive to amiloride. This process is relatively slow. It cannot account for the early response to aldosterone. It could, however, be related to the late mineralocorticoid response, although this has not yet been proven.
2. The second site of action is at the *posttranslational* level. Aldosterone could induce a regulatory protein which would control the assembly of the two subunits, their addressing to the basolateral membrane, or their expression at the cell surface. In other words, aldosterone could regulate the insertion of the sodium pump and/or the activation of silent pumps already present in the membrane. Such a process could explain a number of data which show a rapid effect of aldosterone on $Na^+,K^+$-ATPase activity (Schmidt *et al.*, 1975; Petty *et al.*, 1981; El Mernissi and Doucet, 1982), which appears to be amiloride sensitive (Petty *et al.*, 1981). In this case the effect of aldosterone on $Na^+,K^+$-ATPase could be part of the early response, but secondary to the increased sodium permeability at the apical membrane. In other words, an increased entry of sodium would represent a permissive factor required for the observation of the mineralocorticoid action on $Na^+,K^+$-ATPase.
3. The third site is highly speculative. My colleagues and I term this hypothesis the *sodium pump–channel hypothesis*. It states that the genetic information needed to form a sodium–potassium pump might not be so different from the information required to build a sodium or a potassium channel. In other words, the crucial information needed to synthesize a peptide with the unique property of being selective to sodium and/or potassium could be used in different cells, and in different organelles or different plasma membrane domains of the same cells. Of course, the sodium pump and the sodium channel could be coded for by

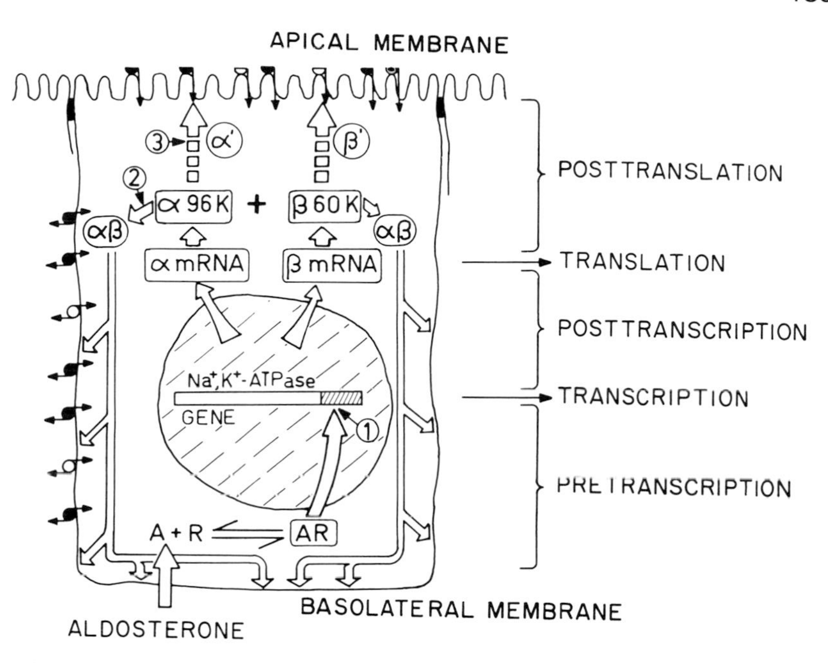


FIG. 6. Possible sites of action of hormones on the control of synthesis and expression of $Na^+,K^+$-ATPase; a model. Three possible sites of action of aldosterone in an epithelial cell are indicated. Site 1. Upon interaction of the aldosterone receptor complex with chromatin acceptor sites located on the $Na^+,K^+$-ATPase gene, the *transcription* of α and β subunit mRNAs is increased, producing an increased level of translatable mRNA and a thereby increased amount of α and β subunits available for insertion and function in the basolateral membrane. The transcriptional effect has a latent period of more than 3 hours. It is amiloride insensitive. $Na^+,K^+$-ATPase is a late aldosterone-induced protein (*late AIP*). Site 2. An aldosterone-induced protein activates the $Na^+,K^+$-ATPase and allows its functional expression at the basolateral membrane. This *posttranslational* effect has a short latent period (60–90 minutes). It is amiloride sensitive, i.e., $Na^+$ entry at the apical membrane has a permissive effect on the action of the aldosterone-induced protein (*early AIP*). Site 3. Sodium channels (α′, β′), derived from the α and/or the β subunits (or from closely related gene products), are controlled by an aldosterone-induced protein. This *posttranslational* effect has a short latent period (60–90 minutes). It is amiloride sensitive and controlled by a regulatory protein which is an *early AIP*.

the *same* gene but processed differently, in order to obtain on the one hand an enzyme addressed to the basolateral membrane, and on the other hand a sodium channel (or part of it) addressed to the apical membrane. Alternatively, the $Na^+,K^+$-ATPase gene and the sodium channel could belong to one gene family, whose gene products would be closely related and immunologically very similar. A similar process could explain the existence of a specific sodium channel in the nerve or in the muscle. The sodium pump–channel hypothesis is not supported by any strong experimental evidence but only by circumstantial evidence from our own experience and from some data taken from the literature. In our experimental model, antigenic determinants were detected by immunocytochemical methods not only (as expected) at the basolateral membrane, but also at the apical membrane (Papermaster *et al.*, 1981). At present, we cannot absolutely rule out the possibility of a contaminant antibody which would detect antigenic sites not yet detected by other immunological techniques described in Section III. We feel however that this is quite unlikely, and we have some tools (for instance, monoclonal antibodies) which will allow us to distinguish between the two possibilities. A few articles recently published seem important to us in this respect, and in no way do they contradict the sodium pump–channel hypothesis. For instance, in the oocyte of *Xenopus laevis,* depolarization can induce the appearance (*within a few minutes*) of specific, tetrodotoxin-sensitive sodium channels, typical of nerve differentiation (Baud *et al.*, 1982). These data are compatible with the idea that the sodium channel might derive (by posttranslational processing) from a precursor pool which could be the $\alpha$ subunit of $Na^+,K^+$-ATPase. Interestingly, during skeletal myogenesis, it appears that the density of $Na^+,K^+$-ATPase and the sodium channel are *coordinately* regulated (Vigne *et al.*, 1982) suggesting a tight and stoichiometric relationship between the ontogeny of the sodium pump and the sodium channel. Finally, two recent papers (Pouysségur *et al.*, 1980, 1982) demonstrate that a fibroblast cell line can express an amiloride-sensitive channel and a tetrodotoxin-sensitive sodium channel, two markers normally absent in this type of cell.

## VII. CONCLUDING REMARKS

The recent development of immunological probes against $Na^+,K^+$-ATPase will certainly enlarge the field of cell and molecular biology of this important plasma membrane protein. Two directions which we can now follow appear equally exciting. One deals with the problem of the assembly of the two subunits, their expression at the cell surface, their turnover, and the regulation of these processes. Of course the comparision between nonepithelial and epithelial cells will be of great interest in order to understand how this enzyme can be selectively sorted out to the basolateral membrane epithelia and thereby become the molecular determinant of epithelial polarity and asymmetry. The other direction deals with the structure and the regulation of the $Na^+,K^+$-ATPase gene. It should ultimately allow a better understanding of how corticosteroid hormone controls sodium and potassium homeostasis.

ACKNOWLEDGMENTS

This work was supported by grants from the Swiss National Foundation (3.646.80) and from the Hoffmann-La Roche Foundation (Grant 156).

This work would not have been possible without a close and fruitful collaboration among a number of scientists: Kathi Geering, Marc Girardet, Hans Peter Gäggeler, Sophie Perret Gentil, and Claude Bron. The idea of the sodium pump–channel emerged from exciting and lively discussions with Jean Pierre Kraehenbühl.

I also thank J. Bonnard for her excellent assistance in preparing the manuscript.

## REFERENCES

Anner, B. M., Lane, L. K., Schwartz, A., and Pitts, B. J. R. (1977). A reconstituted $Na^+ + K^+$ pump in liposomes containing purified $(Na^+ + K^+)$-ATPase from kidney medulla. *Biochim. Biophys. Acta* **467,** 340–345.

Askari, A. (1974). The effects of antibodies to $Na^+,K^+$-ATPase on the reactions catalyzed by the enzyme. *Ann. N.Y. Acad. Sci.* **242,** 372–382.

Askari, A., and Rao, S. N. (1972). $Na^+,K^+$-ATPase complex: Effects of anticomplex antibody on the partial reactions catalyzed by the complex. *Biochem. Biophys. Res. Commun.* **49,** 1323–1328.

Baud, C., Kado, R. T., and Marcher, K. (1982). Sodium channels induced by depolarization of the *Xenopus laevis* oocyte. *Proc. Natl. Acad. Sci. U.S.A.* **79,** 3188–3192.

Caldwell, P. C. (1969). Energy relationships and the active transport of ions. *Curr. Top. Bioenerg.* **3,** 251–278.

Charney, A. N., Silva, P., Besarab, A., and Epstein, F. H. (1974). Separate effects of aldosterone, DOCA, and methylprednisolone on renal Na-K-ATPase. *Am. J. Physiol.* **227,** 345–350.

Chignell, C. F., and Titus, E. (1966). Effect of adrenal steroids on a $Na^+$- and $K^+$-requiring adenosine triphosphatase from rat kidney. *J. Biol. Chem.* **241,** 5083–5089.

Churchill, L., and Hokin, L. E. (1979). Biosynthesis and insertion of $(Na^+ + K^+)$-adenosine triphosphatase subunits into eel electroplax membranes. *J. Biol. Chem.* **254,** 7388–7392.

Crabbé, J. (1977). The mechanism of action of aldosterone. *In* "Receptors and Mechanism of Action of Steroid Hormones" (J. R. Pasqualini, ed.), Part 2, pp. 513–568. Dekker, New York.

Dixon, J. F., and Hokin, L. E. (1974). Studies on the characterization of the sodium-potassium transport adenosine triphosphatase: Purification and properties of the enzyme from the electric organ of *Electrophorus electricus*. *Arch. Biochem. Biophys.* **163,** 749–758.

Edelman, I. S. (1975). Mechanism of action of steroid hormones. *J. Steroid Biochem.* **6,** 147–159.

El Mernissi, G., and Doucet, A. (1982). Sites and mechanism of mineralocorticoid action along the rabbit nephron. *INSERM Symp.* **21,** 269–276.

Esmann, M., Skou, J. C., and Christiansen, C. (1979). Solubilization and molecular weight determination of the $(Na^+ + K^+)$-ATPase from rectal glands of *Squalus acanthias*. *Biochim. Biophys. Acta* **567,** 410–420.

Farley, R. A., Goldman, D. W., and Bayley, H. (1980). Identification of regions of the catalytic subunit of (Na–K)-ATPase embedded within the cell membrane. *J. Biol. Chem.* **255,** 860–864.

Garg, L. C., Knepper, M. A., and Burg, M. B. (1981). Mineralocorticoid effects on Na–K-ATPase in individual nephron segments. *Am. J. Physiol.* **240,** F536–F544.

Geering, K., and Rossier, B. C. (1979). Purification and characterization of $(Na^+ + K^+)$-ATPase from toad kidney. *Biochim. Biophys. Acta* **566,** 157–170.

Geering, K., Girardet, M., Bron, C., Kraehenbühl, J. P., and Rossier, B. C. (1982a). Biosynthesis of the catalytic subunit of $(Na^+,K^+)$-ATPase in toad kidney and toad bladder epithelial cells. *In* "Membranes in Growth and Development" (J. F. Hoffman, G. H. Giebisch, and L. Bolis, eds.), pp. 537–542. Liss, New York.

Geering, K., Girardet, M., Bron, C., Kraehenbühl, J. P., and Rossier, B. C. (1982b). Hormonal regulation of $(Na^+,K^+)$-ATPase biosynthesis in the toad bladder. Effect of aldosterone and 3,5,3′-triiodo-L-thyronine. *J. Biol. Chem.* **257,** 10338–10343.

Geering, K., Girardet, M., Kraehenbühl, J. P., and Rossier, B. C. (1982c). Biosynthesis of $(Na^+,K^+)$-ATPase in epithelial cells of the urinary bladder of *Bufo marinus*. *J. Cell Biol.* **95,** 379a (abstr.).

Geering, K., Claire, M., Gäggeler, H. P., Girardet, M., and Rossier, B. C. (1983). Stimulation of $Na^+$ transport and $(Na^+,K^+)$-ATPase synthesis by aldosterone: Correlation with occupancy of hormone binders. *Experientia* **39,** 663, 1983 (abst.).

Girardet, M. (1982). Etude immunochimique de la Na,K,-ATPase de *Bufo marinus*. Thèse, Université de Lausanne.

Girardet, M., Geering, K., Frantes, J. M., Geser, D., Rossier, B. C., Kraehenbühl, J. P., and Bron, C. (1981). Immunochemical evidence for a transmembrane orientation of both the $(Na^+,K^+)$-ATPase subunits. *Biochemistry* **20,** 6684–6691.

Girardet, M., Geering, K., Rossier, B. C., Kraehenbühl, J. P., and Bron, C. (1983). Hydrophobic labeling of $(Na^+,K^+)$-ATPase: Further evidence that the β subunit is embedded in the bilayer. *Biochemistry* **22,** 2296–2300.

Handler, J. S., Steele, R. E., Sahib, S. K., Wade, J. B., Preston, A. S., Lawson, N. L., and Johnson, J. P. (1979). Toad urinary bladder epithelial cells in culture: Maintenance of epithelial structure, sodium transport, and response to hormones. *Proc. Natl. Acad. Sci. U.S.A.* **76,** 4151–4155.

Hilden, S., and Hokin, L. E. (1975). Active potassium transport coupled to active sodium transport in vesicles reconstituted from purified sodium and potassium ion-activated adenosine triphosphatase from the rectal gland of *Squalus acanthias*. *J. Biol. Chem.* **250,** 6296–6303.

Hill, J. H., Cortas, N., and Walser, M. (1973). Aldosterone action and sodium- and potassium-activated adenosine triphosphatase in toad bladder. *J. Clin. Invest.* **52,** 185–189.

Hokin, L. E., and Dixon, J. F. (1979). Parameters of reconstituted $Na^+$ and $K^+$ transport in liposomes in which purified Na,K-ATPase is incorporated by "freeze-thaw-sonication." *In* "Na,K-ATPase: Structure and Kinetics" (J. C. Skou and J. G. Nørby, eds.), pp. 49–56. Academic Press, New York.

Hokin, L. E., Dahl, J. L., Deupree, J. D., Dixon, J. F., Hakney, J. F., and Perdue, J. F. (1973). Studies on the characterization of the sodium-potassium transport adenosine triphosphatase. *J. Biol. Chem.* **248,** 2593–2605.

Hopkins, B. E., Wagner, H., Jr., and Smith, T. W. (1976). Sodium- and potassium-activated adenosine triphosphatase of the nasal salt gland of the duck (*Anas platyrhynchos*). *J. Biol. Chem.* **251,** 4365–4371.

Ikeda, F., Ikeda, Y., Inagaki, C., and Takaori, S. (1979). Effect of aldosterone on renal Na,K-activated adenosine-triphosphatase activity in non-adrenalectomized rats. *Jpn. J. Pharmacol.* **29,** 138–141.

Jean, D. H., and Albers, R. W. (1976). Immunochemical studies on the large polypeptide of *Electrophorus* electroplax $(Na^+ + K^+)$-ATPase. *Biochim. Biophys. Acta* **452,** 219–226.

Jean, D. H., and Albers, R. W. (1977). Molecular organization of subunits of electroplax (sodium plus potassium)-activated adenosine triphosphatase. *J. Biol. Chem.* **252,** 2450–2451.

Jean, D. H., Albers, R. W., and Koval, G. J. (1975). Sodium-potassium-activated adenosine triphosphatase of *Electrophorus* electric organ. X. Immunochemical properties of the Lubrol-solubilized enzyme and its constituent polypeptides. *J. Biol. Chem.* **250,** 1035–1040.

Jørgensen, P. L. (1972). The role of aldosterone in the regulation of $(Na^+ + K^+)$-ATPase in rat kidney. *J. Steroid Biochem.* **3,** 181–191.

Jørgensen, P. L. (1974a). Purification and characterization of $(Na^+ + K^+)$-ATPase. III. Purification from the outer medulla of mammalian kidney after selective removal of membrane components by sodium dodecylsulphate. *Biochim. Biophys. Acta* **356,** 36–52.

Jørgensen, P. L. (1974b). Purification and characterization of $(Na^+ + K^+)$-ATPase. IV. Estimation of the purity and of the molecular weight and polypeptide content per enzyme unit in preparations from the outer medulla of rabbit kidney. *Biochim. Biophys. Acta* **356,** 53–67.

Jørgensen, P. L. (1975). Isolation and characterization of the components of the sodium pump. *Q. Rev. Biophys.* **7,** 239–274.

Jørgensen, P. L. (1980). Sodium and potassium ion pump in kidney tubules. *Physiol. Rev.* **60,** 864–917.

Jørgensen, P. L., Hansen, O., Glynn, I. M., and Cavieres, J. D. (1973). Antibodies to pig kidney (Na$^+$ + K$^+$)-ATPase inhibit the Na$^+$ pump in human red cells provided they have access to the inner surface of the cell membrane. *Biochim. Biophys. Acta* **291,** 795–800.

Karlish, S. J. D., Jørgensen, P. L., and Gitler, C. (1977). Identification of a membrane-embedded segment of the large polypeptide chain of (Na$^+$,K$^+$)ATPase. *Nature (London)* **269,** 715–717.

Knox, W. H., and Sen, A. K. (1974). Mechanism of action of aldosterone with particular reference to (Na + K)-ATPase. *Ann. N.Y. Acad. Sci.* **242,** 471–488.

Koepsell, H. (1978). Characteristics of antibody inhibition of rat kidney (Na$^+$–K$^+$)-ATPase. *J. Membr. Biol.* **44,** 85–102.

Koepsell, H. (1979). Conformational changes of membrane-bound (Na$^+$–K$^+$)-ATPase as revealed by antibody inhibition. *J. Membr. Biol.* **45,** 1–20.

Kyte, J. (1971). Purification of the sodium- and potassium-dependent adenosine triphosphatase from canine renal medulla. *J. Biol. Chem.* **246,** 4157–4165.

Kyte, J. (1972). Properties of the two polypeptides of sodium- and potassium-dependent adenosine triphosphatase. *J. Biol. Chem.* **247,** 7642–7649.

Kyte, J. (1974). The reactions of sodium and potassium ion-activated adenosine triphosphatase with specific antibodies. *J. Biol. Chem.* **249,** 3652–3660.

Kyte, J. (1976a). Immunoferritin determination of the distribution of (Na$^+$ + K$^+$) ATPase over the plasma membranes of renal convoluted tubules. I. Distal segment. *J. Cell Biol.* **68,** 287–303.

Kyte, J. (1976b). Immunoferritin determination of the distribution of (Na$^+$ + K$^+$) ATPase over the plasma membranes of renal convoluted tubules. II. Proximal segment. *J. Cell Biol.* **68,** 304–318.

Lane, L. K., Copenhaver, J. H., Lindenmayer, G. E., and Schwartz, A. (1973). Purification and characterization of an [$^3$H]ouabain binding to the transport adenosine triphosphatase from outer medulla of canine kidney. *J. Biol. Chem.* **248,** 7197–7200.

Lane, L. K., Potter, J. D., and Collins, J. H. (1979). Large-scale purification of Na,K-ATPase and its protein subunits from lamb kidney medulla. *Prep. Biochem.* **9,** 157–170.

Leaf, A., Anderson, J., and Page, L. B. (1958). Active sodium transport by the isolated toad bladder. *J. Gen. Physiol.* **41,** 657–668.

Le Bouffant, F., Hus-Citharel, A., and Morel, F. (1982). In vitro $^{14}CO_2$ production by single pieces of rat cortical thick ascending limbs and its coupling to active salt transport. *INSERM Symp.* **21,** 363–370.

Levitt, D. G. (1980). The mechanism of the sodium pump. *Biochim. Biophys. Acta* **604,** 321–345.

Lo, C. S., and Edelman, I. S. (1976). Effect of triiodothyronine on the synthesis and degradation of renal cortical (Na$^+$ + K$^+$)-adenosine triphosphatase. *J. Biol. Chem.* **251,** 7834–7840.

Lo, C. S., and Lo, T. N. (1980). Effect of triiodothyronine on the synthesis and degradation of the small subunit of renal cortical (Na$^+$ + K$^+$)-adenosine triphosphatase. *J. Biol. Chem.* **255,** 2131–2136.

Lo, C. S., August, T. R., Liberman, U. A., and Edelman, I. S. (1976). Dependence of renal (Na$^+$ + K$^+$)-adenosine triphosphatase activity on thyroid status. *J. Biol. Chem.* **251,** 7826–7833.

Ludens, J. H., and Fanestil, D. D. (1976). The mechanism of aldosterone function. *Pharmacol. Ther., Part B* **2,** 371–412.

Marver, D. (1980). Aldosterone action in target epithelia. *Vitam. Horm.* **38,** 55–117.

McDonough, A. A., Hiatt, A., and Edelman, I. S. (1982). Characteristics of antibodies to guinea pig (Na$^+$ + K$^+$)-adenosine triphosphatase and their use in cell-free synthesis studies. *J. Membr. Biol.* **69,** 13–22.

Michael, L., Wallick, E. T., and Schwartz, A. (1977). Modification of (Na$^+$,K$^+$)-ATPase function by purified antibodies to the holoenzyme. Effects on enzyme activity and [$^3$H]ouabain binding. *J. Biol. Chem.* **252,** 8476–8480.

Nakao, T., Nakao, M., Nagai, F., Kawai, K., Fugihira, Y., Hara, Y., and Fujita, M. (1973). Purification and some properties of Na,K-transport ATPase. II. Preparations with high specific activity obtained using aminoethyl cellulose chromatography. *J. Biochem. (Tokyo)* **73,** 781–791.

Papermaster, D. S., Lyman, D., Schneider, B. G., Labienic, L., and Kraehenbühl, J. P. (1981). Immunocytochemical localization of the catalytic subunit of ($Na^+$ + $K^+$)ATPase in *Bufo marinus* kidney bladder and retina with biotinyl-antibody and streptavidin-gold complexes. *J. Cell Biol.* **91,** 273a (abstr.).

Perrone, J. R., Hackney, J. F., Dixon, J. F., and Hokin, L. E. (1975). Molecular properties of purified (sodium + potassium)-activated adenosine triphosphatases and their subunits from the rectal gland of *Squalus acanthias* and the electric organ of *Electrophorus electricus*. *J. Biol. Chem.* **250,** 4178–4184.

Peterson, G. L., and Hokin, L. E. (1980). Improved purification of brine-shrimp (*Artemia saline*) ($Na^+$ + $K^+$)-activated adenosine triphosphatase and amino-acid and carbohydrate analyses of the isolated subunits. *Biochem. J.* **192,** 107–118.

Peterson, G. L., Ewing, R. D., and Conte, F. P. (1978). Membrane differentiation and *de novo* synthesis of the ($Na^+$ + $K^+$)-activated adenosine triphosphatase during development of *Artemia salina nauplii*. *Dev. Biol.* **67,** 90–98.

Petty, K. J., Kokko, J. P., and Marver, D. (1981). Secondary effect of aldosterone on Na-K ATPase activity in the rabbit cortical collecting tubule. *J. Clin. Invest.* **68,** 1514–1521.

Pollack, L. R., Tate, E. H., and Cook, J. S. (1981). Turnover and regulation of Na-K-ATPase in HeLa cells. *Am. J. Physiol: Cell Physiol.* **241,** C173–C183.

Pouysségur, J., Jacques, Y., and Lazdunski, M. (1980). Identification of a tetrodotoxin-sensitive $Na^+$ channel in a variety of fibroblast lines. *Nature (London)* **286,** 162–164.

Pouysségur, J., Chambard, J. C., Franchi, A., Paris, S., and Van Obberghen-Schilling, E. (1982). Growth factor activation of an amiloride-sensitive $Na^+/H^+$ exchange system in quiescent fibroblasts: Coupling to ribosomal protein S6 phosphorylation. *Proc. Natl. Acad. Sci. U.S.A.* **79,** 3935–3939.

Rhee, H. M., and Hokin, L. E. (1975). Inhibition of the purified sodium-potassium activated adenosinetriphosphatase from the rectal gland of *Squalus acanthias* by antibody against the glycoprotein subunit. *Biochem. Biophys. Res. Commun.* **63,** 1139–1145.

Rhee, H. M., and Hokin, L. E. (1979). Inhibition of ouabain-binding to ($Na^+$ + $K^+$)ATPase by antibody against the catalytic subunit but not by antibody against the glycoprotein subunit. *Biochim. Biophys. Acta* **558,** 108–112.

Robinson, J. D., and Flashner, M. S. (1979). The ($Na^+$ + $K^+$)-activate ATPase. Enzymatic and transport properties. *Biochim. Biophys. Acta* **549,** 145–176.

Rossier, B. C., Gäggeler, H. P., and Rossier, M. (1978). Effects of 3′deoxyadenosine and actinomycin D on RNA synthesis in toad bladder: Analysis of response to aldosterone. *J. Membr. Biol.* **41,** 149–166.

Rossier, B. C., Rossier, M., and Lo, C. S. (1979a). Thyroxine and $Na^+$ transport in toad: Role in transition from poikilo- to homeothermy. *Am. J. Physiol: Cell Physiol.* **236,** C117–C124.

Rossier, B. C., Gäggeler, H. P., Brunner, D. B., Keller, I., and Rossier, M. (1979b). Thyroid hormone-aldosterone interaction on $Na^+$ transport in toad bladder. *Am. J. Physiol: Cell Physiol.* **236,** C125–C131.

Rossier, B. C., Truscello, A., and Geering, K. (1982). Antimineralocorticoid effect of sodium butyrate and triiodothyronine: Evidence for two pathways in the action of aldosterone on $Na^+$ transport in the toad bladder. *INSERM Symp.* **21,** 225–232.

Sabatini, D. D., Kreibich, G., Morimoto, T., and Adesnik, M. (1982). Mechanisms for the incorporation of proteins in membranes and organelles. *J. Cell Biol.* **92,** 1–22.

Schmidt, U., Schmid, J., Schmid, H., and Dubach, U. C. (1975). Sodium- and potassium-activated ATPase. A possible target of aldosterone. *J. Clin. Invest.* **55,** 655–660.

Sen, A. K., and Post, R. L. (1964). Stoichiometry and localization of adenosine triphosphate-dependent sodium and potassium transport in the erythrocyte. *J. Biol. Chem.* **239,** 345–352.

Skou, J. C. (1957). The influence of some cations on an adenosine triphosphatase from peripheral nerves. *Biochim. Biophys. Acta* **23,** 394–401.

Smith, T. J., and Edelman, I. S. (1979). The role of sodium transport in thyroid thermogenesis. *Fed. Proc. Fed. Am. Soc. Exp. Biol.* **38,** 2150–2153.

Smith, T. W., and Wagner, H. (1975). Effects of ($Na^+$ + $K^+$)-ATPase-specific antibodies on enzymatic activity and monovalent cation transport. *J. Membr. Biol.* **25,** 341–360.

Spooner, P. M., and Edelman, I. S. (1975). Further studies on the effect of aldosterone on electrical resistance of toad bladder. *Biochim. Biophys. Acta* **406,** 304–314.

Stewart, D. J., Semple, E. W., Swart, G. T., and Sen, A. K. (1976). Induction of the catalytic protein of ($Na^+$ + $K^+$)-ATPase in the salt gland of the duck. *Biochim. Biophys. Acta* **419,** 150–163.

Sweadner, K. J. (1979). Two molecular forms of ($Na^+$ + $K^+$)-stimulated ATPase in brain. Separation and difference in affinity for strophanthidin. *J. Biol. Chem.* **254,** 6060–6067.

Truscello, A., Geering, K., Gäggeler, H. P., and Rossier, B. C. (1983). Effects of butyrate on histone deacetylation and aldosterone-dependent $Na^+$ transport in the toad bladder. *J. Biol. Chem.* **258,** 3388–3395.

Uesugi, S., Dulak, N. C., Dixon, J. F., Hexum, T. D., Dahl, J. L., Perdue, J. F., and Hokin, L. E. (1971). Studies on the characterization of the sodium-potassium transport adenosine triphosphatase. IV. Large scale partial purification and properties of a Lubrol-solubilized bovine brain enzyme. *J. Biol. Chem.* **246,** 531–543.

Ussing, H. H., Erlij, D., and Lassen, U. (1974). Transport pathways in biological membranes. *Annu. Rev. Physiol.* **36,** 17–49.

Vigne, P., Frelin, C., and Lazdunski, M. (1982). Ontogeny of the ($Na^+,K^+$)-ATPase during chick skeletal myogenesis. *J. Biol. Chem.* **257,** 5380–5384.

Wallick, E. T., Lane, L. K., and Schwartz, A. (1979). Biochemical mechanism of the sodium pump. *Annu. Rev. Physiol.* **41,** 397–411.

Winter, C. J., and Moss, A. J., Jr. (1979). Ultracentrifugal analysis of the enzymatically active fragments produced by digitonin action on Na,K-ATPase. *In* "Na, K-ATPase, Structure and Kinetics" (J. C. Skou and J. G. Nørby, eds.), pp. 25–32. Academic Press, New York.

# Chapter 8

# Use of Antibodies in the Study of $Na^+,K^+$-ATPase Biosynthesis and Structure

*ALICIA A. MCDONOUGH*

*Department of Physiology and Biophysics*
*University of Southern California School of Medicine*
*Los Angeles, California*



## I. INTRODUCTION

The sodium pump (EC 3.6.1.3) is a plasma membrane-bound oligomer ($\alpha$ catalytic and $\beta$ glycoprotein subunits in a 1:1 ratio) with $Na^+$- and $K^+$-activated ATPase activity ($Na^+,K^+$-ATPase). This ubiquitous enzyme transduces the chemical energy of ATP into the forced exchange of $Na^+$ (out) for $K^+$ (in), an action that maintains cellular ion gradients and membrane electrical potentials. $Na^+,K^+$-ATPase is an ion translocator as well as an ATPase, and these functions are subject to regulation by substrates and various hormones. Its structure, biosynthesis, and turnover are not well understood and are being actively investigated. Antibodies to $Na^+,K^+$-ATPase have proved to be crucial in these studies because they are the most useful specific high-affinity probes for the sodium pump. These probes can recognize and bind specifically to the enzyme subunits




ISBN 0-12-153320-4

under conditions in which the subunits are in low abundance in a background of cellular proteins.

Many of the original studies employing antibodies to $Na^+,K^+$-ATPase produced variable and contradictory results (discussed in Lauf, 1978, and Girardet *et al.*, 1981). In these early studies, the methodologies available to assay and characterize antibodies to the sodium pump (such as complement fixation, functional assays, and immunoprecipitation) were limited in their ability to detect low-level contaminants in the antigen preparation and to characterize the specificity of the antisera for separated α and β subunits.

Two methodological advances have improved and simplified the study of antigen–antibody interactions. The first was the discovery that protein A from *Staphylococcus aureus* binds specifically to the Fc region of many IgG subclasses and can be used to simplify the isolation or identification of immune complexes (Kessler, 1975). The second advance was the development of solid phase assays for antibody–antigen interactions, in which antigen-containing samples are resolved by SDS–polyacrylamide gel electrophoresis (SDS–PAGE) and blotted onto a nitrocellulose sheet (Towbin *et al.*, 1979) or covalently attached diazotized paper (Renart *et al.*, 1979) for subsequent detection of antigen with antibodies. Both the protein A and blotting techniques have improved and simplified the studies of $Na^+,K^+$-ATPase structure and synthesis that depend on antibodies. This article discusses applications of these techniques.

## II. METHODOLOGY: IMMUNE PRECIPITATION AND SOLID PHASE ASSAY

Two approaches that can be used to identify and characterize a low-level protein in a background mixture of proteins are discussed. Both methods depend upon the detection of a radiolabeled protein. The first method is the immunoprecipitation of a labeled antigen, and the second method is a solid phase assay of antigen–antibody complexes with labeled protein A.

The immunoprecipitation approach has proved very useful in the study of the biosynthesis of $Na^+,K^+$-ATPase subunits in cells and tissues labeled in culture (Geering *et al.*, 1982; McDonough, 1983). The labeled cellular proteins are solubilized, incubated overnight at 4°C with specific antiserum, and incubated with formalin-fixed *S. aureus* or protein A bound to agarose beads in order to precipitate the immune complexes. The immune complexes are then solubilized and subjected to SDS–PAGE and fluorography for identification and characterization of the antigen.

The solid phase assay has proved useful in the identification of $Na^+,K^+$-ATPase proteolytic fragments and also in characterizing unlabeled $Na^+,K^+$-ATPase from various tissues or subcellular membrane fractions (McDonough *et al.*, 1982; Schellenberg *et al.*, 1981). In the method developed by Renart *et al.* (1979), the protein mixture is resolved by SDS–PAGE and covalently attached to diazotized paper by blotting. The resolved peptides coupled to diazophenylthioether (DPT) paper are incubated with diluted antiserum, washed, incubated with iodinated protein A, and then washed again. Immune complexes labeled with protein A are visualized by autoradiography. In this manner, α and β subunits have been detected in crude homogenates of various tissues, and proteolytic fragments of the subunits have been defined.

$Na^+,K^+$-ATPase is purified in a membrane-bound form as an oligomer of α (94,000 MW) and β (50,000 MW) subunits (Fig. 1). When this material is injected into a rabbit to generate antibodies, a mixture of anti-α and anti-β antibodies is obtained. The solid phase assay has been adapted to separate subunit- or proteolytic fragment-specific antibodies (McDonough *et al.*, 1982). This method uses the diazotized paper with coupled α or β subunits or coupled

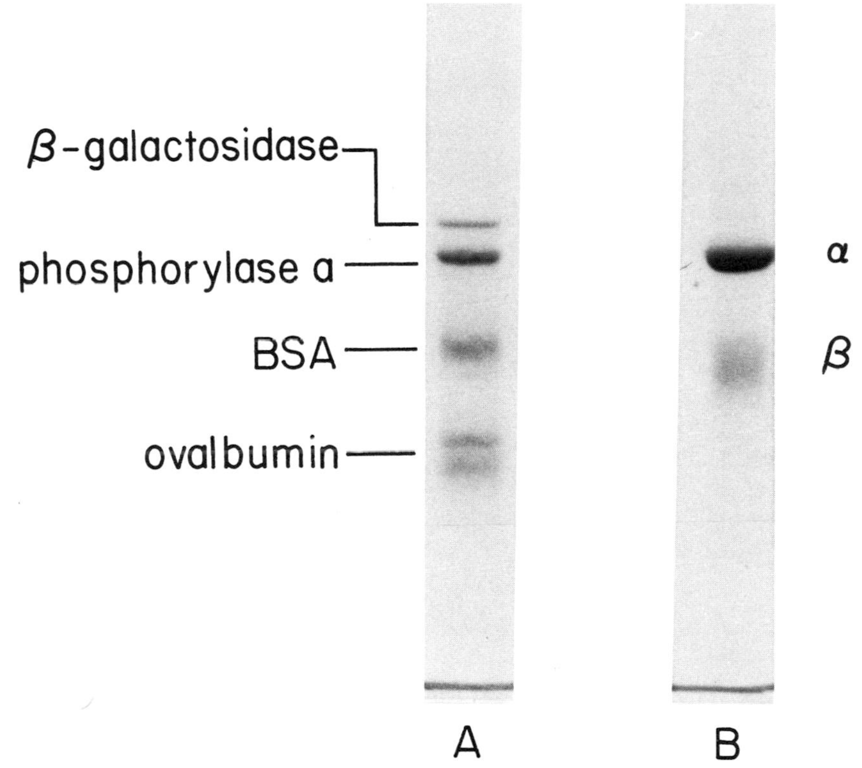


FIG. 1. SDS–PAGE of guinea pig kidney $Na^+,K^+$-ATPase. (A) Molecular weight standards: β-galactosidase (135,000), phosphorylase *a* (94,000), BSA (68,000), and ovalbumin (44,000). (B) $Na^+,K^+$-ATPase: α, catalytic subunit; β, glycoprotein subunit.

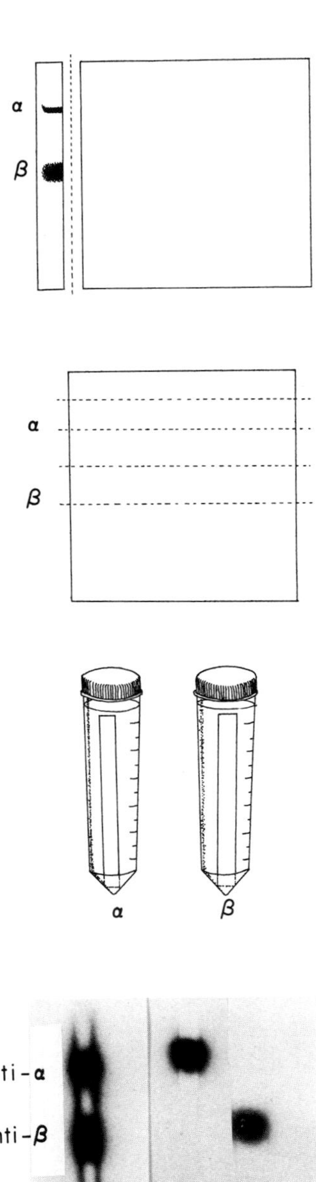
α
β
α
β
α
β

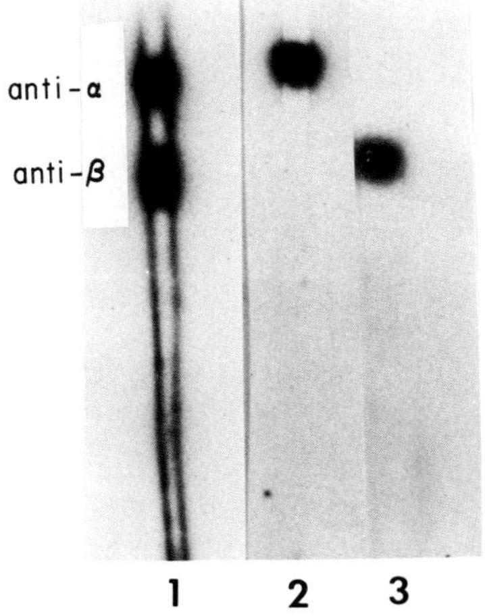
anti-α
anti-β
1
2
3

proteolytic fragments as an affinity ligand. In brief, as illustrated in Fig. 2, purified $Na^+,K^+$-ATPase is solubilized, applied along the top of an SDS slab gel, subjected to electrophoresis, and blotted as described (Renart *et al.*, 1979). Strips containing only α subunits and only β subunits are cut from the paper, incubated with antiserum, and washed to remove nonspecifically bound antibodies. The specific antibodies are eluted from the strips in low-pH buffer and then neutralized. Figure 2 illustrates the results obtained when crude antiserum, anti-α, and anti-β antibodies are tested against strips that have both α and β subunits attached. The separated α and β antibodies have been crucial for the definition of *in vitro* translation products and immunoprecipitates of $Na^+,K^+$-ATPase subunits. Proteolytic fragment-specific antibodies have been used to define antibody-binding regions on the sodium pump subunits (Farley *et al.*, 1983).

## III. APPLICATIONS: BIOSYNTHESIS AND STRUCTURE

In order to illustrate applications of the methodologies outlined above, results and discussion of experiments on biosynthesis, detection, and proteolytic fragmentation of the α catalytic subunit of $Na^+,K^+$-ATPase are given below. Where applicable, anti-α antibodies used in these studies were prepared using the method outlined in Fig. 2.

The mechanism of assembly of the α and β subunits into a functional membrane-bound ATPase and ion pump is not yet understood. Preliminary evidence suggests that, while the β subunit is made on membrane-bound ribosomes, the α subunit is translated on free cytosolic ribosomes (Sherman *et al.*, 1980; Hiatt, 1983). In order to verify this finding regarding the α subunit, an immunoprecipitation of α subunits from cells labeled in culture was conducted. Madin–Darby canine kidney (MDCK) cells, a dog kidney epithelial cell line, were labeled for 20 minutes with $[^{35}S]$methionine. The cells were homogenized and spun at 2,000 *g*. The low-speed pellet (unbroken cells, etc.) was retained and

---

FIG. 2. Preparation of subunit- or fragment-specific antibodies using diazotized paper as an affinity ligand. (1) Resolve peptides by SDS–PAGE. (2) Transfer to diazotized paper. (3) Cut off side strip and test with antiserum to be fractionated. (4) Using the side strip as a guide, cut out perpendicular strips corresponding to the subunit or fragment of interest. (5) Incubate the strips 8–18 hours with antiserum diluted 1:20 to 1:50 in Buffer A (150 m*M* NaCl/5 m*M* EDTA/50 m*M* Tris-HCl pH 7.4/0.25% gelatin/0.05% Nonidet P-40). (6) Wash the strips 8–18 hours in Buffer A to remove nonspecifically bound antibodies. (7) Elute subunit or fragment specific antibodies with 0.2 *M* glycine pH 2.3. for 30 minutes at 37°C. Titrate to pH 7 with 1 *M* Tris-HCl pH 9.0. (8) Repeat to obtain more antibodies. (9) Test separated antibodies against strips of the diazotized paper prepared as in (1) above. 1, Unfractionated antiserum; 2, anti-α; 3, anti-β.

FIG. 3. Immunoprecipitation of $Na^+,K^+$-ATPase α subunit with monospecific antisera. MDCK cells were labeled for 20 minutes with [$^{35}$S]methionine (70 μCi/ml) in methionine-deficient media. The low-speed pellet, cytosol, and membrane fractions were prepared, and the α subunit immunoprecipitated from each with anti-α serum, as described in Section II. The figure shows immunoprecipitated α subunit from C, cytosol; M, membranes; and P, low-speed pellet.

the supernatant was spun at 200,000 *g* for 90 minutes in order to generate cytosol (supernatant) and membrane (pellet) fractions. These fractions were solubilized according to Maccecchini *et al.* (1979) prior to addition of anti-α antibodies. The immune complexes were precipitated with the addition of protein A coupled to Sepharose and processed as described in Section II. The results as shown in Fig. 3 demonstrate that the 94,000 MW α subunit can be immunoprecipitated from both membrane and cytosolic fractions, supporting the finding that the subunit α is translated on free ribosomes. It is, however, possible that the cytosolic α subunits were derived from α subunits loosely associated with membranes, and solubilized during preparation.

If the α subunit is synthesized on free ribosomes, it must become integrated in the membrane somewhere along the endoplasmic reticulum, Golgi, or plasma membrane. With the goal of determining the location of integration, studies were begun in which the α subunit was detected in membranes from MDCK cells fractionated by density gradient centrifugation (McDonough and Mircheff, 1983). The preliminary results presented here demonstrate the feasibility of applying the solid phase assay to studies of protein assembly. Unlabeled MDCK cells were fractionated on a sorbitol gradient and pooled into eight fractions. These fractions were subjected to SDS–PAGE, blotted onto diazotized paper, and assayed for α subunits as described in Section II. Although α subunits cannot be identified in the stained gel because of the presence of hundreds of bands in each fraction, they can be detected in the blots with the anti-α antibodies (Fig. 4). Most of the α subunits are present in fractions 3–5. Enzyme marker studies show that these fractions are enriched in $Na^+,K^+$-ATPase as well as alkaline

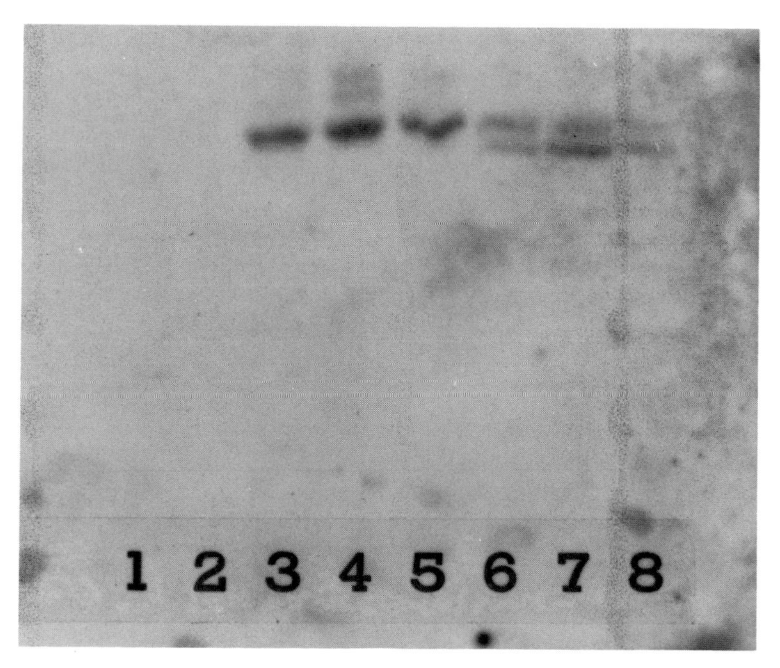


FIG. 4. Detection of $Na^+,K^+$-ATPase α subunit in fractionated membranes of MDCK cells. Membranes from MDCK cells were fractionated by density gradient centrifugation on a 30–65% sorbitol gradient (w/v). The gradient was separated into eight fractions, and the membranes were pelleted and subjected to SDS–PAGE. The resolved peptides were transferred to diazotized paper and probed with anti-α antibodies as described in Section II. The resulting autoradiogram is shown.

phosphatase activity, both markers of plasma membranes. At higher densities there are two bands that bind anti-α antibodies. These fractions are enriched in thiamin pyrophosphatase and glucosaminidase activity, markers of Golgi and lysosomal membranes respectively, and contain a small amount of $Na^+,K^+$-ATPase activity. Pollack *et al.* (1981) have evidence suggesting that tritiated ouabain bound to $Na^+,K^+$-ATPase in the plasma membrane is internalized into a lysosomal compartment as a consequence of enzyme turnover. Our preliminary results showing slightly degraded α subunits in the same fractions as the lysosomal enzyme marker further support their idea. Further purification studies are now under way to separate Golgi and lysosomal membranes and to define the distribution of endoplasmic reticulum in the gradient fractions. These refinements may enable us to identify the site(s) of assembly and turnover of $Na^+,K^+$-ATPase molecules, using the solid phase assay for α and β subunits.

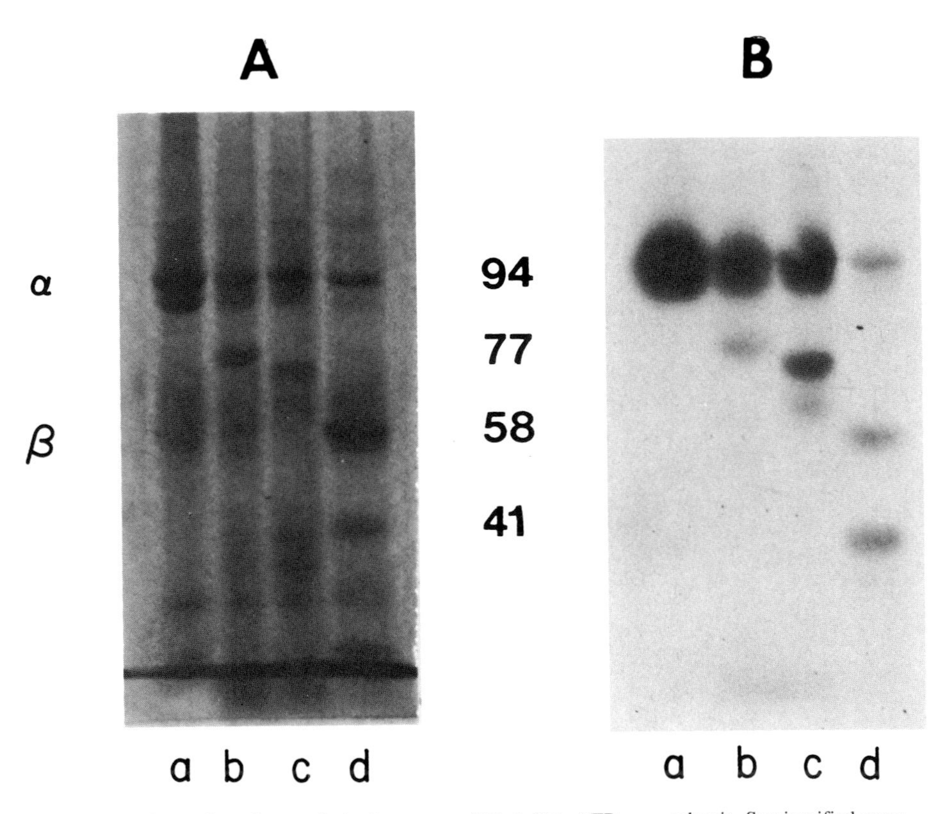


FIG. 5. Detection of proteolytic fragments of $Na^+,K^+$-ATPase α subunit. Semipurified membrane-bound $Na^+,K^+$-ATPase from guinea pig kidney (a) was digested with chymotrypsin (b), chymotrypsin plus ouabain (c), and trypsin plus 150 m*M* KCl (d), all as described by Farley *et al.* (1980). The samples were subjected to SDS–PAGE (A), transferred to diazotized paper, and probed with anti-α antibodies. (B) The resulting autoradiogram.

Antibodies to the sodium pump are also useful in the study of $Na^+,K^+$-ATPase structure. A simple example of the structural application of antibodies is the identification of antigenic proteolytic fragments of the $\alpha$ subunit using the solid phase assay. As shown in Fig. 5, membrane-bound guinea pig $Na^+,K^+$-ATPase (lane a) was partially digested according to Farley *et al.* (1980) with chymotrypsin (lane b), chymotrypsin plus ouabain (lane c), and trypsin (lane d). The samples were resolved by SDS–PAGE, blotted, and tested with anti-$\alpha$ antibodies as described. The antibodies reacted with the 77,000 (chymotrypsin), 58,000, and 41,000 (trypsin) MW fragments. They did not react with the 35,000 and 40,000 MW fragments seen by Farley in the chymotrypsin plus ouabain digests. Proteolytic fragment-specific antibodies have been prepared using the method described in Fig. 2. These fragment-specific antibodies can be used as probes to investigate the structure of $Na^+,K^+$-ATPase in the membrane. Membrane vesicles oriented right side or inside out may be incubated with fragment-specific antibodies and then with labeled protein A, washed, and assayed for bound label in order to detemine whether the fragment is on the outside or inside of the membrane. Experiments of this type could provide information about the asymmetry of the transmembrane sodium pump subunits in the membrane.

## IV. PROBLEMS ASSOCIATED WITH ANTIBODY DETECTION OF MEMBRANE PROTEINS

### A. Solubility

Membrane proteins such as $Na^+,K^+$-ATPase require detergents for solubility in aqueous solutions. Although nonionic detergents such as Triton X-100 or NP-40 usually suffice to solubilize enzymes, the separation of subunits (e.g., $\alpha$ and $\beta$ subunits of the sodium pump) may require ionic detergents such as SDS. Often, membrane proteins must be detergent solubilized during their purification prior to injection as antigen. This solubilization may denature the proteins, causing a change in the antigenic sites. As a result, the antibodies may react poorly with the native protein or subunits unless they too are detergent denatured. Therefore, prior to immunoprecipitation, many investigators boil the labeled protein mixtures in $\sim$2% SDS and then reduce the SDS concentration with a 10-fold dilution in a nonionic detergent buffer (Maccecchini *et al.*, 1979; Geering *et al.*, 1982). In the solid phase assay, the samples are SDS-denatured as a consequence of the SDS–PAGE and need no further treatment prior to assay.

### B. Glycoproteins

Proteins that are synthesized on ribosomes bound to the endoplasmic reticulum and transported through the Golgi apparatus to the plasma membrane are often processed and glycosylated. When processed glycoproteins are injected as antigen they often elicit antibodies to the oligosaccharide residues or processed portions of the molecule. If so, they may not precipitate the nascent unprocessed peptides generated in *in vitro* translation experiments or in biosynthesis experiments. This problem prevented detection of the *in vitro* translation product of the β subunit of the sodium pump (McDonough *et al.*, 1982). To solve this problem, I am purifying nonglycosylated proteolytic fragments of β subunits to use in antibody production.

Antibody cross-reactivity studies suggest that there is more similarity between species in antigenic portions of the α subunit than in the β subunit. As shown in Fig. 6, the anti-α antibodies cross-react with all α subunits tested, albeit with varying intensites. The anti-guinea pig β antibodies, however, recognize only the rat, mouse, and human subunits. The antibody cross-reactivity may reside in the oligosaccharide residues or other processed portions of the β subunit which could vary more from species to species, and even tissue to tissue, than the amino acid sequence of the subunit.

## C. Low Abundance

As has been shown above, antibodies are very useful tools for the study of membrane protein structure and biosynthesis. However, most membrane transport proteins and enzymes are in such low abundance that their purification to homogeneity for antibody production is very difficult. If a protein can be partially purified and identified with a labeled ligand, it is possible to obtain specific antibodies without purification to homogeneity. One approach is to generate a monoclonal antibody to the protein (discussed in Sacktor and Reiss, Chapter 13, this volume). Another approach is to generate polyclonal antibodies (e.g., in a rabbit) to the partially purified membrane protein and to fractionate antibodies specific for the protein of interest, applying the approach outlined in Fig. 2.

## D. Similarity to Other Proteins

The α subunit of $Na^+,K^+$-ATPase is very similar to $Ca^{2+}$-ATPase and $H^+,K^+$-ATPase in molecular weight, amino acid composition, sequence at the phosphorylation site, and tryptic fragmentation patterns (Jørgensen, 1978; Rizzolo and Tanford, 1978; Saccomani *et al.*, 1981). This family of ion-dependent ATPases may contain common antigenic determinants as might families of amino acid, sugar, or ion carriers. Common determinants could be troublesome if antibodies are being used as specific probes of one protein in a tissue that contains a number of similar proteins. Antibodies can be tested for their specificity as shown in Fig. 7. In this case, antibodies to the $Na^+,K^+$-ATPase α subunit

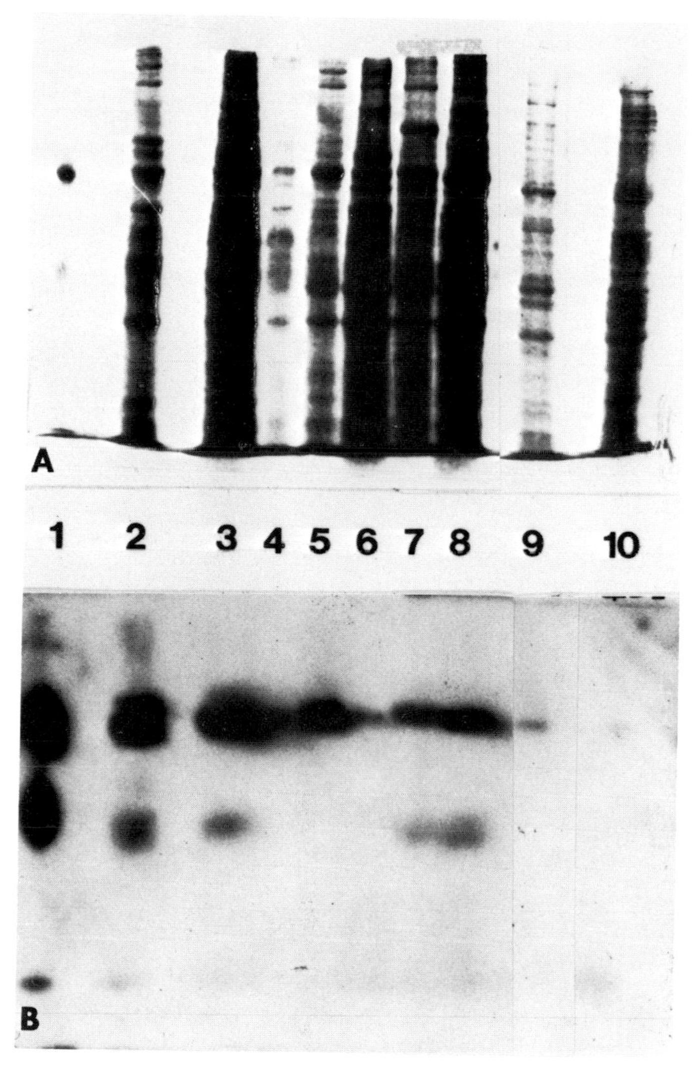


FIG. 6. Cross-reactivity of anti-guinea pig Na$^+$,K$^+$-ATPase antibodies with Na$^+$,K$^+$-ATPase from various species. Membrane fractions from various animals were prepared (McDonough *et al.*, 1982) and subjected to SDS–PAGE. The same amount of Na$^+$,K$^+$-ATPase activity was applied to each lane—1.0 μmol $P_i$/hour/gel lane. (A) The gel was stained with Coomassie blue. (B) Autoradiogram of a duplicate gel which was transferred to diazotized paper, incubated with antisera diluted 1:100, washed, and incubated with $^{125}$I-labeled protein A. 1, Purified guinea pig Na$^+$,K$^+$-ATPase; 2, guinea pig renal outer medulla membranes; 3, human outer medulla membranes; 4, beef outer medulla membranes; 5, dog outer medulla membranes; 6, rabbit outer medulla membranes; 7, rat outer medulla membranes; 8, mouse kidney microsomes; 9, turtle bladder membranes; 10, toad bladder membranes.

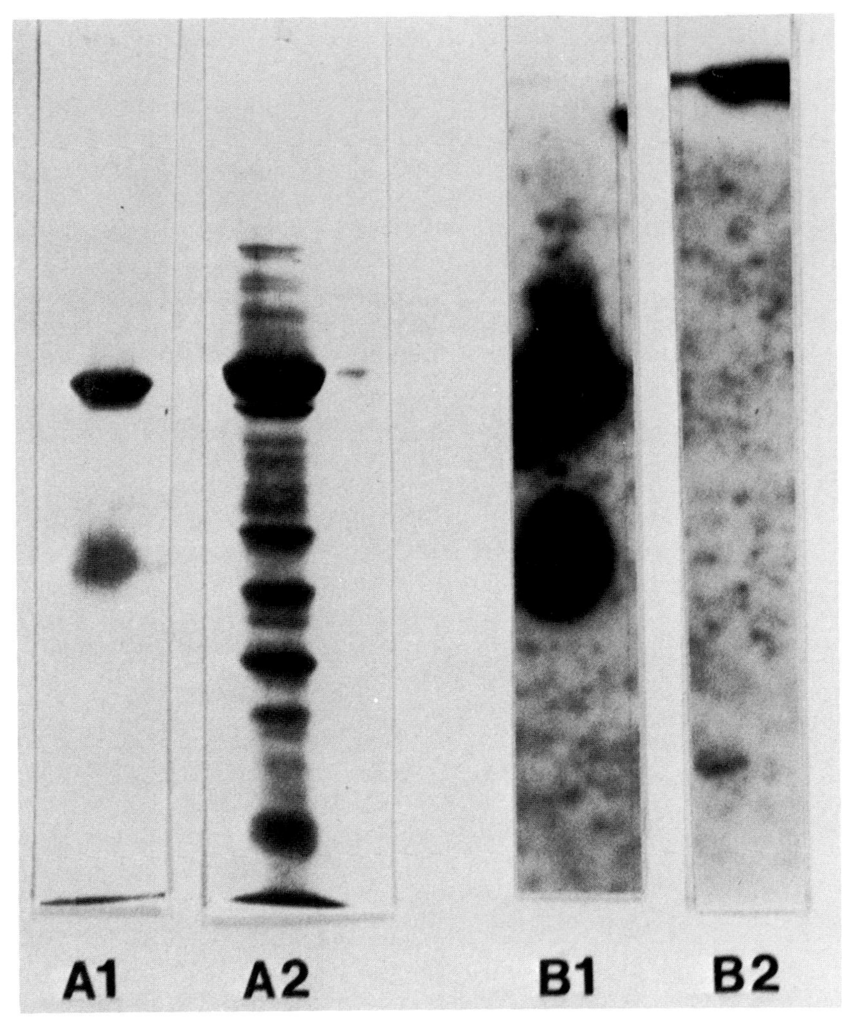


FIG. 7. Cross-reactivity of anti-guinea pig $Na^+,K^+$-ATPase antibodies with guinea pig $Ca^{2+}$-ATPase. SDS–polyacrylamide gel of A1, purified guinea pig $Na^+,K^+$-ATPase, and A2, guinea pig $Ca^{2+}$-ATPase as the major band at 94,000 d. B1, Duplicate gel ($Na^+,K^+$-ATPase) transferred to diazotized paper, incubated with antisera diluted 1:100, incubated with $^{125}I$-labeled protein A, and subjected to autoradiography. B2, Duplicate gel ($Ca^{2+}$-ATPase) treated as for B1.

did not cross-react with $Ca^{2+}$-ATPase from the same species and therefore could be used as specific probes for the $Na^+,K^+$-ATPase. Antibodies that do cross-react with various ATPases would be very interesting and useful probes of the structure of similar proteins.

These are other problems associated with antibody detection of membrane proteins, including the nonspecific binding of antibodies to hydrophobic peptides, proteolysis of antigen during preparation for immunoprecipitation, and the difficulties encountered in raising antibodies to membrane proteins. Despite these problems, antibodies have provided the means of examining aspects of $Na^+,K^+$-ATPase structure, biosynthesis, and regulation that would otherwise have been very difficult.

## REFERENCES

Farley, R. A., Goldman, D. W., and Bayley, H. (1980). Identification of regions of the catalytic subunit of (Na-K)-ATPase embedded within the cell membrane. *J. Biol. Chem.* **255,** 860–864.

Farley, R. A., Tran, C. M., and McDonough, A. A. (1983). Proteolytic fragmentation of the glycoprotein subunit of NaK-ATPase. In preparation.

Geering, K., Girardet, M., Bron, C., Kraehenbühl, J-P., and Rossier, B. (1982). Hormonal regulation of ($Na^+,K^+$)-ATPase biosynthesis in the toad bladder. *J. Biol. Chem.* **257,** 10338–10343.

Girardet, M., Geering, K., Frantes, J. M., Geser, D., Rossier, B. C., Kraehenbühl, J-P., and Bron, C. (1981). Immunochemical evidence for a transmembrane orientation of both the ($Na^+,K^+$)-ATPase subunits. *Biochemistry* **20,** 6684–6692.

Hiatt, A. (1983). In preparation.

Jørgensen, P. L. (1978). Isolation and characterization of the components of the sodium pump. *Q. Rev. Biophys.* **7,** 239–274.

Kessler, S. W. (1975). Rapid isolation of antigens from cells with a staphylococcal protein-A-antibody adsorbent. *J. Immunol.* **115,** 1617–1624.

Lauf, P. K. (1978). Membrane immunological reactions and transport. *In* "Membrane Transport in Biology" (D. C. Tosteson, H. H. Ussing, and G. Giebisch, eds.), p. 291. Springer-Verlag, Berlin and New York.

Maccecchini, M-L., Rudin, Y., Blobel, G., and Schatz, G. (1979). Import of proteins into mitochondria. Precusor forms of the extramitochondrially made F1,-ATPase subunits in yeast. *Proc. Natl. Acad. Sci. U.S.A.* **76,** 343–347.

McDonough, A. A., Hiatt, A., and Edelman, I. S. (1982). Characteristics of antibodies to guinea pig ($Na^+ + K^+$)-adenosine triphosphatase and their use in cell-free synthesis studies. *J. Membr. Biol.* **69,** 13–22.

McDonough, A. A., and Mircheff, A. (1983). NaK-ATPase biosynthesis in MDCK cells. In preparation.

Pollack, L. R., Tate, E. H., and Cook, J. S. (1981). Turnover and regulation of NaK-ATPase in HeLa cells. *Am. J. Physiol.* **241,** C173–C183.

Renart, J., Reiser, J., and Stark, G. R. (1979). Transfer of proteins from gels to diazobenzyloxymethyl-paper and detection with antisera. *Proc. Natl. Acad. Sci. U.S.A.* **76,** 3116–3120.

Rizzolo, L. J., and Tanford, C. (1978). Behavior of fragmented calcium(II) adenosine triphosphatase from sarcoplasmic reticulum in detergent solution. *Biochemistry* **17,** 4049–4055.

Saccomani, G., Sachs, G., Cuppoletti, J., and Jung, C. Y. (1981). Target molecular weight of the gastric ($H^+ + K^+$)-ATPase functional and structural molecular size. *J. Biol. Chem.* **256,** 7727–7729.

Schellenberg, G. D., Pech, I. V., and Stahl, W. L. (1981). Immunoreactivity of subunits of the ($Na^+ + K^+$)-ATPase. *Biochim. Biophys. Acta.* **649,** 691–700.

Sherman, J. M., Sabatini, D. D., and Morimoto, T. (1980). Studies on the biosynthesis of the polypeptide subunits of the Na,K-ATPase of kidney cells. *J. Cell Biol.* **87,** 307a.

Towbin, H., Staehelin, T., and Gordon, J. (1979). Electrophoretic transfer of proteins from polyacrylamide gels to nitrocellulose sheets. *Proc. Natl. Acad. Sci. U.S.A.* **76,** 4350–4354.

CURRENT TOPICS IN MEMBRANES AND TRANSPORT, VOLUME 20

# Chapter 9

# Encounters with Monoclonal Antibodies to $Na^+,K^+$-ATPase

*MICHAEL KASHGARIAN,* **
*DANIEL BIEMESDERFER,*† *AND*
*BLISS FORBUSH III*†

*Departments of *Pathology and †Physiology*
*Yale University School of Medicine*
*New Haven, Connecticut*



## I. INTRODUCTION

Physiologic studies of membrane transport have generally utilized either electrical or chemical methods to trace the movement of ions across apical, basolateral, and intercellular pathways. In a different approach, biochemical studies of the structure of plasma membranes have demonstrated that specific membrane-associated proteins can be identified as either specific ion channels or active ion pumps. Morphological studies, although usually only descriptive, have proved useful where structural changes accompany different functional states. Recently, the introduction of monoclonal antibody technology by Kohler and Milstein (1975) has provided a unique opportunity to unify these three different approaches to the study of transport with the definition of cell surface components and their specific functions. The major advantage inherent in this technology is that a monoclonal antibody produced by an isolated hybridoma




ISBN 0-12-153320-4

clone is a well-defined chemical reagent with which different domains of functional proteins can be reproducibly and positively identified. Furthermore, this technique is ideally suited for the preparation of specific antibodies from a nonpurified antigen. It is this latter advantage that has stimulated us to pursue the development of such antibodies to $Na^+,K^+$-ATPase so that they may be utilized as specific morphologic markers.

Although monoclonal antibodies have great advantages it must be remembered that they are not substitutes for well-absorbed high-titer polyvalent antisera or even affinity-purified polyclonal antibody. Polyvalent antisera and polyclonal antibodies are a mixture of different antibodies with different antigenic specificities and biologic functions which thus can define an antigen at multiple determinants. While the major advantage of monoclonal antibodies is the high degree of specificity, such specificity can be equally a disadvantage if the antigenic determinant is a peptide common to more than one protein. This can easily occur if the antigenic determinant is small, perhaps in the range of 8 or 10 amino acid residues. The resultant antibody, although specific for the determinant, becomes polyspecific because the sequence is common to a number of proteins. This disadvantage can be overcome by utilization of a library of monoclonal antibodies directed against different determinants of the same molecules rather than a single monoclonal antibody to a single determinant. Since each clone synthesizes only a single antibody, that antibody will be of a single class of immunoglobulin. It may or may not fix complement or bind protein A, or its affinity may be so low in its binding to antigen as to negate its usefulness. Furthermore, once a reagent is identified it may not be available to perpetuity. Individual clones may lose antibody activity or may vary in characteristics following subcloning. This requires periodic verification of the specificity and affinity of the antibodies produced.

Despite these caveats there is no doubt that the careful and rigorously controlled utilization of monoclonal antibodies will yield information about transport processes that could not otherwise be obtained. While the utilization of this technique is of considerable value, problems are encountered which may be either inherent in the technology or specific to its application to transport proteins. This review is an attempt to describe our own encounter with these problems and the approaches that we have used to solve them.

## II. IMMUNIZATION PROCEDURES

The development of monoclonal antibodies begins with the initial immunization procedure. Although hybridomas can be generated across species, it is generally simpler to generate them within a single species, the mouse being the species which is best defined. The further advantage in using the mouse as a primary host is the availability of genetically well-defined inbred strains. Since genetic factors influence antigen recognition and the production of antibodies, different strains will yield different families of antibodies to different antigenic determinants. In our experiments, three different strains including BALB/c, NZB, and CB6 were immunized with preparations of $Na^+,K^+$-ATPase. As can be seen in Table I, even within a single strain the immune response varied not only quantitatively but also qualitatively in relation to cross-reactivity with other sources of the same antigen.

The method of immunization has been found to be of importance in determining the success of the hybridization procedure. In our experiments three different immunization procedures were utilized. In the first, mice were immunized with untreated antigen combined with complete Freund's adjuvant in the foot pad and peritoneum. In the second group, mice were immunized with untreated antigen and complete Freund's adjuvant in the peritoneum alone. In a third group, mice were primed with *Bordetella pertussis* organisms and injected intraperitoneally several hours before immunization, and the antigen was aggregated by adsorption with potassium alum before injecting intraperitoneally (Lerner, 1981). Mice immunized in the peritoneum alone developed high antibody titers, but macrophages so dominated the cell population of the spleen that hybridization of these spleens failed. Mice immunized in the foot pad developed an acceptable immune response and hybridization was successful; however, antibody titers and the number of clones produced were limited. In mice treated with *B. pertussis* and immunized with adsorbed antigen, high antibody titers were identified and the number of successful fusions were optimized. The aggregation of the antigen along with the adjuvant effect of the *B. pertussis* is probably responsible for this degree of success. The affinity of antibodies produced in this way also appeared to be greater than that of antibodies produced from fusions of spleen cells from animals immunized in the foot pads.

TABLE I
ANTIBODY RESPONSES OF DIFFERENT MOUSE STRAINS TO DOG KIDNEY $Na^+,K^+$-ATPase

| Strain | Serum antibody (mg/ml) | Cross-reactivity |
|---|---|---|
| 1. BALB/c | 2.5 | — |
| 2. BALB/c | 5 | — |
| 3. BALB/c | 7.5 | + Rat, + human |
| 4. BALB/c | 5 | + Rat |
| 1. NZB | 5 | — |
| 2. NZB | 10 | — |
| 1. CB6 | 5 | — |
| 2. CB6 | 0 | Not applicable |

While possibly the greatest advantage of this technology in the study of membrane proteins is the fact that an impure antigen can be utilized to produce a large number of antibodies which will then be screened for specificity, the screening of antibodies to this impure antigen may become a formidable task and specificity may be difficult to determine. Methods of further purifying the antigen are therefore advantageous. In our experiments $Na^+$,$K^+$-ATPase from the dog was prepared using the Jørgensen method (1974) which enriched it to a concentration comprising 80–90% of the total protein. Such a relatively pure antigen will generate a library of antibodies which will more likely be directed at the molecule in question and will likely comprise antibodies to different antigenic determinants as well. The $Na^+$,$K^+$-ATPase purified from the rat kidney by the Jørgensen method had concentrations of enzyme comprising only 20–30% of total protein, since our preparation of rat outer medulla was more heavily contaminated with cortex.

Two fusions after immunization with dog antigen and one fusion after immunization with rat antigen were carried out. The procedures for fusion and generation of the hybridomas are adequately described elsewhere (Lerner, 1981) and will not be described in detail here. As an example of the vagaries of the method, the clones developed against dog ATPase from our first hybridization ceased to produce antibody after an ascites tumor was generated. Five clones were selected from the second fusion with dog antigen for further characterization, and these have remained stable producers of antibody. Only one clone was selected for further characterization from hybridomas made against rat antigen.

## III. SCREENING PROCEDURES

Screening procedures are necessary initially to identify the presence of antibody and specificity of its reactivity during the immunization and hybridization protocols, and later to characterize the antibody in relation to the antigenic determinant at which it is directed and how it may be related to the structure of the protein under investigation. The most efficient and rapid method for screening of antibody activity in the serum of immunized animals and the supernatant fluid of hybridoma cultures employs either a radioimmunoassay or an enzyme-linked immunoassay. The assay must be rapid and reproducible as well as sensitive. The most popular radioimmunoassay utilizes a radiolabeled second reagent which is usually $^{125}$I-labeled protein A (Kennett *et al.*, 1980). This binds to some classes of mouse immunoglobulin but not all, and therefore may require as an intermediate step a second antibody that does bind protein A directed against mouse immunoglobulin. Alternatives include the use of $^{125}$I-labeled rabbit anti-mouse Ig instead of protein A, or the use of an enzyme-linked immunosorbent assay (ELISA) instead of the radioimmunoassay. In this technique the antigen is bound to microtiter plates, is then reacted with the serum or supernatant fluid to be tested, and after washing is reacted with an enzyme-conjugated anti-mouse Ig (Kennet *et al.*, 1980). The enzyme most frequently used is peroxidase and a color reaction is generated with *o*-phenylenediamine. This method has the added advantage of amplification both by the secondary antibody reaction and by the enzymatic reaction of the conjugate.

While this method is rapid and sensitive, the ability to isolate a clone-producing antibody of desired specificity depends entirely on the purity of the antigen bound to the plates. Thus, although antibody presence can be detected in whatever preparation was used for immunization, the possibility cannot be ruled out that the antibody is reacting with an impurity in the antigen rather than with the protein of interest. For this reason, additional screening techniques are needed to characterize the specificity of the antibody.

A variety of methods have been developed for the detection of specific proteins from crude extracts using gel electrophoresis in the presence of sodium dodecyl sulfate (Laemmli, 1970) combined with the antigen–antibody reaction. These methods combine the high resolution of gel electrophoresis with the specificity of the immune reaction. With the simplest method, either membrane fragments or solubilized membrane proteins are incubated with an excess amount of the antibody under study, to form immune complexes. If membrane fragments are used, they are sedimented, washed, and solubilized in detergent. The immune complexes are removed from solution by binding with either protein A–Sepharose or anti-mouse Ig–agarose. After elution from the beads the proteins are treated with buffered detergent and subjected to SDS–polyacrylamide gel electrophoresis (SDS–PAGE). This method separates all the proteins in solution including the immunoglobulin, and is therefore useful only if the protein or peptide of interest does not migrate with the antibody or its fragments. In a variant of this technique, the proteins are tagged with a radioisotope to facilitate detection and to distinguish them from the unlabeled protein of the antiserum or antibody preparation. With this method, we have utilized the lactoperoxidase-catalyzed radioiodinization of cell surface protein as described by Cone and Brown (1976) to iodinate the membrane proteins. The membrane proteins can also be labeled biosynthetically using [$^{35}$S]methionine (Jackson and Blobel, 1977). The labeled membrane fragments or the solubilized membrane proteins are incubated with the antibody under study to form immune complexes and then processed in the same manner as with the unlabeled method. The gels are stained and dried, and a contact autoradiogram is prepared. The band with radioactivity is then compared with the original gel for identification of the specific protein that binds with the antibody (Kennett *et al.*, 1980). While this method is seemingly simple, there are several points at which negative or anomalous results may be generated. In the radiolabeling procedure, the specific protein in question may not be iodinated during the initial procedure or the iodination may alter the antigenicity of the molecule. Furthermore, during the immunoprecipitation, aggregates containing polypeptides other than the specific protein may be precipitated and detected. In addition, proteolytic degradation of the antigen can sometimes occur during the incubation, precipitation, and elution procedures. Also, the antibody may react differently with different preparations of antigen. For example, if the native holoenzyme is used for the initial immunization, the antibody produced may recognize only the native protein and not a detergent-solubilized form. Even if it does react with the denatured polypeptide, it may have such a low affinity for the subunit as to make immunoprecipitation an inefficient method for detection.

As an alternative approach, one can transfer protein from SDS–PAGE electrophoretograms to nitrocellulose or similar membrane filter paper. Overlaying the paper with the transferred proteins with antibodies has allowed the correlation of the binding activity of the antibody with specific polypeptide bands. The advantages are several, including factors such as ready accessibility of the immunobilized proteins to the antibody and the relatively small amount of reagent necessary for adequate identification. The transfer can be mediated by diffusion or electrophoresis and is generally referred to as the "Western blot" technique (Towbin *et al.*, 1979). In our laboratory we electrophoretically transferred the proteins of our membrane preparations from SDS gels to nitrocellulose paper. Following the transfer, the original gel was stained with Coomassie blue to verify transfer of the proteins, and strips of the nitrocellulose paper were stained with amido black to confirm their presence in the paper. Immunolabeling of the transferred protein was performed by cutting the paper, blocking nonspecific binding with bovine serum albumin and detergent, and incubating with the primary antibody. The strips were then washed in nonionic detergent, incubated with rabbit anti-mouse Ig, washed, and incubated with horseradish peroxidase-labeled goat anti-rabbit Ig (Kennet *et al.*, 1980). The double-labeling technique provides for significant amplification of the signal and reduces the amount of primary antibody and reacting peptide necessary for visualization. The protein bands that were bound to antibody are then visualized by generation of a color reaction using diaminobenzidine in the presence of hydrogen peroxide as a substrate (Graham and Karnovsky, 1966).

This procedure, however, is not without its problems. The protein-binding capacity of the paper may be exceeded, and electrophoresis may actually move the proteins through the paper. Large-molecular-weight proteins may not transfer under the same conditions needed for smaller peptides. Furthermore, the peptides in the gels are denatured, and thus may not react with the antibody if the reactivity of the antibody is determined by the configuration of the protein rather than its sequence.

In the case of antibodies raised to proteins eluted from gel slices, the above techniques are of little value in demonstrating the proper specificity, since any contaminating immunogens were selected to have the same apparent molecular weight as the desired immunogen. Thus it is to be expected that a monoclonal antibody raised to proteins from gel slices will react with a peptide of the "correct" molecular weight, and this is not definitive evidence that the desired antibody has indeed been isolated. Particularly in this situation, the use of two-dimensional gel electrophoresis is indicated.

Thus far we have subjected a few clones to screening in this manner. With a clone generated from rat antigen, we have been able to demonstrate that with both immunoprecipitation and Western blot techniques the antibody reacts against a protein that migrates as a peptide with a molecular weight of approximately 96K, which could correspond to the $\alpha$ subunit of $Na^+,K^+$-ATPase.

Screening of immunoprecipitation of the antibodies raised against native dog ATPase has demonstrated that these antibodies do not appear to react with solubilized dog kidney membrane fragments, while incubation with intact membrane fragments enriched for active holoenzyme resulted in the formation of immune complexes. In contrast, solubilized membrane fragments from cultured MDCK (Madin–Darby canine kidney) cells readily formed immune complexes and simplified the immunoprecipitation.

Another screening method of value is to assess whether there is any functional activity of the antibody such as inhibition of $Na^+,K^+$-ATPase activity or of ouabain binding. All of our clones have been tested for functional activity, and only one of the antibodies directed against dog antigen demonstrated any such activity. Its inhibition is obtained with broken membrane preparations, but not in tight right-side-out vesicles. This antibody inhibited $Na^+,K^+$-ATPase activity and sodium–potassium–magnesium-dependent ouabain binding but not magnesium–phosphorus-dependent ouabain binding or potassium-stimulated *p*-nitrophenylphosphatase activity. This antibody is directed at a binding site that localized to the intercellular aspect of the plasmalemma.

Morphologic screening can be a useful and relatively rapid assessment of antibody activity. Screening can be performed on frozen sections of whole tissue, on membrane preparations, or under certain circumstances on fixed and embedded tissue sections.

Membrane preparations offer further advantages since the same material used in immunochemical screening procedures can also be examined with morphologic techniques. With this method, membranes are incubated with the antibody being studied and then incubated with a second antibody that has a morphologic marker such as colloidal gold. Preparations are applied to coated grids and examined ultrastructurally. This technique complements functional screening techniques also by demonstrating that the antibody is directed at a membrane-associated protein and by assisting in localizing the site of binding, provided that tight intact vesicles with known sidedness can be prepared (Figs. 1–3).

Immunofluorescent localization of the antibody on frozen sections can rapidly demonstrate localization to different nephron segments or different domains of the cell membrane. This can be useful to search for a marker to apical or

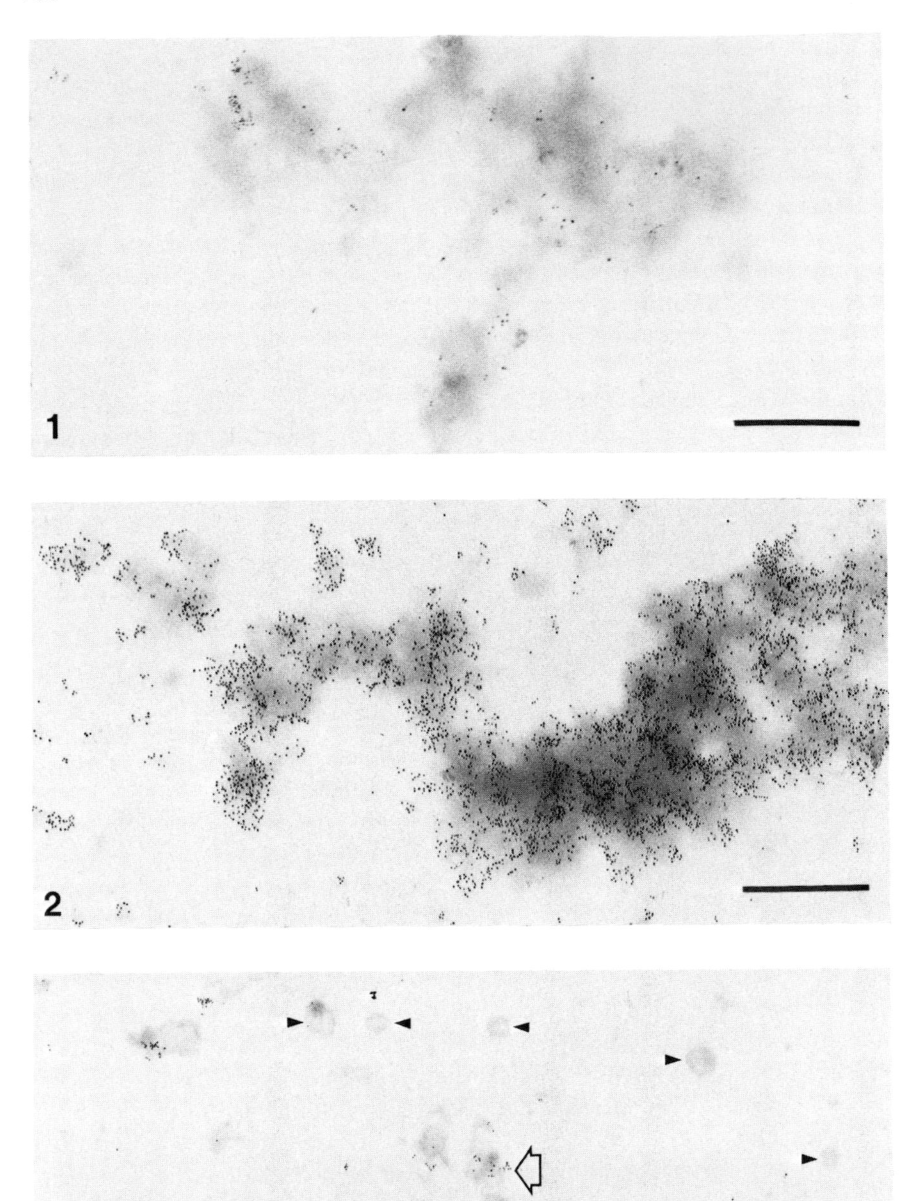
1
2
3

basolateral membranes and, if a specific distribution for a protein is known, to verify antibody activity to it.

It becomes evident from the previous description of these screening procedures that no single method is adequate for complete characterization, and that combination of the various methods is necessary to completely confirm the specific nature of the antibody concerned. As an example, the monoclonal antibody to rat kidney ATPase that we developed reacted well and was easily detected by using ELISA against the antigenic preparation. Characterization by immunoprecipitation and Western blot techniques confirmed that it was an antibody directed against a protein that migrated in the 96K band. This was strongly suggestive that the antibody was directed against the $\alpha$ fragment of $Na^+,K^+$-ATPase. When final screening was done with morphologic localization by both immunofluorescence and electron microscopic immunocytochemical techniques (Fig. 4), it was apparent that the antibody bound primarily to proximal tubular brush border and not to basolateral membrane as might have been anticipated. Several explanations are possible, and the simplest would be that it is directed at a contaminating protein that migrates along with the $\alpha$ fragment on SDS gels. If contamination of the membrane preparation can be excluded, a possibility which should be considered is that there may be common antigenic domains in several membrane proteins that localize to different regions of different cell types. Reactivity of a single antibody to different cell types and domains has been reported by others (Papermaster *et al.*, 1981; Coudrier *et al.*, 1982). Whether such antigenic similarity may be related to any transport function such as ion binding sites raises additional questions which must be explored. To assess whether the antibody is directed at a 96K peptide other than the $\alpha$ subunit, membrane preparations enriched with $Na^+,K^+$-ATPase must be immunoprecipitated with the antibody and the supernatant examined by SDS–PAGE for the presence of a residual 96K protein. Immunoprecipitation of $^{32}P$-labeled $Na^+,K^+$-ATPase would also give a more specific immunoprecipitation reaction than that of iodi-

FIG. 1. Control for immunolabeling of membrane preparations with monoclonal antibodies using colloidal gold as a marker. Gold spheres of 5 nm are coated with rabbit anti-mouse IgG. $Na^+,K^+$-ATPase-enriched membranes from the outer medulla of dog kidney were placed on carbon-coated Formvar grids and incubated with control mouse IgG followed with the gold-conjugated rabbit anti-mouse IgG. Only background labeling is seen. Bar equals 0.5 $\mu$m.

FIG. 2. Immunolabeling of membrane preparations with a monoclonal antibody. $Na^+,K^+$-ATPase-enriched membranes made from the outer medulla of dog kidney were placed on carbon-coated Formvar grids and incubated with the antibody from clone 4. This was followed with gold-conjugated rabbit anti-mouse IgG. Note the heavy gold labeling of the membrane fragments. Bar equals 0.5 $\mu$m.

FIG. 3. This preparation of membranes made from the outer medulla of dog kidney is composed of a large population of intact right-side-out vesicles. Labeling was carried out by incubation with the antibody from clone 4 followed by gold-conjugated rabbit anti-mouse IgG as in the preparation seen in Fig. 2. Intact vesicles (arrows) are unlabeled, while broken membranes (open arrows) exhibit the gold label. Bar equals 0.5 $\mu$m.

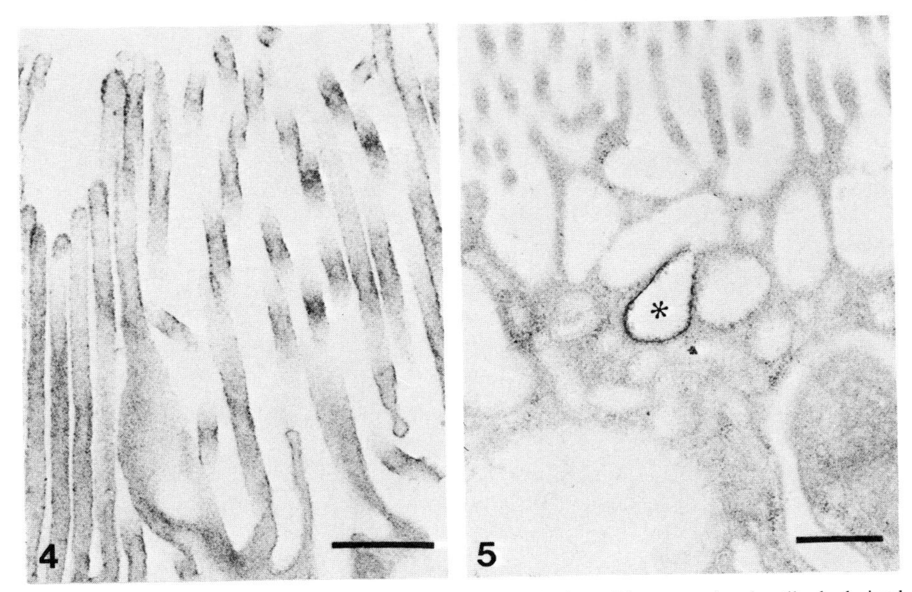


FIG. 4. Perfusion-fixed rat kidney was labeled by incubation with a monoclonal antibody derived following immunization with rat $Na^+,K^+$-ATPase preparation. This antibody reacts with a 96K fragment that migrates with the catalytic subunit of $Na^+,K^+$-ATPase. Following incubation with the monoclonal antibody, a secondary incubation was carried out with a rabbit anti-mouse immunoglobulin antibody. This was followed by peroxidase-conjugated $F(ab')_2$ fragment of goat anti-rabbit IgG. Reaction product was generated by incubation in diaminobenzidine in the presence of hydrogen peroxide. Labeling is restricted to the external surface of the proximal tubular brush border membranes. Bar equals 0.5 μm.

FIG. 5. Electron micrograph of dog proximal tubular brush border following incubation with the monoclonal antibody derived from clone 1 after immunization with dog membranes enriched for $Na^+,K^+$-ATPase. The immunoperoxidase technique as described in Fig. 4 was utilized for generation of the reaction product. Reaction product is confined to coated pits in the apical cytoplasm (*). A similar pattern of staining was seen with the antibody from clone 2. Bar equals 0.5 μm.

nated membranes (Forbush and Hoffman, 1979). Should these confirm that the antibody does indeed react with the α subunit of $Na^+,K^+$-ATPase, then additional studies such as peptide mapping of chymotryptic digests of the immunoprecipitated protein would be necessary to characterize the nature of the cross-reactive sites.

## IV. GENERATION, PURIFICATION, AND CHARACTERIZATION OF MONOCLONAL IMMUNOGLOBULIN

While clones may be expanded into large culture and the supernatant medium used as a source for the monoclonal antibody, there are numerous problems with this method of production. These include all of the difficulties relating to maintenance of large-volume long-term tissue cultures as well as those specific to hybridomas. Sluggish growth or actual cessation of division can suddenly occur, although no definite explanation is available (Lerner, 1981). It is probably due to the presence of suppressor cells or other factors that regulate lymphocyte and plasma cell growth. With our initial fusion following immunization against dog antigen, cell cultures were frozen while the antibody from supernatants was being characterized. After the cultures were thawed, it was discovered that antibody production had ceased. The loss of antibody production in the presence of adequate growth occurs and is probably related to the loss of chromosomes from the hybrid cells. Therefore, the development of ascites tumors as a more stable source of monoclonal antibody production is very useful. BALB/c mice or $F_1$ hybrid mice that match the major histocompatibility type of the fusion partner form ideal hosts for the production of ascites. In practice, the mice are primed with pristane intraperitoneally at least 5 days before injection of the hybrid cells, and then the mice are tapped after ascites develops, usually within 1–3 weeks. We now routinely produce ascites at each cloning step to maintain a constant source of antibody.

While either supernatant fluid or ascites fluid can be used unmodified for many of the screening procedures such as ELISA, spurious results may be generated because of the presence of a large number of other proteins and peptides within the crude preparation. It is therefore necessary to subject the crude antibody source to further purification prior to utilization. Antibody purification can be carried out with various degrees of specificity. Initially, ascites or supernatant fluid can be centrifuged to remove particulate and cellular matter. For an additional degree of purification at least one precipitation using 40% ammonium sulfate should be carried out. Higher concentrations of ammonium sulfate and precipitating agents such as sodium sulfate seem to destroy monoclonal antibody activity and should not be used (Lerner, 1981). To obtain purer samples of monoclonal antibody, the IgG fraction can be further purified by DEAE-cellulose chromatography. The antibody may also be subjected to affinity purification procedures. Protein A–Sepharose columns can be used to separate the antibody, but it must be remembered that not all monoclonal antibodies will bind to protein A. Although antigen affinity purification can be used, it has no major advantages, since such purification does not further refine the specificity of the monoclonal antibody and has the disadvantage of requiring a significant quantity of pure antigen. Affinity purification using a secondary antibody directed against a specific immunoglobulin class of the monoclonal antibody is probably the best final step for purification of monoclonal antibodies. After determining the specific class of the monoclonal antibody by radial immunodiffusion, an affinity-purified antibody to that specific class of mouse immunoglobulin should be obtained and bound to cyanogen bromide-activated Sepharose beads to form an affinity purification column for separation (Gonyea,

1977). The reagent obtained from such a column now has the specificity of the monoclonal antibody combined with purity of the immunoglobulin class.

Determination of these characteristics is necessary if one is to utilize the monoclonal antibody in a quantitative fashion, such as in the assessment of the distribution and density of the transport protein in the plasma membrane. Affinity is defined as a binding strength between the antibody and antigen in the antigen–antibody reaction. This contrasts with avidity, which is the strength of the antibody–antigen bond after the formation of the antigen–antibody complex. Practically, the binding constant can be determined using a solid-phase radioimmunoassay (Kennett *et al.*, 1980). This is carried out by fixing a known amount of antigen to the wells of the microtiter plate followed by a two-step incubation, first with serial dilutions of the monoclonal antibody and then with a radiolabeled secondary antibody. The amount of antibody in each well is determined by comparing the amount of the radioactivity bound to a standard curve developed with an antibody of known concentrations. The total amount of antibody per well is determined by a binding curve with increasing antigen concentrations as the maximum antibody bound at maximum antigen concentration.

## V. MORPHOLOGIC APPLICATION OF MONOCLONAL ANTIBODIES TO MEMBRANE SURFACE PROTEINS

After monoclonal antibodies have been developed, characterized, and purified, they can be used as probes in a wide variety of cell biology studies. Our own interest has been the use of these antibodies as a specific morphological probe in the study of epithelial transport. Using immunocytochemical techniques, localization of the transport protein to different domains of the cell surface membrane as well as to intracellular compartments can be assessed. Furthermore, control of the synthesis and insertion of these proteins in adaptive states can be traced. In tissue culture systems, the development of polarity and differentiation can be examined. Regardless of the question asked, such studies depend on (1) the ability to visualize the antibody with electron optical techniques, (2) preparations of the specimen so as to preserve antigenicity of the protein under study while maintaining structural integrity, and (3) ready access of the antibody to the antigen, particularly at intracellular sites.

A wide variety of markers linked to antibody as a ligand have been used as labels for electron microscopic immunocytochemistry. These include ferritin, horseradish peroxidase, hemocyanin, colloidal gold, and other electron-visible markers. With purified antibodies the most popular have included peroxidase labeling and ferritin labeling (Kraehenbühl and Jamieson, 1974). These have been applied both directly by labeling of the primary antibody and indirectly using a labeled secondary antibody. Peroxidase labeling has the major disadvantage of lacking clear definition, because of the possible spread of reaction product particularly along membrane surfaces. Ferritin molecules are small and not very electron dense and therefore difficult to visualize. Amplification of the ferritin image using ferritin–antiferritin complexes (Willingham, 1980) amplifies the density but does not give labeling close to the antibody binding site. With all of these electron procedures nonspecific background labeling frequently makes accurate application even more difficult. Colloidal gold linked to immunoglobulin offers the advantage of the clear definition of a punctate label which can be easily visualized (DeMey *et al.*, 1981). Furthermore, when directly linked to purified monoclonal antibodies of known affinity, the label may be utilized as an indicator to quantitate antigen binding sites. It also has the advantage of being applicable for localization of the specific protein in freeze-fracture replicas as well as in traditional thin-section transmission electron microscopy. In our own laboratory, we have generally used peroxidase-labeled secondary antibody for initial screening procedures and are concentrating on the use of colloidal gold-labeled primary antibody for definitive localization. An advantage inherent in the use of a label on purified primary antibody rather than a labeled secondary antibody is that it allows for the determination of specificity through the use of unlabeled antibody as a blocking agent and generally results in lower nonspecific background adsorption.

In order to use a labeled antibody quantitatively, one must keep in mind the amount of antibody labeled as well as its affinity and valency. At best, with colloidal gold as an example, even if only one molecule of antibody were bound to each particle there would be at least two antigen binding sites available. If the antigenic sites are properly placed, one particle could be equivalent to two sites or more. If a F(ab′) fragment is utilized, one particle might ideally be the equivalent of a single site. But even such an ideal state would require saturation of the antigenic sites with the antibody, if the incidence of label is to reflect the true number of antigenic sites. In practice, when the membrane to be labeled has a large number of antigenic sites it is unlikely that saturation can be achieved, because of steric hindrance of antibody binding by the presence of a relatively large label. Thus, for several reasons, the number of colloidal gold particles observed by electron microscopy will be an underestimate of the incidence of antigenic sites. It is therefore necessary to define as clearly as possible the ratio of visualized sites to the total number of sites. It is possible to use various concentrations of antigen and antibody to obtain a Scatchard analysis of antigenic sites by radioimmunoassay or ELISA techniques. This information can then be compared to a quantitative morphologic assessment of the binding of directly labeled purified monoclonal antibody to membrane preparations. Following incubation of membrane preparations of $Na^+,K^+$-ATPase of known specific activity with gold-labeled antibody, the number of gold particles per unit area of membrane is measured electron microscopically as an index of the amount of antibody bound at each dilution. After determining the amount of unbound labeled antibody in the supernatant, the number of antigenic sites can be estimated by Scatchard analysis or direct linear or double reciprocal plots. The ratio of labeled to unlabeled antibody may also be defined in such a system using a competitive binding assay. With this information it should be possible to generate a morphologic estimate of antigenic sites in different conditions.

Having defined the immunocytochemical reagent, one must now turn to the problems of application of that reagent in ultrastructural studies. Ideally, the tissue must be fixed, in order to stabilize the cellular structure and prevent loss or relocation of the antigen in question and at the same time not alter the antigenic determinant in question. These two goals are at apparent odds with one another, since fixation usually involves either cross-linking or denaturation which may alter the antigenic determinant, and so techniques must be selected that will provide a reasonable compromise between both goals.

In general, tissue fixation for structural preservation has utilized the reaction of aldehydes with amino and sulfhydryl groups to form intra- and intermolecular bridges. The effect of these fixatives on antigenicity has been studied (Habeeb, 1969) and it can be generally said that fixation with formaldehyde appears to have less of an effect on antigenicity than glutaraldehyde. Careful studies by Kraehenbühl and Jamieson (1974) have suggested that the ideal concentration of fixatives that would preserve both antigenicity and cellular fine structure would be in the range of the 1–2% formaldehyde and 0.25–0.5% glutaraldehyde. Using a fixative with concentrations of 1% formaldehyde, 0.25% glutaraldehyde, and 0.1 *M* cacodylate buffer with osmolarity adjusted with 0.1 *M* sucrose and 3% dextran T40, we have achieved both good fine structural preservation and good preservation of antigenicity of $Na^+,K^+$-ATPase as defined by ELISA as well as by morphologic methods. Indeed, the background adsorption of ELISA was reduced by mild fixation. Probably the most important problem to be faced in immunocytochemical techniques is the access of the antigen to the labeled antibody. Two main types of immunocytochemical techniques are presently being utilized. In diffusion localization techniques, labeled antibodies are incubated with fixed tissues, cells, fragments of cells, or membrane preparations prior to embedding and sectioning. The advantage of this technique is the relatively minimal alteration of the antigenicity of the proteins in question by a mild fixation before exposure to antibody, without the additional insult encountered by dehydration by organic solvents and embedding. Its major disadvantage is that there are barriers to diffusion of particle-tagged antibodies to intracellular locations. An approach that attempts to circumvent this disadvantage has been proposed and applied by Pinto da Silva *et al.* (1981). With this technique, fixed tissues are washed in phosphate-buffered saline and incubated with the cryoprotectant, 30% glycerol. The tissues are then frozen and fractured mechanically with a glass pestle. The fragments are deglycerinated by the addition of glycylglycine and brought to room temperature. Following a wash with additional glycine to occupy any unbound aldehyde groups, the tissue is then incubated with antibody, washed, dehydrated, and embedded for thick section and examination in the electron microscope. Results with this technique have been promising but have the continuing disadvantage that only a small and random portion of the intracellular compartments is available for interaction with the antibody.

We have examined the localization of five different monoclonal antibodies raised against a highly enriched $Na^+,K^+$-ATPase from dog renal outer medulla. Two different patterns of localization were identified using this technique (Figs. 5–11). Antibodies from clones 1, 2, and 5 labeled the external aspect of basolateral membranes as well as some coated pits in the apical cytoplasm but not the apical membranes themselves. Antibodies from clones 3 and 4 on the other hand labeled only the exposed internal aspect of basolateral membrane. It should be noted that the antibody from clone 4 inhibits $Na^+,K^+$-ATPase. The sidedness of the antibody has been confirmed by morphologic examination of membrane preparations, which demonstrated for antibodies from clone 4 that the antibody reacted with $Na^+,K^+$-ATPase enriched membranes that were broken but that labeling was not seen when the antibody was incubated with intact right-side-out vesicles. Examination of frozen sections of dog kidney incubated with the antibodies and stained with fluorescein-labeled anti-mouse Ig confirmed the cytologic localization and also demonstrated differences in the distribution of antibody reactivity to different segments of the tubule. Antibodies from clones 1 and 2 stained only outer medullary and cortical collecting tubules, while the antibody from clone 3 stained the entire collecting duct including the papillary segments. The antibody from clone 5 stained the entire collecting duct as with clone 3 but also stained the thick ascending limb of Henle's loops. The antibody from clone 4 which inhibits $Na^+,K^+$-ATPase had a unique pattern of reactivity which included staining of proximal tubules, distal tubules, and collecting ducts, but it did not appreciably stain the thick limbs of Henle's loop. Staining of the thick limb was better appreciated by immunoelectron microscopy of frozen sections of dog kidney. Thus the ancillary morphologic studies using immunofluorescence and membrane preparations confirmed the validity of the findings using the fractured cellular preparation and diffusion localization techniques.

At the same time the immunofluorescence studies also raise new questions. The heterogeneity of distribution of immunofluorescein labeling confirms that each monoclonal antibody is unique. In preliminary results with the antibodies tested by immunoprecipitation, one reacted with a peptide fragment that migrated with the $\alpha$ subunit of $Na^+,K^+$-ATPase in gel electrophoresis and two others reacted with a peptide migrating as the $\beta$ subunit. These localized in different morphologic patterns. This observation raises doubts as to whether the antibodies are indeed directed against $Na^+,K^+$-ATPase and not other peptides

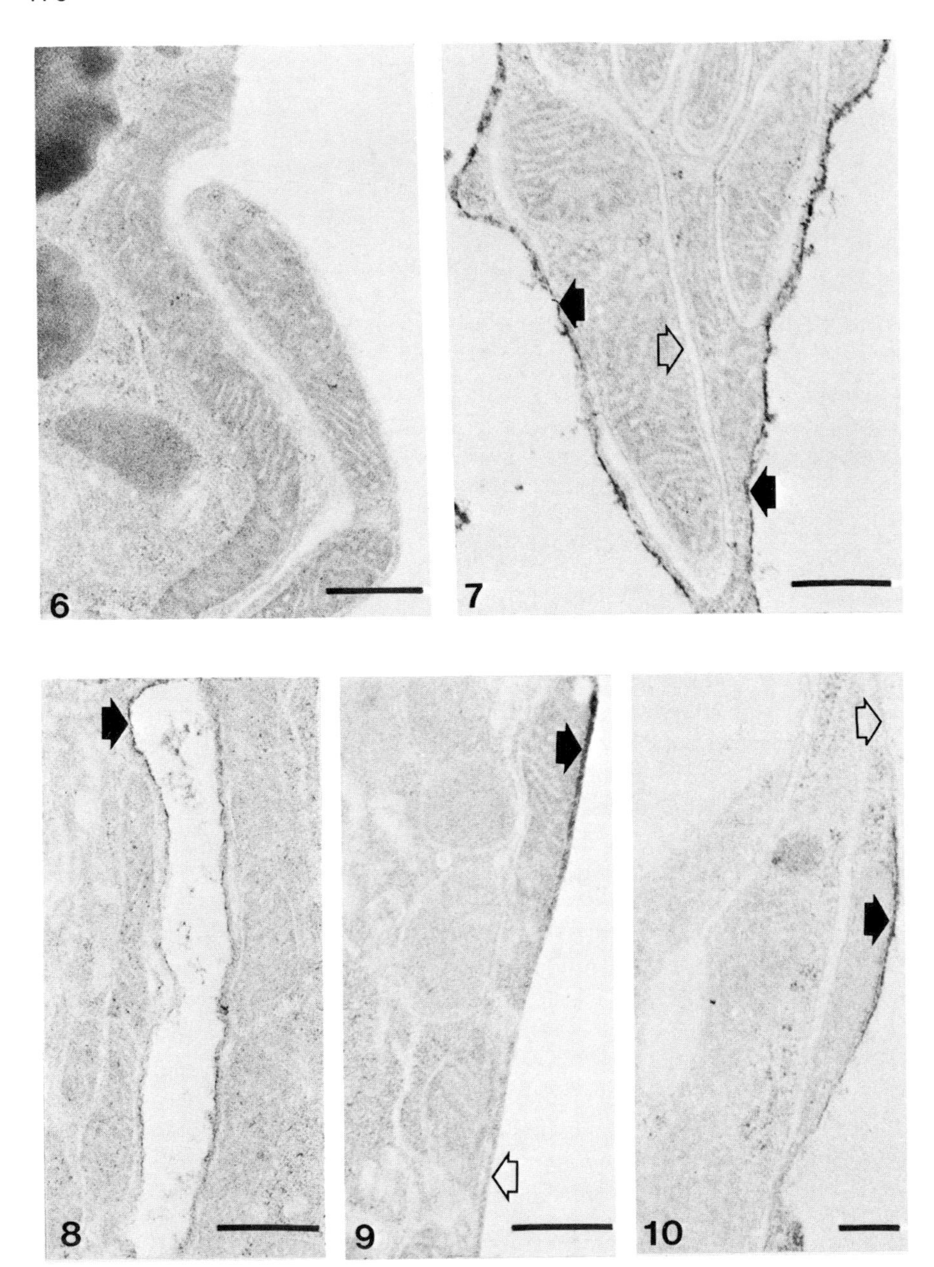
6
7
8
9
10

of similar molecular weight, and it emphasizes the difficulties in absolutely characterizing the specificity of the antibodies by one method alone. Functional inhibition of the enzyme as seen with the antibody raised from clone 4 certainly confirms the specificity yet the lack of prominent staining of the thick ascending limb suggests that there may even be heterogeneity in the enzyme itself, with enzymes derived from different nephron segments possibly being antigenically different. One possibility is that, depending on local conditions at the time of fixation, the enzyme could be in different conformational states in the catalytic cycle resulting in different reactivity of the antibody.

With the surface localization techniques, thin sections of fixed and embedded tissues are prepared prior to incubation with the labeled antibody. The major advantage of surface localization procedures is that there are no diffusion restrictions, because the immunologic reagents may be applied to the entire open surface of the tissue in thin sections. Its disadvantages are related to the relative availability of antigenic sites which may be small in number, and, in addition, to the alteration of the antigenicity of the proteins in question by the dehydration and embedding procedure which requires organic solvents and relatively high temperatures which can easily denature proteins. In one approach, ultramicrocryotomy can be utilized to prepare ultrathin frozen sections. Our initial

---

FIG. 6. Immunocytochemical localization of membrane proteins in the basolateral membrane of the dog proximal tubule. The specimen was prepared by the method of Pinto da Silva prior to incubation with control mouse IgG (nonimmune). Peroxidase labeling was carried out with the same method as in Figs. 4 and 5. Note the lack of reaction product. Bar equals 0.5 μm.

FIG. 7. Immunocytochemical localization of membrane proteins in the basolateral membrane of the dog proximal tubule. This preparation was identical to that in Fig. 6 except the initial incubation was with an antibody from clone 1 generated against dog outer medullary membrane preparations enriched for $Na^+,K^+$-ATPase. The antibody labels the basolateral membranes exposed during the fracturing process (closed arrow). Basolateral membranes not exposed during processing (open arrow) are inaccessible to the antibody and are thus not labeled. Bar equals 0.5 μm.

FIG. 8. Immunocytochemical localization of membrane proteins in the basolateral membrane of the dog proximal tubule. The tissue was prepared in a manner identical to that of the previous figure except that it was incubated with antibody generated from clone 2. This antibody also labels the exposed basolateral membrane (arrow). Bar equals 0.5 μm.

FIG. 9. Immunocytochemical localization of membrane proteins in the basolateral membrane of the dog proximal tubule. The specimen was prepared in the same fashion as the previous figures except that it was incubated with an antibody generated against clone 3. This antibody labeled only the internal aspect of membranes that are fractured (closed arrow). Internal and external aspects of membranes that are not fractured are unlabeled even when clearly exposed (open arrow). Bar equals 0.5 μm.

FIG. 10. Immunocytochemical localization of membrane proteins in the basolateral membrane of the dog proximal tubule. The tissue was prepared as previously described but in this instance was incubated with the antibodies from clone 4, which has been shown to inhibit $Na^+,K^+$-ATPase activity. This antibody only labels selected areas of fractured basolateral membranes on their internal aspects (arrow). Areas of membranes that are unlabeled (open arrow) appear to be unfractured regions of membrane. Bar equals 0.5 μm.

attempts to utilize this method have demonstrated that the ultrastructural resolution of membranes in the electron microscope is relatively poor, making this method less desirable for the localization of membrane proteins. For this reason we are presently pursuing the use of polar low-temperature embedding media as an alternative. This technique has the advantage in that infiltration at low temperature may considerably reduce the extraction of lipids and proteins, and may decrease any denaturation that may occur at higher temperatures. Furthermore, since the medium is polar the degree of dehydration is limited, minimizing the effects of organic solvent denaturation as well. This method has been extensively described in the literature (Carlemalm *et al.*, 1980), and initial studies by Dr. James Wade using polyclonal antibodies to luminal membrane proteins of the toad bladder have demonstrated the feasibility of using such a procedure for immunostaining of membrane-associated proteins.

## VI. SUMMARY

As discussed here, the application of monoclonal and immunocytochemical technology to the problem of localization of membrane-associated transport proteins is not yet thoroughly solved. At this point one can be certain that monoclonal antibodies directed against some transport proteins such as the $Na^+,K^+$-ATPase pump can be generated and that the techniques described can be utilized for proper characterization. With the use of the screening procedures outlined, which include immunochemical, electrophoretic, and functional techniques, complete characterization can be accomplished regarding specificity and the physicochemical characteristics of the antibodies themselves. Even when immunizing with a well-characterized protein such as $Na^+,K^+$-ATPase, it is difficult to establish with complete confidence that the monoclonal antibodies raised bind to that protein and only that protein. The problem of accurate immunocytochemical localization is greatly alleviated by the use of labeled monoclonal antibodies but still is not reduced to a trivial task. Optimization of specific labeling with minimal nonspecific labeling must be achieved and factors considered that may influence this, including the affinity and avidity of the antibody, the preservation and availability of the antigenic site, and the effectiveness of the labeling technique to be used for electron microscopic localization. It must be emphasized that when using monoclonal antibodies several different approaches must be used morphologically in order to confirm the true localization of membrane proteins.

Note: Recently, there has been a published report of isolation and characterization of monoclonal antibodies to $Na^+,K^+$-ATPase (Ball *et al.*, 1982). The experimence of Ball and co-workers using both electrophoretically purified subunits and holoenzyme is similar to our own.

## REFERENCES

Ball, W. J., Schwartz, A., and Lessard, J. L. (1982). Isolation and characterization of monoclonal antibodies to (Na+K)-ATPase. *Biochim. Biophys. Acta* **719,** 413–423.

Carlemalm, E., Filliger, W., and Acetarin, J. D. (1980). Advances in specimen preparation for electron microscopy. 1. Novel low temperature embedding resin. *Experientia* **36,** 6.

Cone, R. E., and Brown, W. (1976). Isolation of membrane associated immunoglobulins from T lymphocytes by non-ionic detergents. *Immunochemistry* **13,** 571.

Coudrier, E., Reggio, H., and Louvard, D. (1982). The cytoskeleton of intestinal microvilli contains two polypeptides immunologically related to proteins of striated muscle. *Cold Spring Harbor Symp. Quant. Biol.* **46,** 881–92.

DeMey, J., Moeremans, M., Gevens, G., Nuydens, R., and DeBrabander, M. (1981). High resolution light and electron microscopic localization of tubulin with the IGS (immunogold staining) method. *Cell Biol. Int. Rep.* **5,** 889–899.

Forbush, B., and Hoffman, J. F. (1979). Evidence that ouabain binds to the same large polypeptide chain of dimeric Na-K ATPase that is phosphorylated from Pi. *Biochemistry* **18,** 2308–2315.

Gonyea, L. M. (1977). Purification and iodination of antibody for use in an immunoradiometric assay for ferritin. *Clin. Chem.* **23,** 234–236.

Graham, R. C., and Karnovsky, M. J. (1966). The early stages of absorption of an infected horseradish peroxidase in the proximal tubules of mouse kidney. *J. Histochem. Cytochem.* **14,** 291–302.

Habeeb, A. F. S. A. (1969). A study of the antigenicity of formaldehyde and gluteraldehyde treated bovine serum albumin and ovalbumin and bovine serum albumin complex. *J. Immunol.* **102,** 457–465.

Jackson, R. C., and Blobel, G. (1977). Post-transitional cleavage of presecretory proteins with an extract of rough microsomes from dog pancreas containing signal peptidase activity. *Proc. Natl. Acad. Sci. U.S.A.* **74,** 5598–5602.

Jørgensen, P. C. (1974). Purification and characterization of Na+ K-ATPase. III. Purification from the outer medulla of mammalian kidney after selective removal of membrane components by sodium dodecyl sulfate. *Biochim. Biophys. Acta* **356,** 36–52.

Kennett, R. H., McKearn, T. J., and Bechtol, K. B. (1980). "Monoclonal Antibodies." Plenum, New York.

Kohler, G., and Milstein, C. (1975). Continuous cultures of fused cells secreting antibody of predefined specificity. *Nature (London)* **256,** 495.

Kraehenbühl, J. P., and Jamieson, J. D. (1974). Localization of intracellular antigens by immunoelectron microscopy. *Int. Rev. Exp. Pathol.* **13,** 53.

Laemmli, U. K. (1970). Cleavage of structural proteins during the assembly of the head of bacteriophage T4. *Nature (London)* **227,** 680–685.

Lerner, E. A. (1981). How to make a hybridoma. *Yale J. Biol. Med.* **54,** 387–402.

Papermaster, D. S., Lyman, D., Schneider, B. G., Labienic, L., and Kraehenbühl, J. P. (1981). Immunocytochemical localization of the catalytic subunit of Na-K ATPase in *Bufo marinus* kidney, bladder and retina with biotinyl-antibody-strepavidin gold complexes. *J. Cell Biol.* **91,** 273A.

Pinto da Silva, P., Kachar, B., Torrisi, M. R., Brown, C., and Parkison, C. (1981). Freeze fracture cytochemistry: Replicas of critical point dried cells and tissues. *Science* **213,** 230–233.

Towbin, H., Stachelin, T., and Gordon, J. (1979). Electrophoretic tranfer of proteins from polyacrylamide gels to nitrocellulose sheets. *Proc. Natl. Acad. Sci. U.S.A.* **76,** 4350–4354.

Willingham, M. (1980). Electron microscopic immunocytochemical localization of intracellular antigens in cultured cells. The EGS and ferritin bridge procedures. *Histochem. J.* **12,** 419–434.

# Chapter 10

# Monoclonal Antibodies as Probes of Epithelial Cell Polarity

*GEORGE K. OJAKIAN AND DORIS A. HERZLINGER*

*Department of Anatomy and Cell Biology*
*Downstate Medical Center*
*State University of New York*
*Brooklyn, New York*



The tubules of the mammalian nephron are composed of a variety of epithelial cell types that have their own distinct structural, biochemical, and physiological properties. This epithelial heterogeneity allows each portion of the nephron to exhibit tubule segment-specific functions, and renal physiologists and biochemists have been able to describe unique differences in the transport of ions, sugars, and amino acids along the nephron. Until recently, this functional segmentation has been studied primarily by utilizing micropuncture analysis and microdissected nephron segments (21, 29, 44). To obtain more precise information about renal function at both the cellular and molecular levels, investigators have been using the following approaches: (1) the isolation of homogeneous cell populations from both kidney cortex and medulla (14, 26), (2) the establishment of kidney epithelial primary cell cultures (63), and (3) the utilization of epithelial cell lines that have retained the differentiated functions of transporting epithelia in culture (23). One of these cell lines, Madin–Darby canine kidney (MDCK) cells, has been studied in considerable detail due to the retention of many differentiated epithelial functions in culture. MDCK cells are morphologically polarized, possessing apical microvilli and basolateral membranes with lateral inter-




ISBN 0-12-153320-4

digitations (8, 9, 39, 42, 47, 49) that are typical of transporting epithelia (4). Biochemically, MDCK cells have an asymmetric distribution of membrane proteins, with leucine aminopeptidase found primarily on the apical cell membrane and $Na^+$,$K^+$-ATPase on the basolateral membrane (40). The lateral space is sealed by tight junctions and, when grown on permeable substrates, MDCK cells can generate transepithelial electrical resistances of 100–300 $\Omega$ $cm^2$ (8, 42, 47, 49). Low-passage MDCK cells have been reported to have resistances of 4000 $\Omega$ $cm^2$ (51). These properties give rise to a functional polarity that allows a net transepithelial movement of both sodium and water across the MDCK monolayer in the apical-to-basal direction (8, 39, 42). We have utilized MDCK cells to study the polarity of epithelial cells and the differential expression of cell surface proteins along the nephron.

## I. PRODUCTION OF MONOCLONAL ANTIBODIES AGAINST MDCK CELL SURFACE PROTEINS

For these studies, an immunological approach has been used to produce antibodies against MDCK cells that can be utilized as high-affinity ligands to study the distribution of membrane proteins on the surfaces of MDCK and kidney epithelial cells. Until recently, the conventional approach was to isolate and purify the membrane protein of interest, inject it into an animal, and collect an immune serum against the molecule, providing the investigator with a polyclonal antibody to use. These antisera have proved to be extremely powerful tools for studying membrane protein structure, function, and distribution (see Rossier, Chapter 7, and McDonough, Chapter 8, this volume). One disadvantage in using such antisera is that large quantities of purified protein must be obtained for the immunizations, which usually requires working with major membrane proteins or using very large quantities of membranes to purify the quantities of minor membrane components sufficient for immunization. Another disadvantage is that the antisera are complex mixtures which contain antibodies directed against different antigenic determinants on the same molecules as well as antibodies that bind with different specificities to the same antigenic site. The advantages and disadvantages of polyclonal antibodies have been discussed in an excellent review (35).

To overcome these difficulties and work with less complex antibodies, we and other investigators have been utilizing the hybridoma technique developed by Kohler and Milstein (33, 34) to produce monoclonal antibodies against membrane components. These procedures take advantage of the observation that lymphocytes from the spleen of an immunized animal (usually a mouse) making a unique species of antibody against a single antigenic determinant can be isolated and grown in culture (33, 34). Monoclonal antibodies can be produced in large quantities (up to 10 mg/ml in ascites fluid; see Refs. 35 and 58) and have been successfully utilized to study membrane components from both pure (50, 64) and impure antigens (2, 3, 15, 24, 50, 62, 64, 65). For an extensive overview of the literature, refer to the recent review by Kennett (31).

For our studies on epithelial cell surface proteins, mice were immunized with intact, confluent MDCK cells, and hybridoma cell lines were generated using the fusion protocol of Gefter *et al.* (20). Hybridomas secreting monoclonal antibodies directed against antigenic determinants on MDCK cells were selected by an indirect radioimmunoassay (RIA) that employs $^{125}$I-labeled goat anti-mouse IgG (GAM) to detect levels of monoclonal antibody bound to the MDCK cell surface (27). After MDCK antibodies were obtained, the hybridoma supernatants were screened simultaneously on MDCK cells and fibroblasts (Table I) and only those hybridomas secreting epithelia-specific monoclonal antibodies were retained for further cloning in soft agar. The data presented in Table I depict an example of a hybridoma cell line that is secreting monoclonal antibody directed against a membrane component present on MDCK cells but not fibroblasts.

Membrane proteins recognized by the monoclonal antibodies were identified by immunoprecipitation and sodium dodecyl sulfate–polyacrylamide gel electrophoresis (SDS–PAGE). MDCK cells were grown either on [$^{35}$S]methionine- or [$^{3}$H]glucosamine-containing medium to isotopically label membrane proteins and glycoproteins, respectively. The labeled cells were detergent solubilized and the cell lysate incubated with either purified monoclonal antibody coupled di-

TABLE I
BINDING OF ANTIBODY 11B8 DETERMINED BY INDIRECT RIA[a,b]

| | First incubation | | |
|---|---|---|---|
| Cell type | 11B8 | Anti-myosin | PBS |
| MDCK | 3685 ± 306 | 354 ± 20 | 150 ± 8 |
| Rat embryo fibroblasts | 119 ± 18 | 129 ± 7 | — |
| Dog kidney primary fibroblasts | 169 ± 6 | 173 ± 10 | 147 ± 4 |

[a] Reproduced from *The Journal of Cell Biology*, 1982, **93**, 269–277 by copyright permission of The Rockefeller University Press.

[b] Cells (2 × 10$^{4}$) in microtiter wells were incubated with 11B8 hybridoma supernatant for 1 hour at 4°C, washed, and incubated with $^{125}$I-labeled GAM (10$^{5}$ cpm) as described in Herzlinger *et al.* (27). Cells were solubilized and the bound radioactivity was determined. As controls, an anti-myosin monoclonal antibody or washing buffer (phosphate-buffered saline, PBS) was substituted for antibody 11B8. Cell-associated radioactivity was determined for six replicate samples and expressed as the average cpm bound ± the standard error of the mean.

rectly to Sepharose 4B or sequentially with hybridoma supernatant and then GAM F(ab$'$)$_2$–Sepharose 4B. Proteins were eluted from the Sepharose beads and identified by SDS–PAGE (37) and fluorography (5). With these procedures, we have isolated three MDCK cell surface polypeptides recognized by monoclonal antibodies. They include a 20,000-d (20K) glycoprotein rocognized by antibodies from five different clones, a 35K glycoprotein rocognized by hybridoma clone 11B8 (27), and a 52K protein recognized by clone H6. We have not yet determined whether the 52K protein is glycosylated.

## II. DISTRIBUTION OF POLARIZED CELL SURFACE PROTEINS

Polarity of epithelial cell surface domains has been determined by biochemical studies on purified apical and basolateral membrane preparations. Each membrane domain has its own distinct polypeptide composition (17, 52), and enzymatic analysis has demonstrated the presence of alkaline phosphatase, disaccharidases, and aminopeptidases in apical membranes, while $Na^+$,$K^+$-ATPase and hormone receptors are localized preferentially in the basolateral membranes (6, 12, 32, 45, 59, 60). Supporting structural studies have demonstrated polarized distributions of leucine aminopeptidase and $Na^+$,$K^+$-ATPase (13, 36, 40).

We have been utilizing monoclonal antibodies to determine the polarity of cell surface proteins on subconfluent MDCK cells by both light and electron microscopy. Immunofluorescence studies were done as described previously (27). For electron microscopy, monoclonal antibodies bound to proteins on the MDCK cell surface were localized by GAM coupled to horseradish peroxidase (HRP–GAM) using diaminobenzidine as the substrate.

Immunofluorescence microscopy has demonstrated that both the 35K glycoprotein (27) and the 52K protein are homogeneously distributed on all cells in growing, subconfluent MDCK cultures. Careful examination of the fluorescence images indicates that these membrane proteins are present on apical surface microvilli, on the lateral membranes between subconfluent cells in regions of cell contact, and, by examining cells at different focal levels, on the basal membrane adjacent to the substratum. These observations were confirmed for the 35K glycoprotein ultrastructurally by the detection with HRP–GAM of the presence of this polypeptide on both the apical and basal plasma membrane of subconfluent MDCK cells (Fig. 1). In addition, intense staining of the 35K glycoprotein was observed on the lateral membranes of subconfluent cells in contact (data not shown), supporting the observations made by immunofluorescence microscopy.

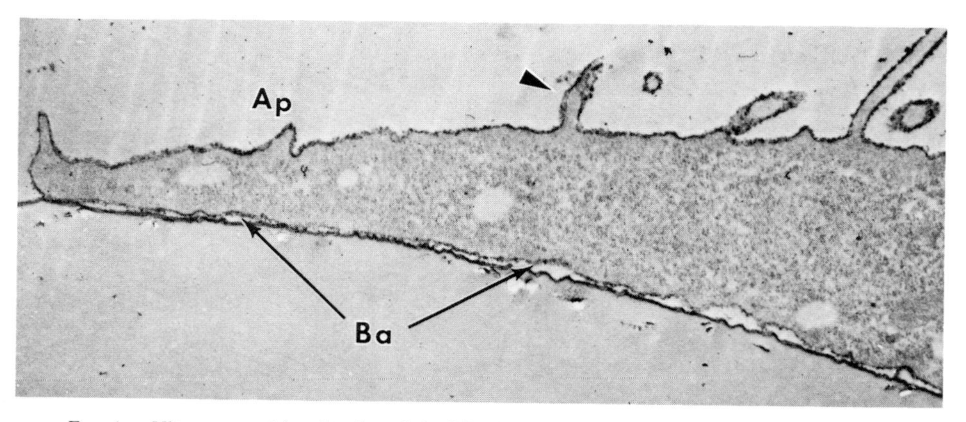


FIG. 1. Ultrastructural localization of the 35K glycoprotein on subconfluent MDCK cells. Low-density cultures were fixed in 4% paraformaldehyde–0.1% glutaraldehyde and then incubated in hybridoma supernatant 11B8 followed by HRP–GAM. The electron-dense reaction product is clearly visualized on the apical (Ap) and basal (Ba) cell surfaces as well as apical microvilli (arrowhead). The uniformity in staining suggests that the 35K glycoprotein is evenly distributed over the entire cell surface. ×12,600.

After growth of MDCK cells into a confluent monolayer of tightly packed cells, immunofluorescence observations demonstrated that the 35K and the 52K proteins were no longer distributed randomly over the entire cell surface but were now asymmetrically distributed and could only be localized on the basolateral membrane. When confluent monolayers (Fig. 2a) were stained with antibody H6, little, if any, 52K protein could be detected on the apical cell surface (Fig. 2b). Since these monolayers are sealed by tight junctions (7, 8, 42, 49), it is reasonable to assume that the antibodies did not have access to the lateral space and were prevented from detecting proteins residing on the basolateral membrane. To demonstrate the presence of the 52K protein on the basolateral membrane, experimentally induced tears were made in the MDCK monolayer after fixation, removing some cells and allowing the antibodies access to the basolateral membrane by diffusion under the monolayer. Immunofluorescent microscopy demonstrated the presence of the 52K protein on the basolateral membranes of the cells adjacent to the tear (Fig. 2b). Similar results were obtained for the 35K glycoprotein. The basolateral localization of the 35K glycoprotein was confirmed ultrastructurally by scraping fixed, confluent MDCK monolayers from the culture dish and incubating them with antibody 11B8 followed by HRP–GAM to allow the antibodies access to both the apical and basolateral domains. These results are not presented here as they are identical to those obtained in the kidney for the 35K glycoprotein (see Section IV, Fig. 6).

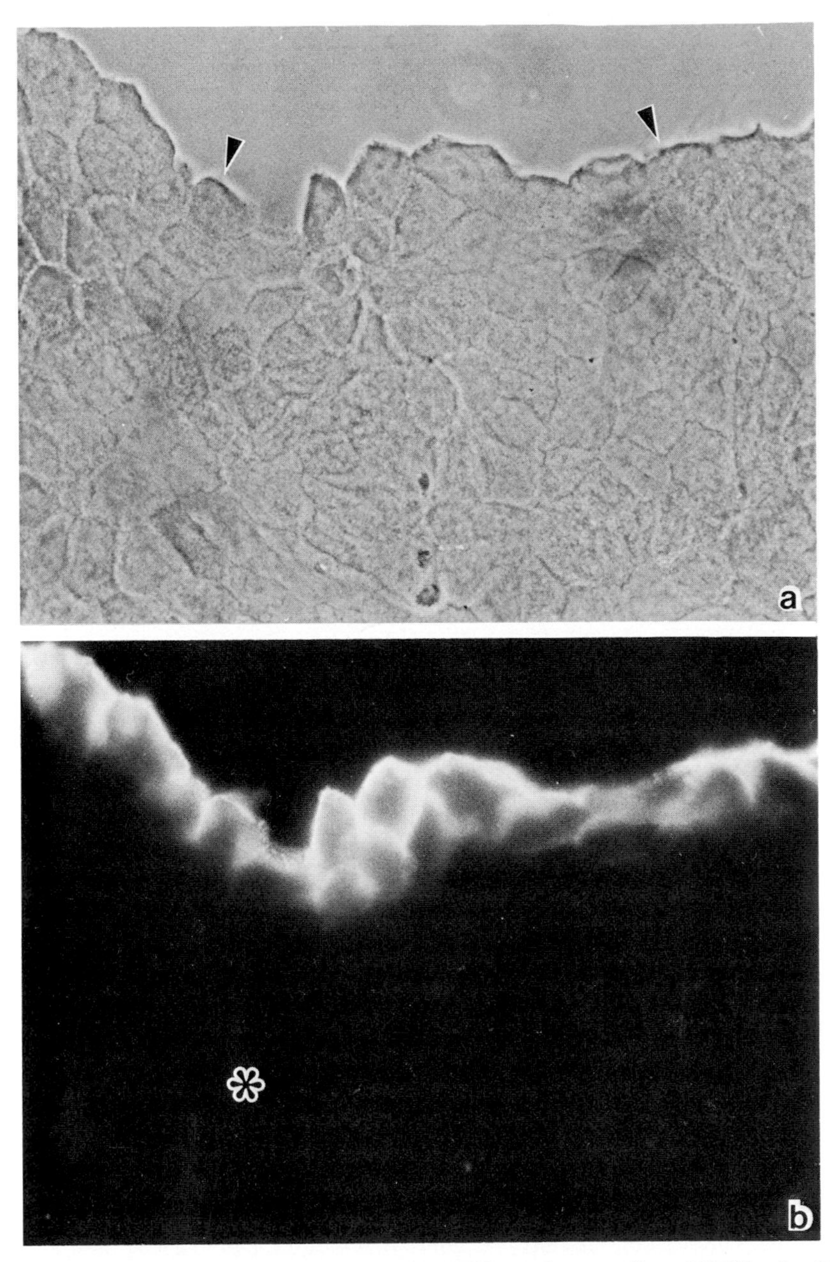


FIG. 2. Immunofluorescence localization of the 52K protein on confluent MDCK cells. Cells were fixed in 4% paraformaldehyde–0.05% glutaraldehyde and tears were induced in the monolayer to allow antibodies access to the basolateral as well as the apical cell surfaces. With phase-contrast microscopy (a) the edge of the tear (arrowheads) can be visualized in the upper portion of the field. The 52K protein was localized by incubating the monolayer in hybridoma supernatant H6 followed by GAM coupled to rhodamine. In the corresponding immunofluorescence micrograph (b), the 52K protein was not detected on the apical cell surface (*); however, antibodies can diffuse under the cells at the edge of the tear, and cells one to four deep were outlined by fluorescence staining, demonstrating that the 52K protein is localized to the basolateral membrane. ×800.

## III. DEVELOPMENT AND MAINTENANCE OF EPITHELIAL CELL POLARITY

Because the 35K and 52K proteins had a random distribution during MDCK cell growth but assumed a polarized distribution after reaching confluency, experiments were designed to follow the development of cell polarity. MDCK cells were plated on both Millipore filters (42) and cover glasses at subconfluent densities and allowed to grow into a confluent monolayer, so that the formation of tight junctions could be monitored by transepithelial electrical resistance measurements (8, 42) simultaneously with monitoring the redistribution of the 35K and 52K proteins by immunofluorescence microscopy. Prior to the development of electrical resistance, both proteins could be detected on the apical and basolateral surfaces (Fig. 1). After 2–3 days in culture, transepithelial electrical resistances of up to 300 $\Omega$ $cm^2$ had developed, demonstrating the formation of functional tight junctions (16), and immunofluorescence microscopy demonstrated that both proteins could now be localized primarily on the basolateral membrane (see Fig. 2). These results suggest that as soon as functional tight junctions are present in an MDCK monolayer, a regulatory mechanism becomes operational that can detect the presence of basolateral cell surface proteins in the apical surface, remove them, and allow cell polarity to become established. To test this hypothesis further, the distribution of cell surface proteins was studied after disassembly of MDCK tight junctions by EGTA (8, 18, 28, 41, 48). Previous studies have demonstrated that asymmetrically distributed cell surface components can redistribute when tight junctions are disrupted (10, 18, 48, 66), providing evidence that these junctions can function as intramembrane barriers to the movement of epithelial plasma membrane proteins from one membrane domain to the other. These constraints were removed by incubating confluent MDCK monolayers in EGTA to produce cell rounding, break tight junctions (Fig. 3a), and allow the movement of mobile basolateral membrane proteins (27) into the apical cell surface (Fig. 3b). The EGTA was removed and replaced by medium containing calcium, allowing the cells to spread and reform cell-to-cell contacts (Fig. 3c). One hour after the addition of calcium, a considerable amount of both 35K glycoprotein (Fig. 3d) or 52K protein could be detected on the apical surface by immunofluorescence; however, after 4 hours in calcium the amount of apical protein was reduced considerably (Fig. 3f) and by 24 hours (Fig. 3h) was qualitatively similar to intial levels (Fig. 2b).

At the present time, it is not clear how MDCK cells are able to remove basolateral membrane proteins from their apical surface and what the fate of these proteins is. One possibility is that membrane proteins that are present in the incorrect membrane domain are detected by the cell, removed by endocytosis, and either degraded or returned to the proper membrane domain. The latter suggestion is supported by the observation that in MDCK cells, leucine ami-

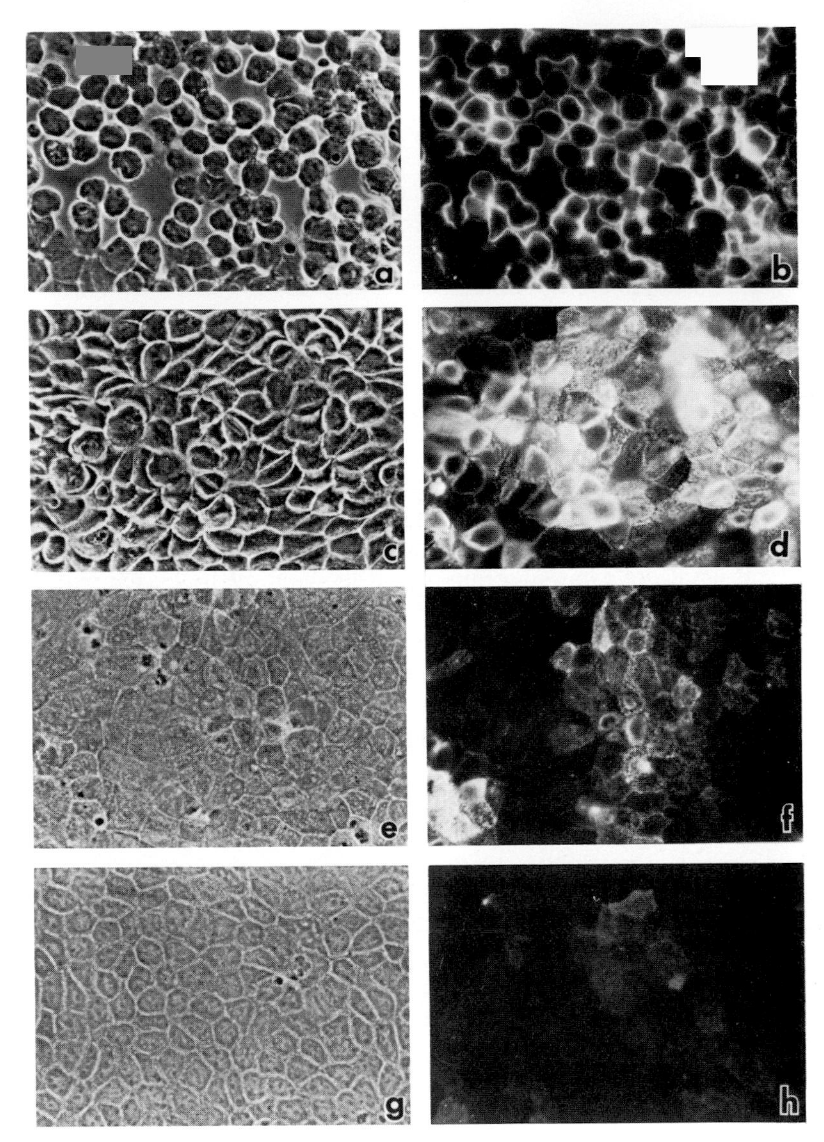


FIG. 3. Removal of the 35K glycoprotein from the MDCK cell surface after EGTA treatment. Phase-contrast, a, c, e, and g; immunofluorescent micrograph, b, d, f, and h. Incubation of confluent MDCK monolayers in 2.5 m*M* EGTA for 45 minutes at 37°C produces considerable rounding (a); increased levels of fluorescent staining (b) demonstrates that the majority of the 35K glycoprotein is distributed on the basolateral membrane. The EGTA was then removed, calcium-containing medium was added, and the cells were fixed and processed for immunofluorescence microscopy after 1 hour (c,d), 4 hours (e,f), and 24 hours (g,h). Phase-contrast microscopy demonstrates that the MDCK cells progress from a rounded configuration (a) back to a flatter monolayer (c,e,g) identical to that observed for untreated MDCK (Fig. 2a). One hour after return to calcium, a considerable amount of 35K glycoprotein can be observed on the apical cell surface (d). After 4 hours (f), apical surface staining is reduced considerably, and, after 24 hours (h), the amount of apical 35K glycoprotein is back to initial levels (Fig. 2a). ×280.

nopeptidase can be taken up in vesicles and then recycled back to the apical cell surface within 30 minutes (40). Another possibility is that a normal membrane protein turnover mechanism becomes operational and a loss of basolateral protein from the apical surface occurs during the normal degradation process. Membrane proteins would be removed from both membrane domains and newly synthesized protein inserted into only the basolateral membrane after "sorting out" within the cell takes place (54, 57), allowing the establishment of cell polarity.

Of particular interest are the differences between the observations presented here on MDCK cell surface protein polarity and those reported by Rodriguez-Boulan and Sabatini (54) for the polarized budding of lipid envelope viruses from MDCK cells. Further studies have demonstrated that the viral membrane glycoproteins are inserted into the proper membrane domain of confluent MDCK cells (55) and that polarized viral budding is maintained in single MDCK cells attached to a collagen substrate (56), suggesting that intact tight junctions are not required. These observations are in marked contrast to those reported here for the 35K and 52K MDCK cell surface proteins, which are distributed symmetrically on both the apical and basal membrane of subconfluent MDCK cells (Fig. 1). One possibility for these observed differences is that viral membrane proteins are "sorted out" (54–57) by a mechanism different from that of MDCK membrane proteins and take a different intracellular route to the cell surface. Another is that the viral nucleocapsid associates with the inner plasma membrane surface of subconfluent MDCK in a polarized manner and the symmetrically distributed viral membrane proteins move laterally in the membrane plane to form a complete virion. Confirmation of this suggestion will be determined in studies of the distribution of viral membrane proteins immediately after their insertion into the membrane of subconfluent MDCK cells. At the present time, it is not known whether MDCK membrane proteins are inserted into the cell surface of subconfluent MDCK cells in a nonpolarized manner, but from the data presented here, it seems certain that tight junctions are necessary for the maintenance of MDCK cell polarity, after the asymmetric distribution of membrane proteins has been established.

## IV. DEMONSTRATION OF NEPHRON SEGMENT-SPECIFIC PROTEINS

Because MDCK cells were originally isolated from dog kidney cortex (19), studies were initiated first to determine whether proteins recognized by monoclonal antibodies of MDCK culture cells were expressed in the kidney and then to localize these proteins within the nephron. Even though monoclonal antibodies have an extremely high specificity, there have been reports describing monoclonal antibodies that cross-react with proteins of different identity (11, 38). To be certain that the MDCK proteins recognized by monoclonal antibodies were the same in the kidney, dog kidney medulla plasma membranes (3) were detergent solubilized and passed over an affinity column made up of monoclonal antibody directly coupled to Sepharose 4B, and the antigenic proteins were radioiodinated in the presence of chloramine-T (22). The $^{125}$I-labeled proteins were then eluted from the column and their molecular weights were determined by SDS-PAGE and autoradiography. Monoclonal antibody 11B8 recognizes a protein of 35,000 MW in kidney membranes (data not shown), the same molecular weight as that of the glycoprotein that was isolated from MDCK cells by this antibody (see Section I). Based on this evidence, it seems reasonable to assume that we are probably studying the same protein in both the MDCK cell line and kidney membranes. Further molecular characterization such as peptide mapping and the determination that the 35K kidney protein is glycosylated would certainly strengthen this argument.

The distribution of the 35K glycoprotein was studied in frozen sections of dog kidney by immunofluorescence microscopy (27). In the cortex, all epithelial cells of the thick ascending limb (TAL), macula densa, and distal convoluted tubule (DCT) have the 35K glycoprotein, whereas fluorescent staining was not observed in glomeruli, the proximal convoluted tubule, renal vasculature, or interstitium (Fig. 4). In addition, some cells of the cortical collecting tubule also stained with antibody 11B8. In the medulla, the fluorescence staining pattern was especially striking since all epithelial cells of the TAL have the 35K glycoprotein (Fig. 5). Staining was not observed on the medullary collecting tubules, descending thick limb, or the thin loops and vas recta of the vascular bundles (Fig. 5). Careful observation of both cortical and medullary sections suggested that this protein is localized only on the basolateral membrane (27). These observations were confirmed by immunoelectron microscopic studies. Subcellular localization of kidney proteins was done on 40-μm-thick cryostat sections that were incubated with monoclonal antibodies and HRP–GAM, processed, and embedded for electron microscopy according to published methods (61). Ultrastructural observations of unstained thin sections demonstrated that the 35K glycoprotein was present on the basolateral plasma membrane of epithelial cells in the cortical and medullary TAL and the DCT and could not be detected on their apical cell surface (Fig. 6). The distribution of the 35K glycoprotein in the nephron is illustrated diagrammatically in Fig. 7.

We have also identified and mapped the distribution of the 20K glycoprotein in the kidney, using the same biochemical and morphological methods described above. One interesting result is that we have isolated hybridoma clones that secrete monoclonal antibodies to at least three different antigenic sites on the 20K glycoprotein. This was determined by immunofluorescence staining of dog, rabbit, and rat kidney frozen sections with antibodies from the five hybridoma clones mentioned previously (see Section I). Of these, two clones were specific

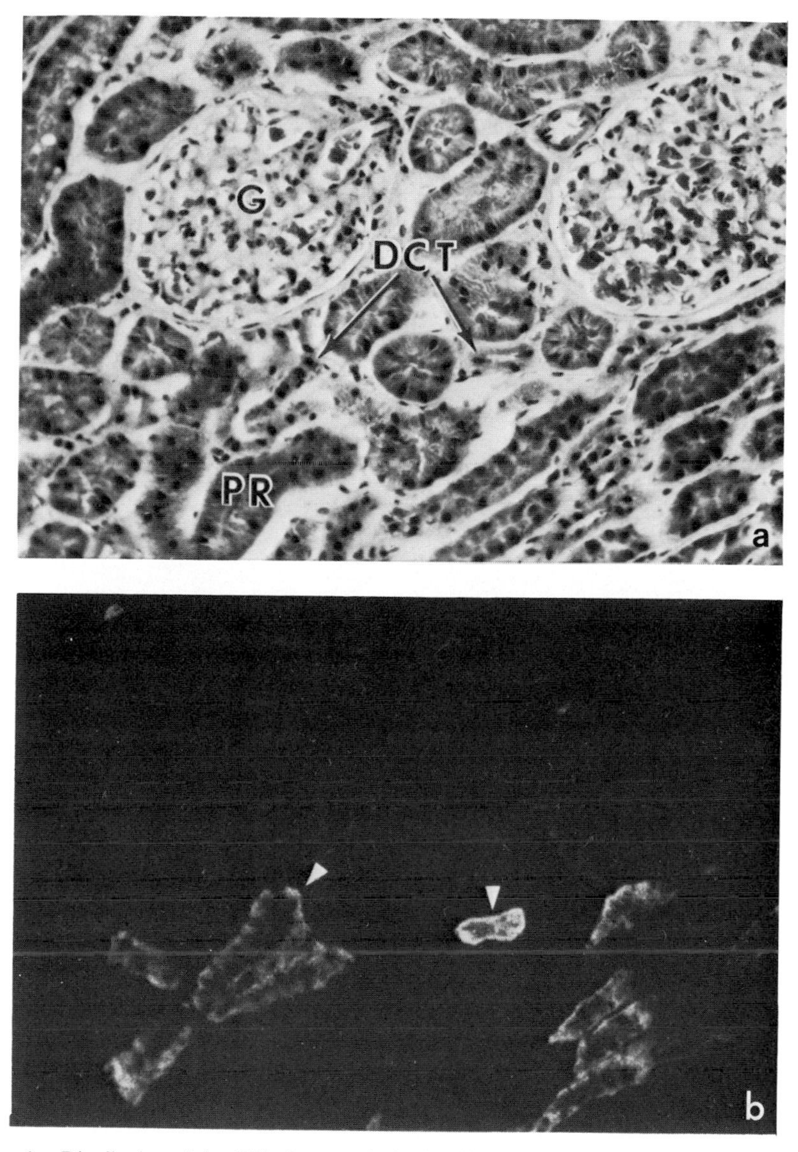


FIG. 4. Distribution of the 35K glycoprotein in dog kidney cortex. Serial frozen sections were processed for immunofluorescence and conventional hematoxylin and eosin (H/E) staining to verify the phase-contrast identification of renal tubules that bind antibody 11B8 (Fig. 3). Micrograph of H/E stained cortex (a) and fluorescence micrograph (b) of the identical field on an adjacent section demonstrate that the antibody binds to epithelial cells of the distal convoluted tubules (DCT) but not to the proximal tubules (PR) or glomeruli (G). Reproduced from *The Journal of Cell Biology*, 1982, **93,** 269–277 by copyright permission of The Rockefeller University Press.

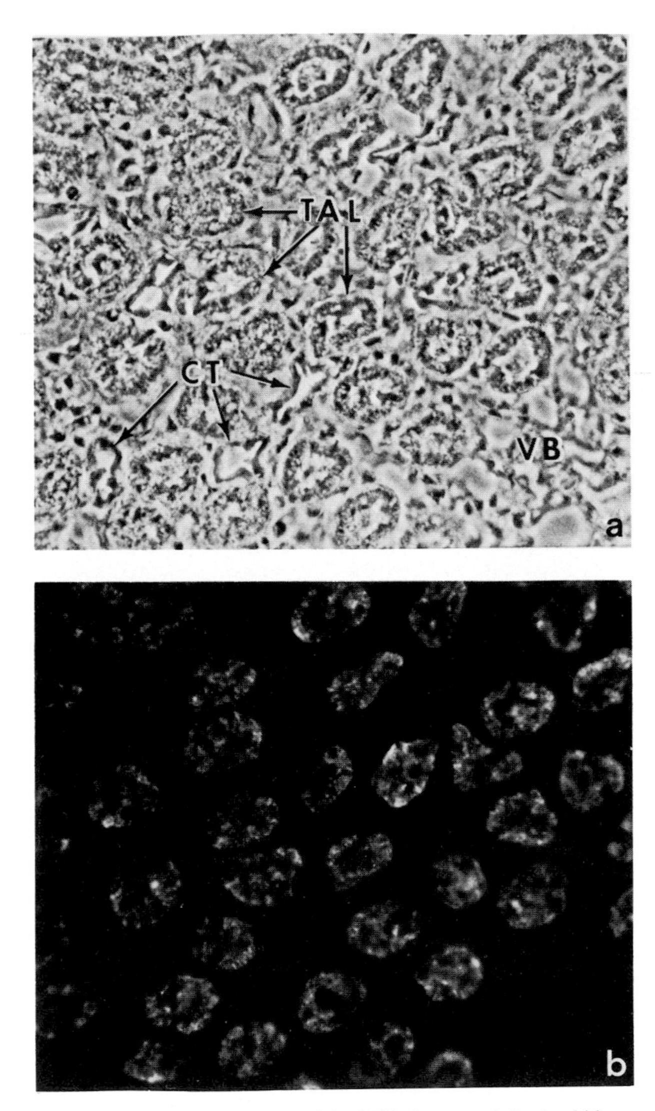


FIG. 5. Immunofluorescence localization of the 35K glycoprotein in dog kidney medulla. Corresponding phase-contrast (a) and immunofluorescence (b) micrographs demonstrate that only the epithelial cells of the thick ascending limb (TAL) express the 35K glycoprotein. Collecting tubules (CT) and elements of the vascular bundles (VB) do not bind detectable levels of antibodies. Reproduced from *The Journal of Cell Biology*, 1982, **93,** 269–277 by copyright permission of The Rockefeller University Press.

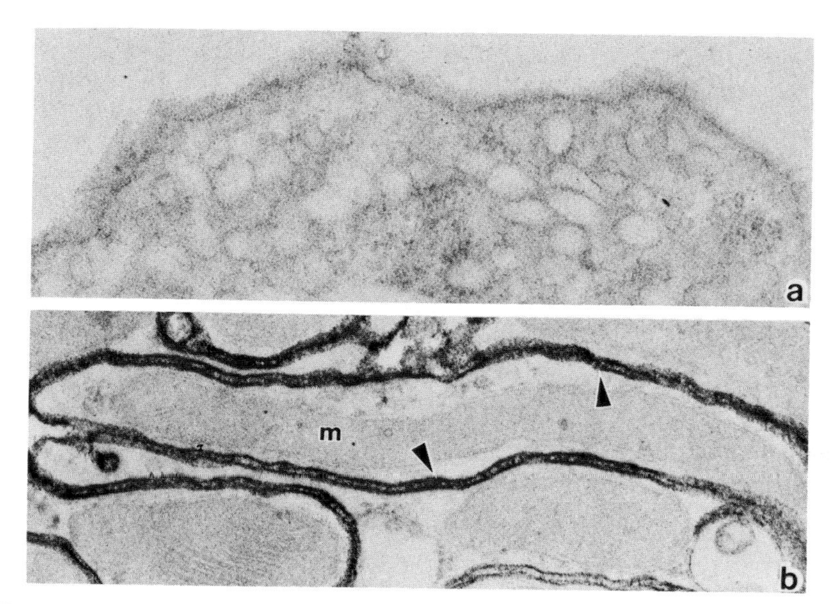


FIG. 6. Ultrastructural localization of the 35K glycoprotein in dog kidney cortex. Thick cryostat sections (40 μm) were incubated in monoclonal antibody 11B8 followed by HRP–GAM. In unstained thin sections of the distal convoluted tubule, the 35K glycoprotein was not detected on the apical cell surface (a) and was localized to the basolateral membrane (b: see arrowheads). Mitochondria (m) are present between the basal membrane infoldings. ×23,500.

for dog only, two were dog and rabbit specific and one was dog, rabbit, and rat specific. These results provide strong evidence that the biochemical composition of the glycoprotein varies between these animal species, but do not indicate whether these differences reside in the primary sequence, carbohydrate groups, or the conformation of the molecule. Immunofluorescence microscopy has demonstrated that the 20K glycoprotein is present on all epithelial cells of the cortical TAL (but not medullary), the DCT, and both cortical and medullary collecting tubules, whereas ultrastructural studies have localized this protein only on the basolateral plasma membrane. Of particular interest is the observation that the 20K glycoprotein could be detected only on the cortical TAL, since recent studies have demonstrated physiological differences between the cortical and medullary TAL (25; see Ref. 29 for a review of the literature).

The results reported here along with those published previously (27) illustrate both the strengths and weaknesses of utilizing monoclonal antibodies. These highly specific ligands are excellent probes for identifying, isolating, and studying the distribution of proteins available either in small quantities, in impure membrane fractions (2, 3, 15, 24, 62, 65), or on subpopulations of cells from tissues containing many cell types (1, 27, 67). A major disadvantage is that the

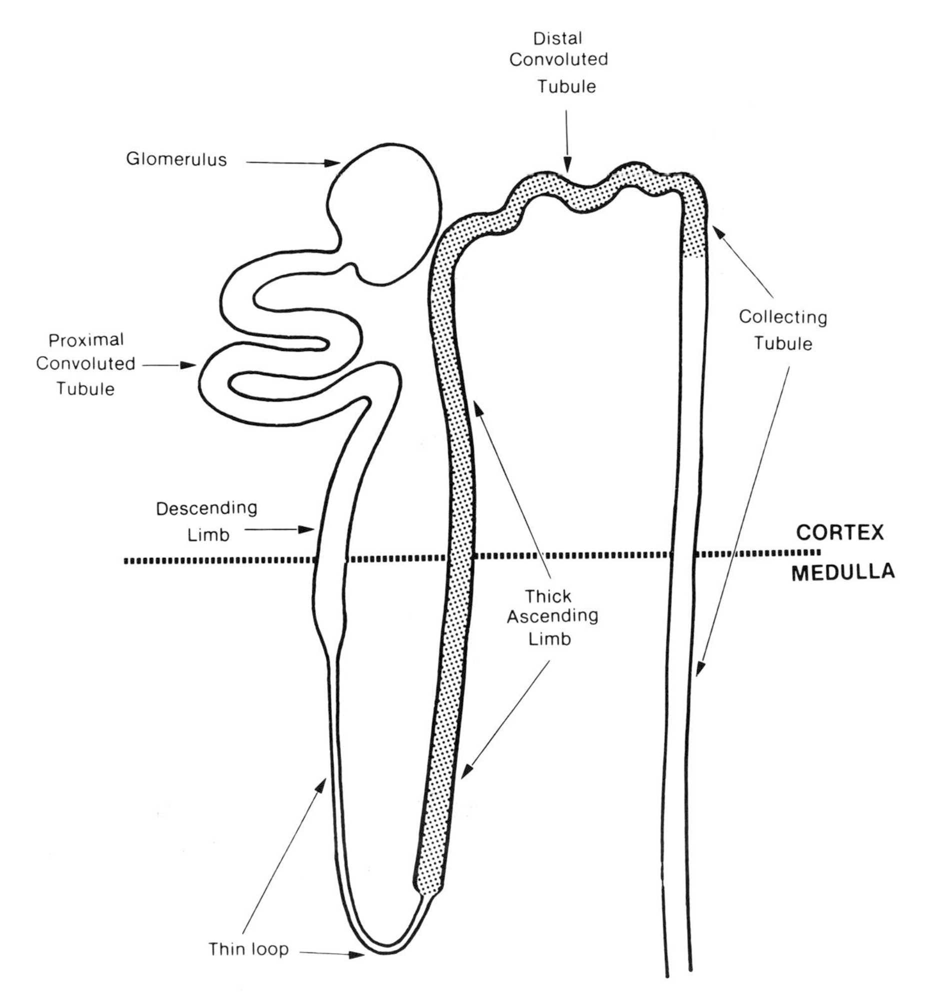


FIG. 7. The distribution of the 35K glycoprotein in the nephron. The 35K glycoprotein is localized to the cortical and medullary thick ascending limb, distal convoluted tubule, and some cells of the cortical collecting tubule (see stippled areas).

function of the protein(s) being investigated may not be known. In our studies, we have not as yet been able to determine in what physiological processes the 20K and 35K glycoproteins are involved. In view of the fact that they are localized on the basolateral membrane, it would be reasonable to assume that these proteins could either be involved in transepithelial transport or function as hormone receptors. For example, the distribution of the 35K glycoprotein in the nephron is identical to that mapped for calcitonin-stimulated cyclase activity on microdissected nephron segments, and is similar to that for isoproterenol-stimulated activity (44). Because MDCK cells exhibit an increase in adenylate cyclase activity in response to calcitonin, isoproterenol, vasopressin, and a variety of other hormones (43, 53), it should now be possible to take advantage of this simple model system to study the physiological function of membrane proteins to which we have monoclonal antibodies.

## REFERENCES


1. Barnstable, C. J. (1980). Monoclonal antibodies which recognize different cell types in the rat retina. *Nature (London)* **286,** 231–235.
2. Barnstable, C. J., Bodmer, W. E., Brown, G., Galfre, G., Milstein, C., Williams, A. F., and Ziegler, A. (1978). Production of monoclonal antibodies to group A erythrocytes, HLA and other human cell surface antigens—New tools for genetic analysis. *Cell* **14,** 9–20.
3. Beisiesel, U., Schneider, W. J., Goldstein, J. L., Anerson, R. G., and Brown, M. S. (1981). Monoclonal antibodies to the low density lipoprotein receptor as probes for study of receptor mediated endocytosis and the genetics of familial hypercholesterolemia. *J. Biol. Chem.* **256,** 11923–11931.
4. Berridge, M. J., and Oschman, J. L. (1972). "Transporting Epithelia." Academic Press, New York.
5. Bonner, W. M., and Laskey, R. A. (1974). A film detection method for tritium-labeled proteins and nucleic acids in polyacrylamide gels. *Eur. J. Biochem.* **46,** 83–88.
6. Booth, A. G., and Kenny, A. J. (1974). A rapid method for presentation of microvilli from rabbit kidney. *Biochem. J.* **142,** 575–581.
7. Cereijido, M., Ehrenfeld, J., Meza, I., and Martinez-Palomo, A. (1980). Structural and functional membrane polarity in cultured monolayers of MDCK cells. *J. Membr. Biol.* **52,** 147–159.
8. Cereijido, M., Robbins, E. S., Dolan, W. J., Rotunno, C. A., and Sabatini, D. D. (1978). Polarized monolayers formed by epithelial cells on a permeable and translucent support. *J. Cell Biol.* **77,** 853–880.
9. Cramer, E. B., Milks, L. C., and Ojakian, G. K. (1980). Transepithelial migration of human neutrophils: An *in vitro* model system. *Proc. Natl. Acad. Sci. U.S.A.* **77,** 4069–4073.
10. Dragsten, P. R., Blumenthal, R., and Handler, J. S. (1981). Membrane asymmetry in epithelia: Is the tight junction a barrier to diffusion in the plasma membrane? *Nature (London)* **294,** 718–722.
11. Dulbecco, R., Unger, M., Bologna, H., Battifora, P., P. Syka, and Okada, S. (1981). Cross-reactivity between Thy 1 and a component of intermediate filaments demonstrated using a monoclonal antibody. *Nature (London)* **292,** 772–774.
12. Ebel, H., Aulbert, E., and Merker, H. J. (1976). Isolation of the basal and lateral plasma membrane of rat kidney tubule cells. *Biochim. Biophys. Acta* **433,** 531–546.
13. Ernst, S. A., and Mills, J. W. (1977). Basolateral plasma membrane localization of ouabain-sensitive sodium transport sites in the secretory epithelium of the avian salt gland. *J. Cell Biol.* **75,** 74–94.
14. Eveloff, J., Haase, W., and Kinne, R. (1980). Separation of renal medullary cells: Isolation of cells from the thick ascending limb of Henle's loop. *J. Cell Biol.* **87,** 672–681.
15. Fraser, C. M., and Venter, J. C. (1980). Monoclonal antibodies to beta-adrenergic receptors: Use in purification and molecular characterization of beta receptors. *Proc. Natl. Acad. Sci. U.S.A.* **77,** 7034–7038.
16. Frömter, E., and Diamond, J. (1972). Route of passive ion permeation in epithelia. *Nature (London) New Biol.* **235,** 9–13.

17. Fujita, M., Kawai, Asane, S., and Nakae, M. (1973). Protein components of two different regions of an intestinal epithelial cell membrane. *Biochim. Biophys. Acta* **307,** 141–151.
18. Galli, P., Brenna, A., DeCamilli, P., and Medolesi, J. (1976). Extracellular calcium and the organization of tight junctions in pancreatic acinar cells. *Exp. Cell Res.* **99,** 178–183.
19. Gausch, C. R., Hard, W. L., and Smith, T. F. (1966). Characterization of an established line of canine kidney cells (MDCK). *Proc. Soc. Exp. Biol. Med.* **122,** 931–935.
20. Gefter, M. L., Margulies, D. H., and Scharff, M. D. (1977). A simple method for polyethylene glycol-promoted hybridization of mouse myeloma cells. *Somatic Cell Genet.* **3,** 231–236.
21. Grantham, J. J., Irish, J. M., and Hall, D. A. (1978). Studies of isolated renal tubules *in vitro*. *Annu. Rev. Physiol.* **40,** 249–277.
22. Greenwood, E. C., Hunter, W. M., and Glover, S. (1963). The preparation of $^{131}$I-labeled human growth hormone of high specificity radioactivity. *J. Biochem.* **89,** 114–122.
23. Handler, J. S., Perkins, F. M., and Johnson, J. P. (1980). Studies of renal cell function using cell culture techniques. *Am. J. Physiol.* **238,** F1–F9.
24. Hauri, H. P., Quaroni, A., and Isselbacher, K. J. (1980). Monoclonal antibodies to sucrase/isomaltase:Probes for the study of postnatal development and biogenesis of the intestinal microvillus membrane. *Proc. Natl. Acad. Sci. U.S.A.* **77,** 6629–6633.
25. Hebert, S. C., Culpepper, M., and Andreoli, T. E. (1981). NaCl transport in mouse thick ascending limbs. I. Functional nephron heterogeneity and ADH-stimulated NaCl cotransport. *Am. J. Physiol.* **241,** F412–431.
26. Heidrich, H. G., and Dew, M. E. (1977). Homogeneous cell populations from rabbit kidney cortex. Proximal, distal tubule and renin-active cells isolated by free-flow electrophoresis. *J. Cell Biol.* **74,** 780–788.
27. Herzlinger, D. A., Easton, T. G., and Ojakian, G. K. (1982). The MDCK epithelial cell line expresses a cell surface antigen of the kidney distal tubule. *J. Cell Biol.* **93,** 269–277.
28. Hoi Sang, U., Saier, M. H., and Ellisman, M. H. (1979). Tight junction formation is closely linked to the polar redistribution of intramembranous particles in aggregating MDCK epithelia. *Exp. Cell Res.* **122,** 384–391.
29. Jacobson, H. R. (1981). Functional segmentation of the mammalian nephron. *Am. J. Physiol.* **241,** F203–F218.
30. Jørgensen, P. L. (1974). Purification and characterization of $Na^+$-$K^+$-ATPase. III. Purification from outer medulla of mammalian kidney after selective removal of membrane components by sodium dodecyl sulfate. *Biochim. Biophys. Acta* **356,** 36–52.
31. Kennett, R. H. (1981). Hybridomas: A new dimension in biological analysis. *In Vitro* **17,** 1036–1050.
32. Kinsella, J. L., Holohan, P. D., Pessah, N. I., and Ross, C. R. (1979). Isolation of luminal and antiluminal membranes from dog kidney cortex. *Biochim. Biophys. Acta* **552,** 468–477.
33. Kohler, G., and Milstein, C. (1975). Continuous culture of fused cells secreting antibody of predefined specificity. *Nature (London)* **256,** 495–497.
34. Kohler, G., and Milstein, C. (1976). Deriviation of specific of antibody-producing tissue culture and tumor lines by cell fusion. *Eur. J. Immunol.* **6,** 511–519.
35. Kwan, S.-P., Yelton, D. E., and Scharff, M. D. (1980). Production of monoclonal antibodies. *Genet. Eng.* **2,** 31–46.
36. Kyte, J. (1976). Immunoferritin determination of the distribution of $Na^+$-$K^+$ ATPase over the plasma membranes of renal convoluted tubules. I. Distal segment. *J. Cell Biol.* **68,** 287–303.
37. Laemmli, U. K. (1970). Cleavage of structural proteins during assembly of the head of bacteriophage T4. *Nature (London)* **277,** 680–685.
38. Lane, D. P., and Hoeffler, W. K. (1980). SV40 large T shares an antigenic determinant with a cellular protein of molecular weight 68,000. *Nature (London)* **288,** 167–170.
39. Leighton, J., Estes, L. W., Mansukhani, S., and Brada, Z. (1970). A cell line derived from dog

kidney (MDCK) exhibiting qualities of papillary adenocarcinoma and of renal tubular epithelium. *Cancer* **26,** 1022–1028.

40. Louvard, D. (1980). Apical membrane aminopeptidase appears at site of cell-cell contact in cultured kidney epithelial cells. *Proc. Natl. Acad. Sci. U.S.A.* **77,** 4132–4136.
41. Meldolesi, J., Castiglioni, G., Parma, R., Nassivera, N., and De Camilli, P. (1978). $Ca^{++}$-dependent disassembly and reassembly of occluding junctions in guinea pig pancreatic acinar cells: Effect of drugs. *J. Cell Biol.* **79,** 156–172.
42. Misfeldt, D. S., Hammamoto, S. T., and Pitelka, D. R. (1976). Transepithelial transport in cell culture. *Proc. Natl. Acad. Sci. U.S.A.* **73,** 1212–1216.
43. Mita, S. M., Takeda, H., Nakane, H., Yasuda, M., and Endo, H. (1980). Response of a dog kidney cell line (MDCK) to antidiuretic hormone (ADH) with special reference to activation of protein kinase. *Exp. Cell Res.* **130,** 169–173.
44. Morel, F. (1981). Sites of hormone action in the mammalian nephron. *Am. J. Physiol.* **240,** F159–F164.
45. Murer, H., and Kinne, R. (1980). The use of isolated membrane vesicles to study epithelial transport processes. *J. Membr. Biol.* **55,** 81–95.
46. O'Farrell, P. H. (1975). High resolution two-dimensional electrophoresis of proteins. *J. Biol. Chem.* **250,** 4007–4021.
47. Ojakian, G. K. (1981). Tumor promoter-induced changes in the permeability of epithelial cell tight junctions. *Cell* **23,** 95–103.
48. Pisam, M., and Ripoche, P. (1976). Redistribution of surface macromolecules in dissociated epithelial cells. *J. Cell Biol.* **71,** 907–920.
49. Rabito, C. A., Tchao, T., Valentich, J., and Leighton, J. (1978). Distribution and characteristics of the occluding junctions in a monolayer cell line (MDCK) derived from dog kidney. *J. Membr. Biol.* **43,** 351–365.
50. Reichman, M. E., Spencer, A. M., and Edelman, I. S. (1981). Generation of monoclonal antibodies directed against sodium potassium ATPase. *In* "Monoclonal Antibodies in Endocrine Research" (R. E. Fellows and G. S. Eisenbarth, eds.), pp. 135–142. Raven, New York.
51. Richardson, J. C. W., Scalera, V., and Simmons, N. L. (1981). Identification of two strains of MDCK cells which resemble separate nephron tubule segments. *Biochim. Biophys. Acta* **673,** 26–36.
52. Richardson, J. C. W., and Simmons, N. L. (1979). Demonstration of protein asymmetries in the plasma membrane of cultured renal (MDCK) epithelial cells by lactoperoxidase-mediated iodination. *FEBS Lett.* **105,** 201–204.
53. Rindler, M. J., Chuman, L. M., Shaffer, L., and Saier, Jr., M. H. (1979). Retention of differentiated properties in an established cell line (MDCK). *J. Cell Biol.* **81,** 635–648.
54. Rodriguez-Boulan, E., and Sabatini, D. D. (1978). Asymmetric budding of viruses in epithelial monolayers: A model system for study of epithelial polarity. *Proc. Natl. Acad. Sci. U.S.A.* **75,** 5071–5075.
55. Rodriguez, Boulan, E., and Pendergast, M. (1980). Polarized distribution of viral envelope proteins in the plasma membrane of infected epithelial cells. *Cell* **20,** 45–54.
56. Rodriguez-Boulan, E., Paskiet, K. T., and Sabatini, D. D. (1983). Assembly of enveloped viruses in MDCK cells: Polarized budding from single attached cells and from clusters of cells in suspension. *J. Cell Biol.* **96,** 866–874.
57. Sabatini, D. D., Kreibich, G., Morimoto, T., and Adesnik, M. (1982). Mechanisms for the incorporation of proteins in membranes and organelles. *J. Cell Biol.* **92,** 1–22.
58. Scharff, M. D., and Roberts, S. (1981). Present status and future prospects for the hybridoma technology. *In Vitro* **17,** 1072–1077.
59. Semenza, G. (1976). Small intestinal disaccharidases: Their properties and role as sugar translocators across natural and artificial membranes. *Enzymes Biol. Membr. 1976* **3,** 349–382.

60. Schlatz, L. J., Schwartz, I. L., Kinne-Saffran, E., and Kinne, R. (1975). Distribution of parathyroid hormone-stimulated adenylate cyclase in plasma membranes of cells of the kidney cortex. *J. Membr. Biol.* **24,** 131–144.
61. Sisson, S. P., and Vernier, R. L. (1980). Methods for immunoelectron microscopy: Localization of antigens in rat kidney. *J. Histochem. Cytochem.* **28,** 441–452.
62. Stern, P. L., Willison, K. R., Lennox, E., Galfre, G., Milstein, C., Secher, D., Ziegler, A., and Springer, T. (1978). Monoclonal antibodies as probes for differentiation and tumor associated antigens: A Forssman specificity on teratocarcinoma stem cells. *Cell* **14,** 775–783.
63. Taub, M., Chuman, L., Saier, M. H., and Sato, G. (1979). Growth of Madin-Darby canine kidney epithelial cell (MDCK) line in hormone-supplemented, serum-free medium. *Proc. Natl. Acad. Sci. U.S.A.* **76,** 3338–3342.
64. Tzartos, S. J., Rand, D. E., Einarson, B. L., and Lindstrom, J. M. (1981). Mapping of surface structures of *Electrophorus* acetylcholine receptor using monoclonal antibodies. *J. Biol. Chem.* **256,** 8635–8645.
65. Williams, A. F., Galfre, G., and Milstein, C. (1977). Analysis of cell surfaces by xenogeneic myeloma-hybrid antibodies: Differentiation antigens of rat lymphocytes. *Cell* **12,** 663–673.
66. Ziomek, C. A., Schulman, S., and Edidin, M. (1980). Redistribution of membrane proteins in isolated mouse intestinal epithelial cells. *J. Cell Biol.* **86,** 849–857.
67. Zipser, B., and McKay, R. (1981). Monoclonal antibodies distinguish identifiable neurons in the leech. *Nature (London)* **289,** 549–554.

# Chapter 11

# Immunolabeling of Frozen Thin Sections and Its Application to the Study of the Biogenesis of Epithelial Cell Plasma Membranes

*IVAN EMANUILOV IVANOV, HEIDE PLESKEN, DAVID D. SABATINI, AND MICHAEL J. RINDLER*

*Department of Cell Biology*
*New York University School of Medicine*
*New York, New York*



## I. INTRODUCTION

The functional and morphological polarization of transporting epithelial cells is largely a consequence of the asymmetric distribution of their membrane proteins into apical or basolateral plasma membranes. The biogenesis of these two domains necessarily involves sorting out processes which are responsible for directing the intracellular transport of membrane proteins from their sites of synthesis to their ultimate destinations in the cell (cf. Sabatini *et al.*, 1982).

We have employed the epithelial tissue culture cell line MDCK (Madin–




ISBN 0-12-153320-4

Darby canine kidney) (Misfeldt, 1976; Cereijido *et al.*, 1978) as a useful *in vitro* model system for studies on the mechanisms of plasma membrane polarization. A striking morphological manifestation of cell polarity is observed when MDCK is infected with certain enveloped viruses (Rodriguez-Boulan and Sabatini, 1978). For example, after the cells are infected with influenza virus, viral particles are formed exclusively at the luminal plasma membrane. Vesicular stomatitis virions (VSV), on the other hand, are preferentially assembled at basal and lateral surfaces of the cell following infection of MDCK with this virus. It has also been demonstrated that this pattern of viral assembly corresponds to the polarized distribution of the newly synthesized envelope glycoproteins of the viruses (Rodriguez-Boulan and Pendergast, 1980). Moreover, when combinations of viruses, such as influenza and VSV, are utilized in a simultaneous double infection, the overall distribution of budding viruses is preserved (Rindler *et al.*, 1982).

In order to identify and characterize the molecular events that control the process of cell polarization, it was of interest to identify the cellular site of sorting out of apical and basolateral viral membrane glycoproteins. From cell fractionation and immune localization studies in nonpolarized cells, it was already known that the viral glycoproteins generally follow the classical pathway of intracellular transport established for secretory proteins—namely, synthesis in the rough endoplasmic reticulum and transport through the Golgi apparatus before insertion into the plasma membrane (Palade, 1975; Compans, 1973; Knipe *et al.*, 1977; Bergmann *et al.*, 1981). To further characterize this pathway in the polarized epithelial cell line, we have employed immunolabeling techniques at the electron microscope (EM) level.

The efficacy of immunolabeling procedures designed for the EM level largely depends on the ability of the immunoreagents to gain access to their corresponding antigens within the tissue or the cell (Kraehenbühl and Jamieson, 1974). Two general strategies for immunolabeling have been devised to overcome this limitation and to exploit the specificity of the antibody–antigen interaction and the ultrastructural resolution of the EM. One group includes those protocols that rely on the partial permeabilization of the cellular membranes by detergents such as saponin to give IgGs and EM markers access to intracellular compartments (Ohtsuki *et al.*, 1978; Willingham *et al.*, 1978), and the other consists of those that employ thin (~100 nm) or thick (5–100 μm) sectioned material to provide direct exposure of intracellular antigens to the immunoreagents. After immunolabeling, permeabilized cells and thick frozen sections must be embedded and processed by conventional techniques to visualize the reaction product, whereas thin sections may be viewed directly in the EM. Immunolabeling of thin sections has been applied with varying levels of success to specimens embedded in methacrylate (Kawarai and Nakane, 1970), polyethylene glycol (PEG) (Mazurkiewicz and Nakane, 1972), Epon (Bendayan *et al.*, 1980), Lowicryl (Roth *et al.*, 1981), serum albumin (McLean and Singer, 1970), and sucrose (Tokuyasu, 1973).

In our studies on the biogenesis of epithelial cell polarity, we have applied the immunolabeling technique to ultrathin frozen sections prepared from virally infected MDCK cells. This method enabled us to localize viral glycoproteins inside the cell as well as on its surface and to obtain information concerning the pathways followed by these proteins from their site of synthesis to their site of assembly into virions. Our experience with this method and the variables associated with each step is discussed in this chapter together with the results of other investigators using similar or alternative approaches.

## II. PRODUCTION OF ULTRATHIN FROZEN SECTIONS

### A. Fixation

Chemical fixation is a prerequisite for the production of ultrathin frozen sections from tissues or from cultured cells for use in immunolabeling studies. The selection of an appropriate fixative must take into account two conflicting requirements. The desire for reasonable ultrastructural preservation of intracellular organelles dictates the use of stronger fixation procedures, which, in turn, may have a negative impact on the need to maintain the integrity of the antigenic determinants and on the accessibility of the antibodies to their binding sites. Thus, it is essential to optimize fixation conditions for a given tissue and antigen and to strike a balance between ultrastructural and antigenic preservation. To summarize the experience of investigators in this field, glutaraldehyde (0.2–2%) or some mixture of formaldehyde and glutaraldehyde is commonly employed (Bourguignon and Butman, 1982; Green *et al.*, 1981; Griffiths *et al.*, 1981; Rindler *et al.*, 1982). The general observation is that the use of glutaraldehyde provides better ultrastructure but can lead to a decrease in the amount of immunolabeling (Kyte, 1976; Tokuyasu, 1980). In some instances it is desirable to partially mitigate the effects of these aldehyde fixatives by pretreating the samples with the monofunctional imidate ethyl acetimidate (Tokuyasu and Singer, 1976; Chen and Singer, 1982). Because this reagent reacts with the same chemical sites as formaldehyde and glutaraldehyde, the approach leads to final formation of a less dense protein meshwork and, as a result, to better antibody penetrability, but at the same time, to general deterioration of the cellular ultrastructure.

In our work with the MDCK cell line infected with enveloped viruses, we have been able to obtain good results, in terms of both ultrastructural determination and specific labeling, using 1–2% glutaraldehyde fixation in phosphate-buffered saline (PBS) for times ranging from 30 to 60 minutes (Fig. 1). For tissues such as

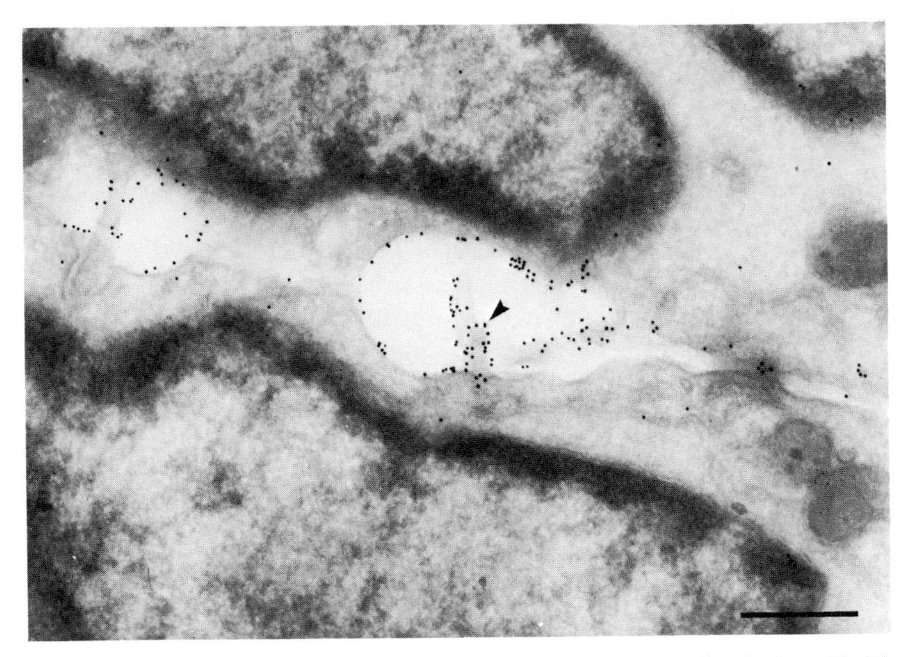

FIG. 1. Frozen thin sections of VSV-infected MDCK cells prepared after fixation with 2% glutaraldehyde and infusion with 0.6 *M* sucrose. The sections were incubated with rabbit antibody against the viral glycoprotein (G) followed by affinity-purified goat anti-rabbit IgG complexed to 18-nm colloidal gold particles. The label is largely confined to the lateral plasma membranes of adjacent cells where the budding virions are also located (arrowhead). Bar, 0.5 μm.

liver, glutaraldehyde is also effective (Fig. 2), although a mixture of 2% formaldehyde and 0.2% glutaraldehyde gives more than satisfactory structural delineation. Thicker frozen sections (500 nm) can be useful for immunofluorescence studies; in this case, paraformaldehyde fixation is preferable. A small amount of glutaraldehyde (0.2%) may also be included, although this fixative contributes a considerable background fluorescence when used at high concentrations (Weber *et al.* , 1978).

## B. Sucrose Infusion

Early attempts to obtain ultrathin frozen sections by workers such as Fernandez-Moran (1952), Bernhard and Leduc (1967), Iglesias *et al.* (1971), and

---

FIG. 2. Frozen thin sections of rat liver produced after fixation in 2% glutaraldehyde and infusion with 1.5 *M* sucrose. They have been labeled sequentially with biotinylated concanavalin A and avidin–ferritin. Among the cellular structures depicted in (a) are the cell nucleus (N), mitochondria (M), Golgi apparatus (GA), and bile caniliculus (BC). The ferritin particles tended to accumulate over lysosomes (*). (b) An enlargement of a segment of the designated lysosome. Bar, 0.25 μm.

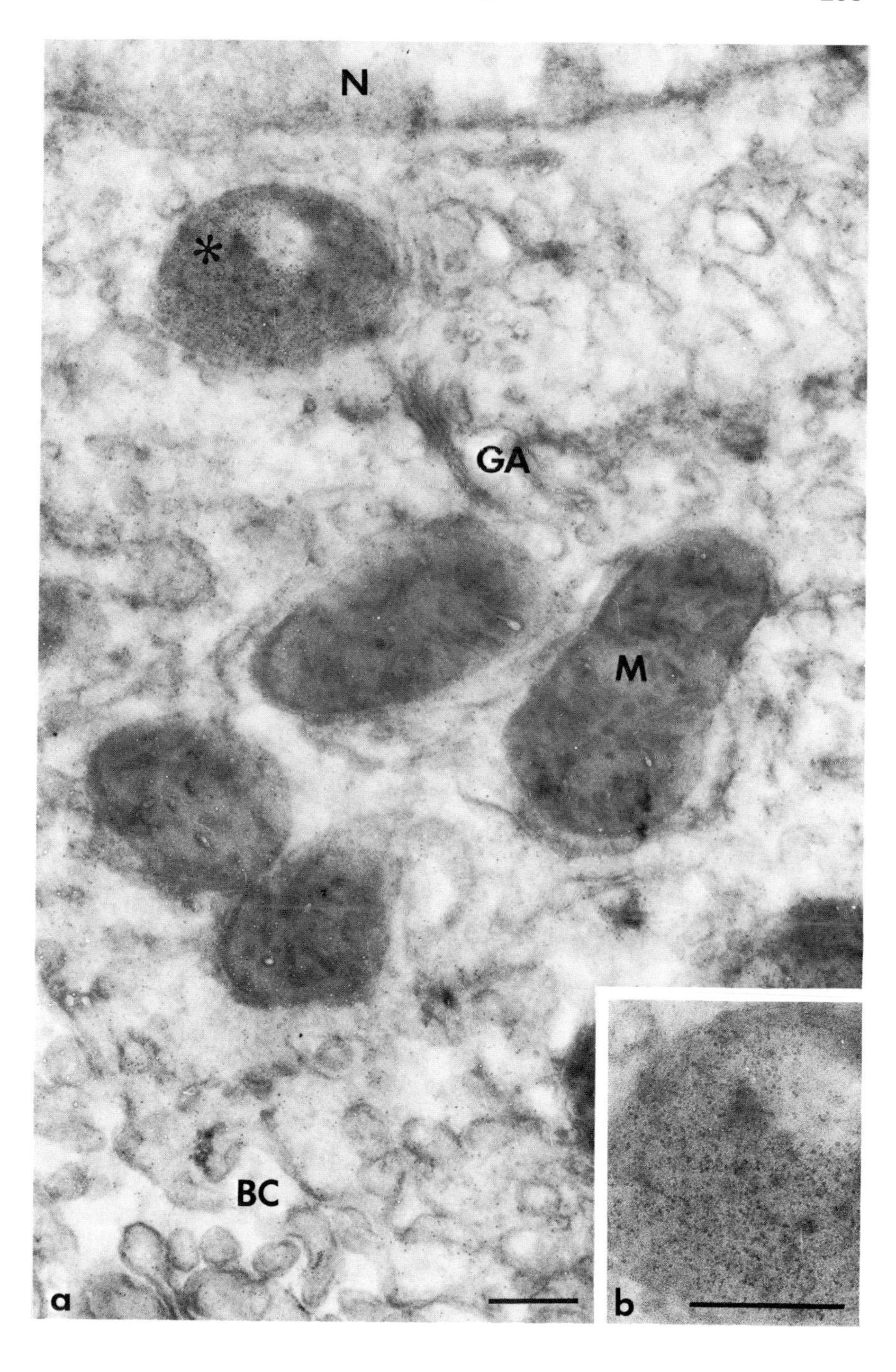
N
*
GA
M
BC
a
b

Christensen (1970, 1971) met with limited success due to technical difficulties. In this respect, the introduction by Tokuyasu (1973) of sucrose infusion of the tissue after fixation represented a significant contribution to the development of this method. Sucrose serves as a cryoprotectant, preventing the formation of large ice crystals which may damage the tissue ultrastructure. It facilitates sectioning of the tissue at thicknesses suitable for the EM (60–100 nm), although low temperatures in the range of −80 to −110°C are required. In addition, sucrose is easily washed out of the sections, thus providing better penetration of the tissue by the labeling reagents.

In practice, small blocks of tissue (~0.5 mm$^3$) are infused with sucrose after fixation. Here again, the optimal conditions for adequate thin sectioning and ultrastructural preservation must be determined empirically depending on the kind of tissue or cell. One overall rule is that the higher the concentration of sucrose employed, the lower the temperature used for sectioning. There are two general tendencies in the field. One is to infuse with 0.6–1.6 *M* sucrose and to section at temperatures of about −80 to −90°C (Tokuyasu, 1973, 1980), and a second is to utilize highly concentrated sucrose (2.3 *M*) and a temperature of sectioning of about −110°C. (Griffiths *et al.*, 1981). We have been successful with the first set of conditions in our work with MDCK cells and rat liver. Well-preserved ultrathin sections could be obtained, for example, from specimens of glutaraldehyde-fixed liver after infusion with sucrose at concentrations ranging from 0.6 to 1.5 *M* in 0.1 *M* phosphate buffer at pH 7.4. However, maximal reproducibility was achieved with 1.3 *M* sucrose infusion for at least 1 hour at room temperature and a temperature of sectioning of about −85°C. When sections of 500 nm are to be produced for immunofluorescence studies, sucrose concentrations of 1.3–1.6 *M* and a higher temperature of sectioning (about −50°C) are employed.

## C. Sectioning at Low Temperatures

The basic procedure introduced by Tokuyasu (1973, 1980) has been generally adopted. Accordingly, after fixation and infusion with sucrose, the small sample (~0.5 mm$^3$) is mounted on the tip of a copper specimen holder which is then immersed in liquid nitrogen or Freon to achieve rapid freezing. The holder with the sample is subsequently transferred to the precooled chamber of the microtome. Ultrathin frozen sections are obtained using special cryoattachments to normal microtomes. (We employ a DuPont/Sorvall FTS cryoattachment to a Dupont/Sorvall MT-2 microtome.) The attachment regulates the overall temperature of the chamber by injecting a stream of vapor from a liquid nitrogen tank. A dry precooled glass knife is used for cutting the sections, which then accumulate on its surface. Here they are picked up by adhesion to small drops of

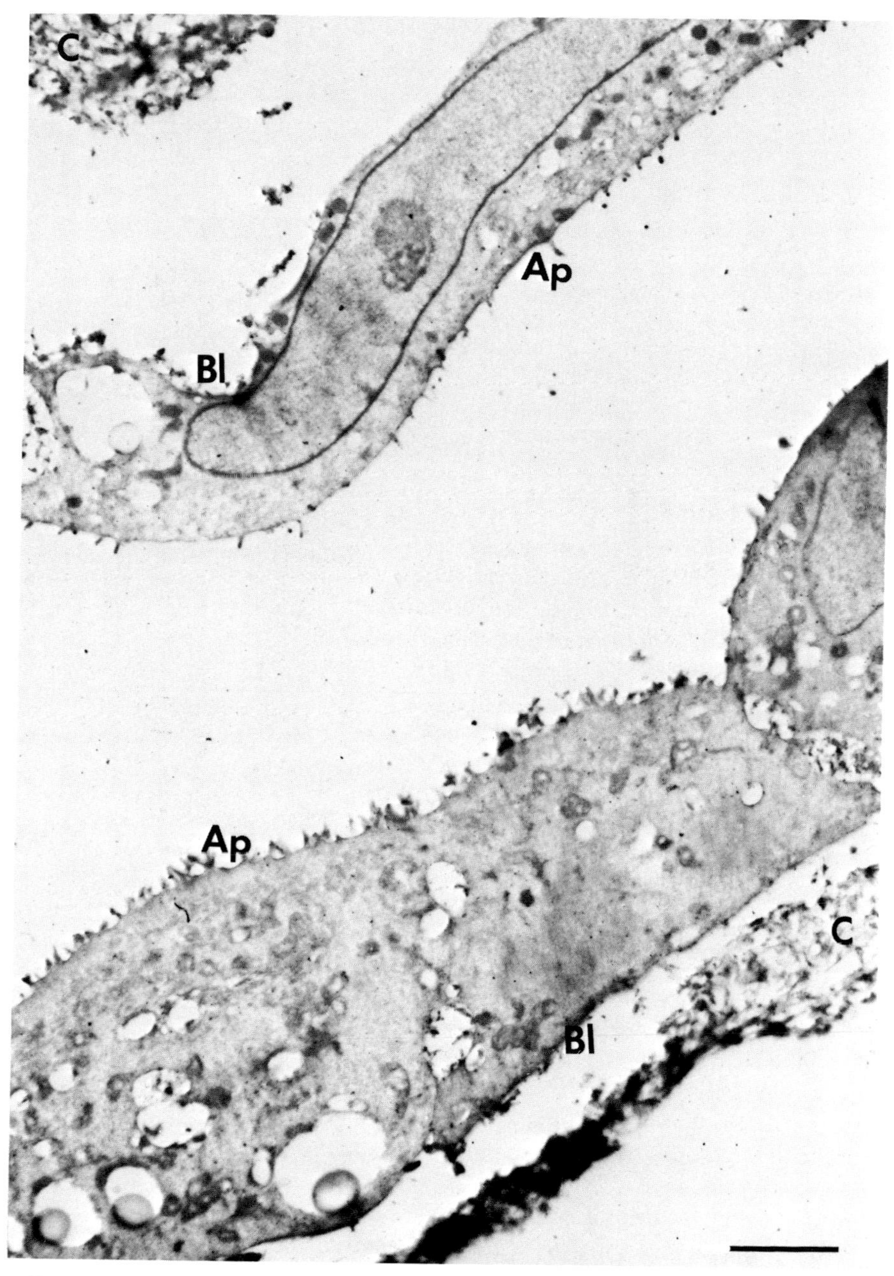


FIG. 3. Strips of MDCK cells derived from material subjected to ultrathin frozen sectioning. Monolayers grown on collagen (C) were first fixed in 2% glutaraldehyde and infused with 0.6 *M* sucrose. Ap, Apical plasma membrane; Bl, basolateral plasma membrane. Bar, 2 μm.

semifrozen sucrose hanging on wire loops, and then transferred to the surface of EM grids.

When the specimen is a tissue sample, the blocks to be cut are preliminarily shaped as small pyramids in such a way that after mounting on a microtome holder no additional trimming of the block is needed. In our work with virus-infected MDCK cells, confluent monolayers grown on thick collagen layers were first peeled off and infused with 0.6 *M* sucrose for 2 hours at room temperature, and then were organized into a pellet on the specimen holder. In this way, we were able to obtain cross sections of longer strips of cells and to provide access of labeling IgGs to both apical and basolateral surfaces of the epithelial cells as well as to their intracellular structures (Fig. 3).

## III. INDIRECT IMMUNOLABELING

### A. First Antibodies

After the frozen thin sections are attached to the EM grids they are subjected to a series of washes with PBS before immunolabeling. These washes serve several purposes. They remove the sucrose, which presumably enhances the penetration of labeling reagents into the thin section. At this stage, the unreacted aldehyde groups are quenched by inclusion in the washing buffer of a free amino acid such as glycine (0.01 *M*). In addition, the sections are treated with 2% gelatin (300 bloom) which serves to decrease nonspecific binding of IgG molecules.

In the first step of indirect immunolabeling, the thin frozen sections are incubated with a specific IgG against the antigen of interest. The choice of a suitable first antibody is perhaps the most critical aspect of the immunolabeling process. Three factors contribute to the selection: the titer of the antiserum, the level of nonspecific binding, and the sensitivity of the antigenic sites to fixation (especially when glutaraldehyde is used). It is preferable that an IgG fraction rather than whole serum be utilized, because the nonspecific binding of serum components may cloud the final image and may increase background labeling. In fact, most investigators employ affinity-purified antibodies at concentrations of 50–200 μg/ml (Tokuyasu, 1980; Green *et al.*, 1981). We have found that affinity-purified antibodies are not essential so long as the initial antiserum has a sufficiently high titer; thus our rabbit anti-vesicular stomatitis virus glycoprotein (G) IgG is used at a concentration of ~1 mg/ml. When using monoclonal antibodies, it is not necessary to purify the IgG from ascites fluid so long as a sufficient dilution (we used ~1:30) still yields satisfactory results. Every antiserum must first be tested by immunofluorescence microscopy on tissue culture cells or on semithick (500 nm) frozen sections attached to glass slides. It should be noted, however, that the concentrations of primary antibodies that provide for satisfactory labeling in indirect immunofluorescence are considerably lower than those required for optimal labeling at the EM level. Consequently, it is necessary to optimize the concentration of each reagent on the ultrathin sections themselves in order to achieve a balance between specific labeling and the nonspecific binding that inevitably results when high concentrations are employed. To minimize this background, many investigators include a competitor protein such as IgG of another species or bovine serum albumin (IgG-free) in the incubation mixture at concentrations of 3–10 mg/ml.

### B. Second Antibody Conjugates

The selection of a competitor protein is dependent upon the labeling procedure chosen as the next step, in which an electron-dense marker must be employed to visualize the antigen–antibody interaction. For example, investigators using a labeling particle complexed to protein A should avoid the use of a competitor IgG. Historically, ferritin has been the major EM marker and is ordinarily coupled covalently to second antibodies recognizing IgG of the initial species (Kishida *et al.*, 1975), although more recently the iron–dextran complex Imposil (Dutton *et al.*, 1979) and colloidal gold particles have been successfully applied. Colloidal gold has some advantages over ferritin including the fact that it can be easily produced in a wide range of sizes (from 3 to greater than 20 nm) (Frens, 1973; Faulk and Taylor, 1971; Horisberger and Rosset, 1977) and that it is much more electron opaque, although in some cases (Papermaster *et al.*, 1978) the apparent density of the ferritin core can be enhanced substantially by applying a bismuth subnitrate treatment (Ainsworth and Karnovsky, 1972). The visibility of the colloidal gold leads to the possibility of using lower EM magnifications while still being able to distinguish the particles (Figs. 1 and 2). It also allows visualization and reasonable interpretation of thicker or more intensely contrasted areas of the section. Gold particles are generally coupled to *Staphyloccoccus* protein A (Romano and Romano, 1977; Roth *et al.*, 1978), to lectins (Horisberger and Rosset, 1977; Roth and Binder, 1978), or to second antibodies (Faulk and Taylor, 1971; Geoghegan and Ackerman, 1977) by electrostatic interactions. Detailed accounts for the production of such particles and their conjugation with protein A have appeared (Slot and Geuze, 1981).

We have elected to use colloidal gold as a marker in our studies on the biogenesis of epithelial cell polarity in virally infected MDCK cells (Rindler *et al.*, 1982). Moreover, we have employed colloidal gold particles of several sizes complexed to affinity-purified second antibodies by procedures similar to those utilized for protein A. In our protocol, however, the coupling is performed at a pH above 7.8 using two or three times the minimal amount necessary for antibody stabilization. To avoid aggregation, which is a serious problem when

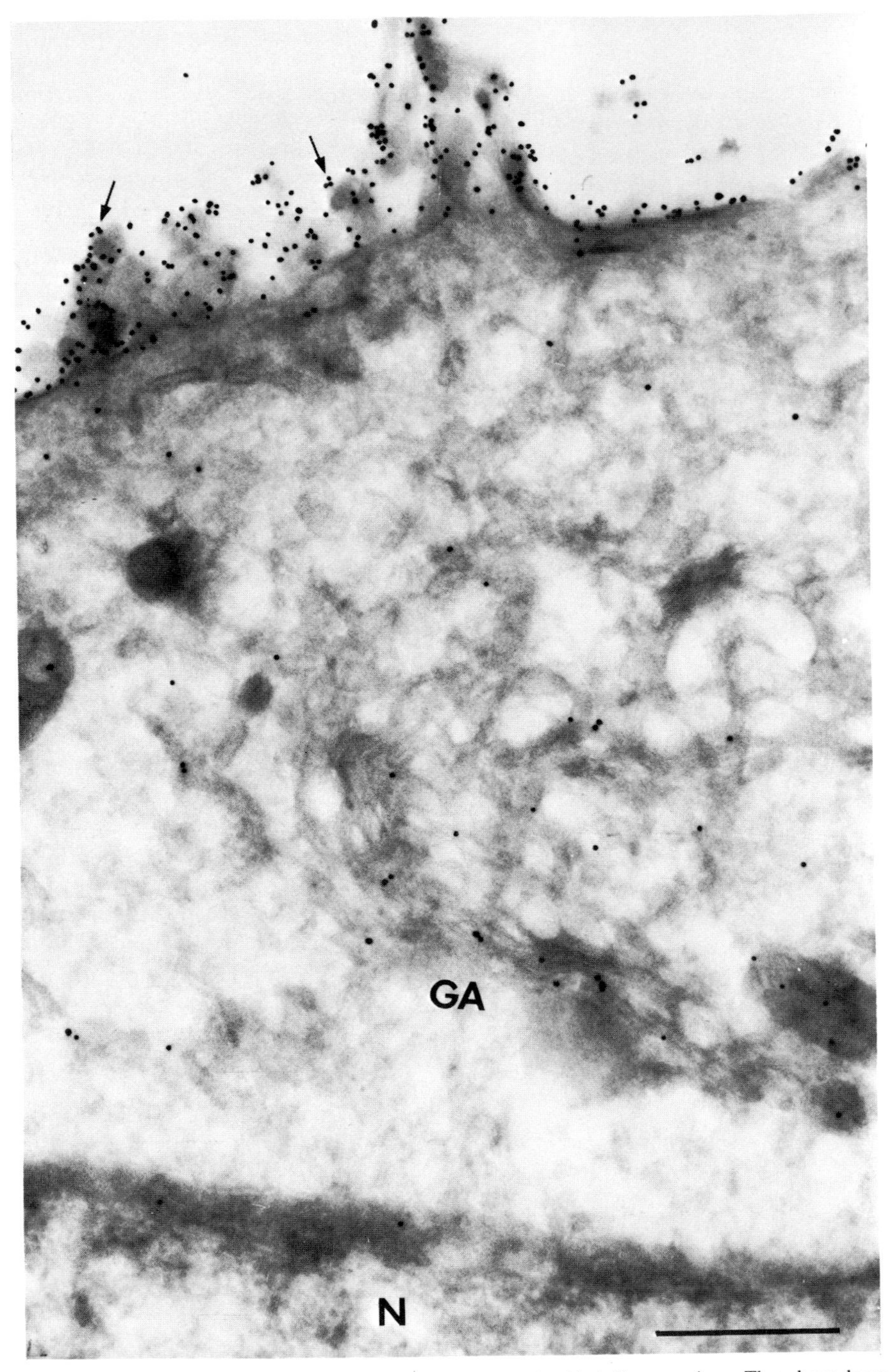


FIG. 4. Frozen thin sections of MDCK cells infected with influenza virus. They have been treated with monoclonal antibodies against the hemagglutinin glycoprotein of influenza virus followed by affinity-purified goat anti-mouse IgG complexed to 18-nm colloidal gold particles. Labeling was observed in the Golgi apparatus (GA) and on the apical plasma membrane where influenza virions (arrows) are found to assemble. N, Nucleus. Bar, 0.5 μm.

working with antibody–gold complexes, the conjugates should be recovered using a minimal sedimentation speed. For example, 18-nm gold particles coupled to IgG are recovered by centrifugation at 12,000 *g* for 1 hour in a Sorvall centrifuge. Subsequently, singlets and doublets can be separated from larger aggregates by performing sucrose gradient centrifugation (Slot and Geuze, 1981). When only a single marker is needed, we have typically applied an indirect procedure using 18-nm gold particles coupled to affinity-purified goat anti-mouse or goat anti-rabbit antibodies. In this manner we were able to examine the intracellular pathway followed by newly synthesized viral glycoproteins in their transport to the surface of MDCK cells. The hemagglutinin of influenza, a virus that assembles exclusively at the apical surface of the polarized cells, was localized to the apical plasma membrane and to the Golgi apparatus of infected cells (Fig. 4). Similarly, the glycoprotein (G) of vesicular stomatitis virus (VSV), which assembles at the basal and lateral surfaces of the cells, accumulated predominantly on the basolateral surfaces (Fig. 1) and was also found intracellularly in the Golgi apparatus. The same viral proteins could be localized in

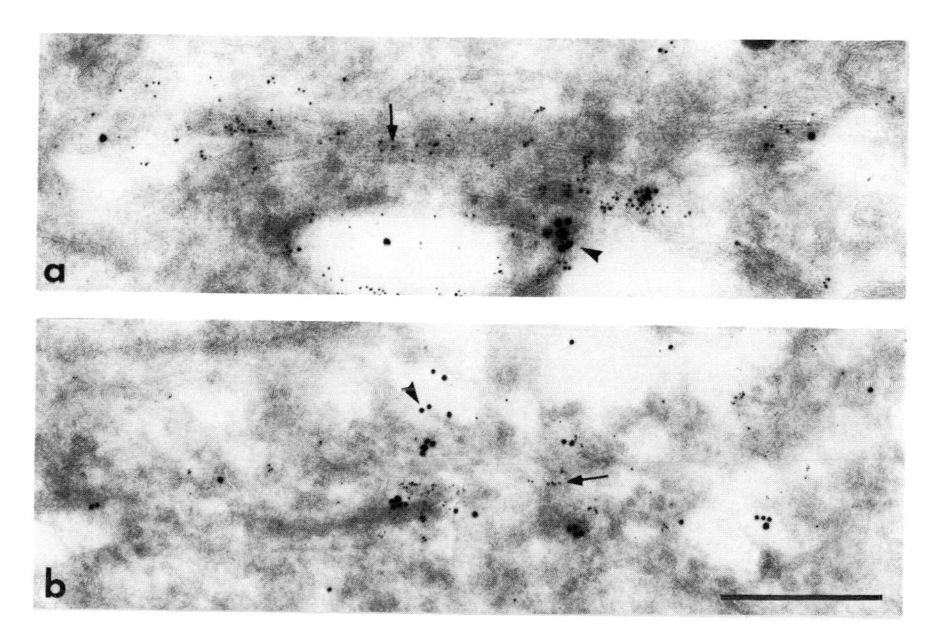


FIG. 5. Frozen thin sections produced from MDCK cell monolayers doubly infected with influenza and VSV. (a and b) Examples of typical Golgi apparatus of these cells manifesting labeling for the influenza hemagglutinin (arrowheads) and the VSV G (arrow) proteins. The sections were treated sequentially with a mouse monoclonal antibody against the influenza HA, affinity-purified goat anti-mouse IgG complexed to 18-nm collodial gold, rabbit anti-VSV G protein, and finally affinity-purified goat anti-rabbit IgG complexed to 5-nm colloidal gold particles. Bar, 0.5 μm.

cells simultaneously infected by the two viruses. To accomplish this, we used to particular advantage the ability to produce colloidal gold of different sizes (Geuze *et al.*, 1981). Colloidal gold conjugates of 5 and 18 nm were complexed with goat anti-rabbit IgG and goat anti-mouse IgG, respectively. Our first antibodies were a rabbit antiserum in the case of the VSV G and a mouse monoclonal antibody against the hemagglutinin of influenza. Thus, we were able to determine that viral glycoproteins specific for different plasma membrane domains passed through the same Golgi apparatus (Fig. 5) en route to their final destinations. These results, made possible by the technique of immunolabeling on ultrathin frozen sections, suggest that the sorting out of apical and basolateral plasma membrane glycoproteins occurs at a very late stage of passage through the Golgi apparatus or takes place after traversal to the Golgi.

## IV. EMBEDDING AND STAINING OF FROZEN THIN SECTIONS

The staining and embedding procedures employed after immunolabeling, which largely determine the level of ultrastructural detail present in electron micrographs of frozen sections, are preceded by a brief treatment with glutaraldehyde (1–2%) in order to stabilize the antigen–antibody linkages. Embedding in a thin film of supporting medium is a necessary step, because sections washed free of sucrose are subject to collapse on the EM grid during the final drying of the specimens. To provide this supporting film, Tokuyasu (1978) introduced methyl cellulose, a medium which is not only water soluble but is also translucent to electrons and relatively resistant to beam damage in the electron microscope. Usually the mounted section is incubated for a variable period of time in a drop of embedding medium which may contain a heavy metal stain solution to enhance contrast. The fluid is then removed by absorption and the section is dried in air. The thickness of the embedding film after drying should be one which yields a silver-golden interference color.

The method of staining chosen depends primarily on the EM marker used and should, of course, allow the unequivocal recognition of the cellular organelles labeled. In general, if ferritin or small (~5 nm) colloidal gold particles are employed, low to medium positive contrast is preferred since a dark background could mask some of the intracellular labeling. If the larger, more dense colloidal gold particles (8–20 nm) are used, then higher positive contrast or even light negative contrast could be utilized. In most recent studies, an absorption method of staining frozen thin sections has been applied (Tokuyasu, 1978). This technique consists of two consecutive steps. The frozen sections are first treated with 2% uranyl acetate at a pH of 7–9, and, after a few short washings, they are incubated again with uranyl acetate under acidic conditions (pH 4) either before or during embedding. The reproducibility of the staining and the nature of the contrast obtained depend critically on the washing step after the uranyl acetate treatment, and are affected by the time of incubation in the drop of embedding medium and by the composition of the medium itself (see below). The results obtained may vary from partial negative contrast (Griffiths *et al.*, 1981, 1982) to positive contrast (Bergmann *et al.*, 1981; Bourguignon and Butman, 1982).

In the course of our work with sections of MDCK cells and rat liver, we introduced some modifications in Tokuyasu's procedures which reproducibly provide a higher level of positive contrast and ultrastructural detail of cellular organelles. An important one is a postfixation step in which sections are incubated with 0.3–1.0% $OsO_4$ in 0.1 *M* sodium cacodylate buffer before embedding. This enhances the positive contrast of the cellular membranes. For very reproducible staining of MDCK cells we have embedded the sections in a mixture of 0.8% methyl cellulose (15 cP), 0.4% PEG (1540), and 0.003% acidic uranyl acetate for 60 minutes before drying. This staining and embedding protocol leads to a very fine appearance of ultrastructure detail (Fig. 6), although the

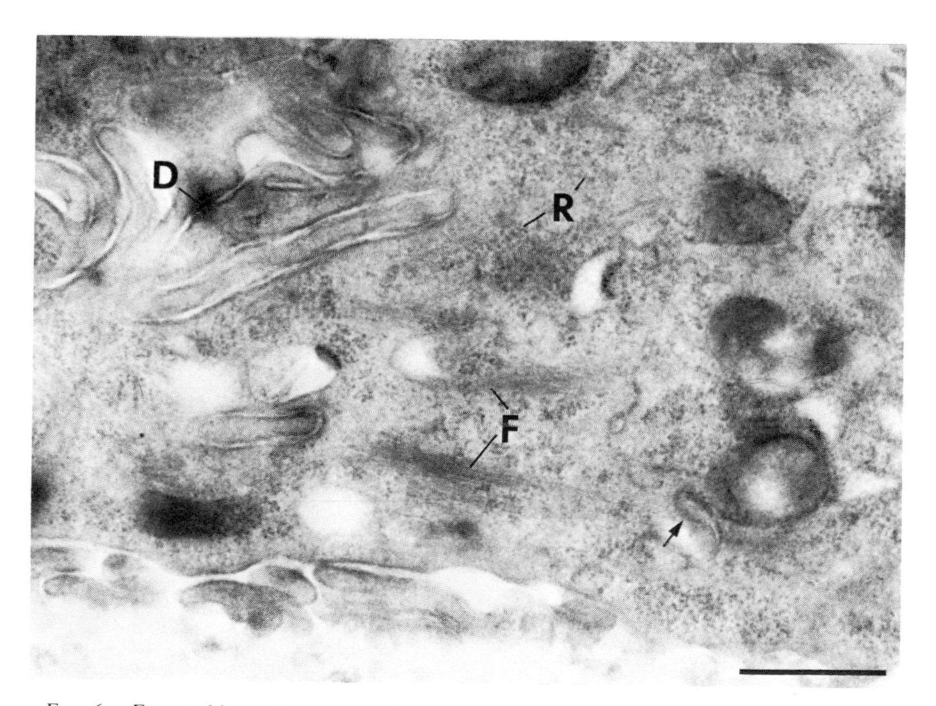


FIG. 6. Frozen thin section of a MDCK cell not exposed to the immunolabeling procedure. Embedding and staining (see text for details) were performed immediately after washing out the sucrose. A segment of the lateral membranes of two adjacent cells is shown to contain a desmosome (D), and filament bundles (F) can be observed in the cytoplasm near the basal surface. Ribosomes can be seen free in the cytoplasm (R) as well as bound to rough endoplasmic reticulum cisternae (arrow). Bar, 0.5 μm.

additional steps required for immunolabeling do tend to partially blur the images of nonmembranous cytoplasmic structures. To achieve the desired level of contrast for a particular tissue, one can regulate the time of incubation and the concentration of uranyl acetate in the embedding medium without being concerned with the variability introduced by the intermediate washing steps in the previously described procedure.

## V. CONCLUDING REMARKS

As mentioned in Section I, intracellular immunolabeling can be carried out not only on ultrathin frozen sections but also on thin sections of fixed material embedded in media such as cross-linked bovine serum albumin (BSA) (McLean and Singer, 1970), polyethylene glycol (Mazurkiewicz and Nakane, 1972), Epon (Bendayan *et al.*, 1980), and methacrylates (Kawarai and Nakane, 1970), including Lowicryl K 4M (Roth *et al.*, 1981). Although the use of plastic embedded material for immunolabeling would be the most convenient approach, there are several caveats associated with it, including the fact that only antigens exposed at the surface of the section or made accessible after etching can be labeled. Dehydration in organic solvents, required for embedding these media, may also affect antigenicity. In this regard, Lowicryl K 4M, a plastic which can be polymerized at low temperature, has been shown in some cases to provide better preservation of antigenicity than Epon (Roth and Berger, 1982). In addition, when metal staining is applied to Epon or Lowicryl sections after immunolabeling, a "negative" appearance of membranes is obtained (Bendayan *et al.*, 1980; Roth *et al.*, 1978), and hence a somewhat more painstaking ultrastructural interpretation may be required.

A singularity that distinguishes the BSA-embedding technique from all of the other procedures is the lack of support for intracellular organelles during drying of the specimen caused by the inability of the albumin to penetrate the cells. As a result, the morphological details of the ultrastructure are difficult to discern (Papermaster *et al.*, 1978). On the other hand, the intracellular antigens within the thickness of the section itself, now not subject to interference from the embedding media, would, in principle, remain more accessible to labeling reagents. However, the harsh fixation conditions required for the cross-linking of extracellular BSA cause a general cross-linking of intracellular materials and in this case also limit antibody access.

The question of whether sucrose-infused cryosections provide for better labeling than sections of specimens embedded in other media may in fact be moot, because preservation of antigenicity and accessibility to labeling probes are greatly affected by the fixation conditions employed, which in turn are a function of the degree of ultrastructural preservation required. A more effective fixation will result in a greater cross-linking of the cellular protein matrix, impairing penetrability to the point that it will not be much greater than in sections of BSA or even plastic-embedded samples. It has been demonstrated that the extent of cross-linking can be reduced by the inclusion of monofunctional imidates in the fixative and that this can lead to a dramatic (severalfold) increase in the amount of observable labeling (Tokuyasu, 1978; Chen and Singer, 1982); however, this procedure also results in a poorer overall ultrastructural appearance.

A recent comparison of immunolabeled specimens obtained from BSA-embedded and frozen sectioned material indicated that more effective labeling was achieved in BSA-embedded sections, but better ultrastructural preservation was obtained in thin frozen sections (Griffiths and Jockusch, 1980). The authors pointed out, however, that the fixation and immunolabeling procedures were not optimized. Preliminary experiments in our laboratory using lectin–colloidal gold conjugates revealed significantly more binding sites in frozen sections of fixed liver tissue than in those obtained from the same material after Lowicryl K 4M embedding (C. DeLemos, personal communication).

In summary, a main advantage of the technique of immunolabeling of thin frozen sections is that it ensures direct accessibility to cellular compartments while still maintaining a reasonably interpretable ultrastructural preservation. There are, of course, drawbacks associated with the technique as well. Admittedly, the quality of the ultrastructural detail of cellular organelles observable in frozen thin sections, especially after the immunolabeling procedure, is not yet comparable to that normally attained with Epon sections, although one might expect the gap to narrow with improvements in the overall technique. Those who wish, therefore, to take advantage of this method must be ready to accept a modicum of deterioration in the ultrastructure for the sake of obtaining valuable functional information from positive immunolabeling. A major drawback of this technique is the difficulty in consistently obtaining sections of high quality. This problem seems to stem from a large thermal gradient between the chamber of the cryoattachment and the rest of the ultramicrotome, a circumstance that creates mechanical instability in the movement of the sample across the knife. Further improvements in available instrumentation may greatly facilitate obtaining good ultrathin frozen sections and contribute to greater use of this approach to intracellular immunolabeling.

ACKNOWLEDGMENTS

We wish to acknowledge the gifts of monoclonal antiserum against HA from Dr. R. G. Webster and to thank Dr. E. Rodriguez-Boulan for the initial participation in the project, Dr. C. DeLemos for providing the biotinylated concanavalin A and helpful comments, Dr. Papadopoulos for reading the text and helpful comments, Mr. G. Davy for technical help, Mr. B. Zietlow and Ms. J. Culkin for photographic assistance, and Ms. S. Martinez for typing the manuscript. This work was supported by NIH Grants GM 20277, AG 01461, AG 00378, and an NIH postdoctoral fellowship to M.J.R.

## REFERENCES

Ainsworth, S. K., and Karnovsky, M. J. (1972). An ultrastructural staining method for enhancing the size and electron opacity of ferritin in thin sections. *J. Histochem. Cytochem.* **20,** 225–229.

Bendayan, M., Roth, J., Perrelet, A., and Orci, L. (1980). Quantitative immunocytochemical localization of pancreatic secretory proteins in subcellular compartments of the rat acinar cell. *J. Histochem. Cytochem.* **28,** 149–160.

Bergmann, J. E., Tokuyasu, K. T., and Singer, S. J. (1981). Passage of an integral membrane protein, the vesicular stomatitis virus glycoprotein through the Golgi apparatus en route to the plasma membrane. *Proc. Natl. Acad. Sci. U.S.A.* **78,** 1746–1750.

Bernhard, W., and Leduc, E. H. (1967). Ultrathin frozen sections. I. Methods and ultrastructural preservation. *J. Cell Biol.* **34,** 757–771.

Bourguignon, L. Y. W., and Butman, B. T. (1982). Intracellular localization of certain membrane glycoproteins in mouse T-Lymphoma cells using immunoferritin staining of ultrathin frozen sections. *J. Cell. Physiol.* **110,** 203–212.

Cereijido, M., Robbins, E. S., Dolan, W. J., Rotunno, C. A., and Sabatini, D. D. (1978). Polarized monolayers formed by epithelial cells on a permeable and translucent support. *J. Cell Biol.* **77,** 853–880.

Chen, W.-T., and Singer, S. J. (1982). Immunoelectronmicroscopic studies of the sites of cell-substratum and cell-cell contacts in cultured fibroblasts. *J. Cell Biol.* **95,** 205–222.

Christensen, A. K. (1970). Preparation of frozen thin sections of fresh tissue for electron microscopy. *Proc. 28th Annu. Meet., Electron Micros. Soc. Am.*, 294.

Christensen, A. K. (1971). Frozen thin sections of fresh tissue for electron microscopy with a description of pancreas and liver. *J. Cell Biol.* **51,** 772–804.

Compans, R. W. (1973). Influenza virus proteins. II. Association with components of cytoplasm. *Virology* **51,** 56–70.

Dutton, A., Tokuyasu, K. T., and Singer, S. J. (1979). Iron-dextran antibody conjugates; A general method for the simultaneous staining of two components in high resolution immunoelectron microscopy. *Proc. Natl. Acad. Sci. U.S.A.* **76,** 3392–3396.

Faulk, W. P., and Taylor, G. M. (1971). An immunocolloid method for the electron microscope. *Immunochemistry* **8,** 1081–1083.

Fernandez-Moran, H. (1952). The submicroscopic organization of vertebrate nerve fibres. An electron microscopic study of myelinated and unmyelinated nerve fibres. *Exp. Cell Res.* **3,** 282–359.

Frens, G. (1973). Controlled nucleation for the regulation of the particle size in monodisperse gold suspensions. *Nature (London), Phys. Sci.* **241,** 20–22.

Geoghegan, W. D., and Ackerman, G. A. (1977). Adsorption of horseradish peroxidase, ovomucoid and anti-immunoglobulin to colloidal gold for the indirect detection of concanavalin A, wheat germ agglutinin and goat antihuman immunoglobulin G on cell surfaces at the electron microscopic level: A new method, theory and application. *J. Histochem. Cytochem.* **25,** 1187–1200.

Geuze, H. J., Slot, J. W., van der Ley, P. A., and Scheffer, R. C. T. (1981). Use of colloidal gold particles in double labelling immunoelectron microscopy of ultrathin frozen sections. *J. Cell Biol.* **89,** 653–665.

Green, J., Griffiths, G., Louvard, D., Quinn, P., and Warren, G. (1981). Passage of viral membrane proteins through the Golgi complex. *J. Mol. Biol.* **152,** 663–698.

Griffiths, G. W., and Jockusch, B. M. (1980). Antibody labeling of thin sections of skeletal muscle with specific antibodies: A comparison of bovine serum albumin (BSA) embedding and ultracryomicrotomy. *J. Histochem. Cytochem.* **28,** 969–978.

Griffiths, G., Warren, G., Stuhlfauth, I., and Jokusch, B. M. (1981). The role of clathrin-coated vesicles in acrosome formation. *Eur. J. Cell Biol.* **26,** 52–60.

Griffiths, G., Brands, R., Burke, B., Louvard, D., and Warren, G. (1982). Viral membrane proteins acquire galactose in trans Golgi cisternae during intracellular transport. *J. Cell Biol.* **95,** 781–792.

Horisberger, M., and Rosset, J. (1977). Colloidal gold, a useful marker for transmission and scanning electronmicroscopy. *J. Histochem. Cytochem.* **25,** 295–305.

Iglesias, R., Bernier, R., and Simard, R. (1971). Ultracryotomy: A routine procedure. *J. Ultrastruc. Res.* **36,** 271–289.

Kawarai, Y., and Nakane, P. K. (1970). Localization of tissue antigens on the ultrathin sections with peroxidase-labeled antibody method. *J. Histochem. Cytochem.* **18,** 161–166.

Kishida, Y., Olsen, B. R., Berg, R. A., and Prockop, D. J. (1975). Two improved methods for preparing ferritin-protein conjugates for electron microscopy. *J. Cell Biol.* **64,** 331–339.

Knipe, D. M., Baltimore, D., and Lodish, H. F. (1977). Separate pathways of maturation of the major structural proteins of vesicular stomatitis virus. *J. Virol.* **21,** 1128–1139.

Kraehenbühl, J. P., and Jamieson, J. D. (1974). Localization of intracellular antigens by immunoelectron microscopy. *Int. Rev. Exp. Pathol.* **13,** 1–53.

Kyte, J. (1976). Immunoferritin determination of the distribution of ($Na^+$ + $K^+$) ATPase over the plasma membranes of renal convoluted tubules. *J. Cell Biol.* **68,** 287–303.

Mazurkiewicz, J. E., and Nakane, P. K. (1972). Light and electron microscopic localization of antigens in tissues embedded in polyethylene glycol with a peroxidase-labeled antibody method. *J. Histochem. Cytochem.* **20,** 969–974.

McLean, J. D., and Singer, S. J. (1970). A general method for the specific staining of intracellular antigens with ferritin-antibody conjugates. *Proc. Natl. Acad. Sci. U.S.A.* **65,** 122–128.

Misfeldt, D. S., Hamamoto, S. T., and Pitelka, D. R. (1976). Transepithelial transport in cell culture. *Proc. Natl. Acad. Sci. U.S.A.* **73,** 1212–1216.

Ohtsuki, I., Manzi, R. M., Palade, G. E., and Jamieson, J. D. (1978). Entry of macromolecular tracers into cells fixed with low concentrations of aldehydes. *Biol. Cell.* **31,** 119–126.

Palade, G. E. (1975). Intracellular aspects of the process of protein synthesis. *Science* **189,** 347–358.

Papermaster, D. S., Schneider, B. G., Zorn, M. A., and Kraehenbühl, J. P. (1978). Immunocytochemical localization of a large intrinsic membrane protein to the incisures and margins of frog rod outer segment disks. *J. Cell Biol.* **78,** 415–425.

Rindler, M. J., Ivanov, I. E., Rodriguez-Boulan, E., and Sabatini, D. D. (1982). Biogenesis of epithelial cell plasma membranes. *Ciba Found. Symp.* **92,** 184–208.

Rodriguez-Boulan, E. J., and Sabatini, D. D. (1978). Asymmetric budding of viruses in epithelial monolayers: A model for the study of epithelial polarity. *Proc. Natl. Acad. Sci. U.S.A.* **75,** 5071–5075.

Rodriguez-Boulan, E. J., and Pendergast, M. (1980). Polarized distribution of viral envelope proteins in the plasma membrane of infected epithelial cells. *Cell* **20,** 45–54.

Romano, E. L., and Romano, M. (1977). Staphylococcal proteins A bound to colloidal gold: A useful reagent to label antigen-antibody sites in electron microscopy. *Immunochemistry* **14,** 711–715.

Roth, J., and Berger, E. G. (1982). Immunocytochemical localization of galactosyl transferase in HeLa cells: Codistribution with thiamine pyrophosphatase in trans-Golgi cisternae. *J. Cell Biol.* **92,** 223–229.

Roth, J., and Binder, M. (1978). Colloidal gold, ferritin, and peroxidase as markers for electron microscopic double labeling lectin techniques. *J. Histochem. Cytochem.* **26,** 163–169.

Roth, J., Bendayan, M., and Orci, L. (1978). Ultrastructural localization of intracellular antigens by the use of the protein A-gold complex. *J. Histochem. Cytochem.* **26,** 1074–1081.

Roth, J., Bendayan, M. Carlemalm, E., Villiger, W., and Garavito, M. (1981). Enhancement of structural preservation and immunocytochemical staining in low temperature embedded pancreatic tissue. *J. Histochem. Cytochem.* **29,** 663–671.

Sabatini, D. D., Kreibich, G., Morimoto, T., and Adesnik, M. (1982). Mechanisms for the incorporation of proteins in membranes and organelles. *J. Cell Biol.* **92,** 1–22.

Slot, J. W., and Geuze, H. J. (1981). Sizing of protein A-colloidal gold probes for immunoelectron microscopy. *J. Cell Biol.* **90,** 533–536.

Tokuyasu, K. T. (1973). A technique for ultracryotomy of cell suspensions and tissues. *J. Cell Biol.* **57,** 551–565.

Tokuyasu, K. T., and Singer, S. J. (1976). Improved procedures for immunoferritin labeling of ultrathin frozen sections. *J. Cell Biol.* **71,** 894–906.

Tokuyasu, K. T. (1978). A study of positive staining of ultrathin frozen sections. *J. Ultrastruct. Res.* **63,** 287–307.

Tokuyasu, K. T. (1980). Immunocytochemistry on ultrathin frozen sections. *Histochem. J.* **12,** 381–403.

Weber, K., Rathke, P. C., and Osborn, M. (1978). Cytoplasmic microtubular images in glutaraldehyde-fixed tissue culture cells by electron microscopy and by immunofluorescence microscopy. *Proc. Natl. Acad. Sci. U.S.A.* **75,** 1820–1824.

Willingham, M. C., Yamada, S. S., and Pastan, I. (1978). Ultrastructural antibody localization of $\alpha_2$-macroglobulin in membrane-limited vesicles in cultured cells. *Proc. Natl. Acad. Sci. U.S.A.* **75,** 4359–4363.

# Chapter 12

# Development of Antibodies to Apical Membrane Constituents Associated with the Action of Vasopressin

*JAMES B. WADE,[1] VICTORIA GUCKIAN,[1] AND INGEBORG KOEPPEN[2]*

*Department of Physiology*
*Yale University School of Medicine*
*New Haven, Connecticut*



[1]Present address: Department of Physiology, University of Maryland School of Medicine, Baltimore, Maryland.

[2]Present address: Anatomisches Institut der Universität Heidelberg, Heidelberg, Federal Republic of Germany.




ISBN 0-12-153320-4

## I. INTRODUCTION

Although immunological characterization has been widely utilized by immunologists and cell biologists to characterize and isolate plasma membrane proteins, application of this approach to epithelial transport problems has been more limited. Recent reports, however, point to some of the potential applications made possible by the technical advances that have been made in immunological methods (5, 11, 12, 20, 33, 36). It is clear that if a specific protein is isolated and purified, antibodies can be raised to localize the protein (8, 14, 39, 43). Alternatively, antibodies specifically directed against a membrane protein of interest can be utilized not only for localization studies but also for affinity isolation of the membrane or specific protein of interest (5, 28, 32). The difficulty in utilizing this approach to study constituents associated with the action of vasopressin is that neither isolated proteins nor specific antibodies are yet available. Indeed, there are a large number of important transport proteins for which this "catch-22" applies. Since in many cases these proteins probably occur in relatively low abundance along with a large number of contaminating constituents, the prospects for identification and purification are extremely bleak unless a specific ligand such as an antibody is developed.

Two strategies have developed in recent years that allow development of specific antibodies even when starting with an impure antigen. The key to the first approach is the use of adsorption steps to separate the desired antibodies from antibodies to contaminants (31). A second approach is to use the hybridoma technique of Kohler and Milstein (29). By fusing spleen cells of an immunized animal with a myeloma line, antibody-producing cell lines can be selected and cloned that produce antibodies with the desired specificity.

With either approach using an impure antigen, the biggest problem is to develop an antibody-screening system that is rapid but will allow selection of those antibodies specific for the transport system of interest. The option of screening for antibodies that inhibit transport is not a suitable approach in most cases for several reasons. First, because a large number of antibody preparations must be screened at multiple dilutions, most transport assays take too much time and material to serve as an initial screening assay. Second, valuable antibodies might be missed which bind to the protein of interest without inhibiting function. It is even possible that noninhibitory antibodies in a polyclonal antiserum may bind in such a way that access by inhibitory antibodies would be blocked so that their presence would remain undetected. For these reasons, initial screening assays that have been used simply measure binding of antibody to antigen. The key to designing assays for binding to specific transport systems is to exploit what is known about the transport system in the design of the assay. Our laboratory has attempted to take advantage of recent insights into the action of vasopressin to design a screening system that could detect antibodies to apical membrane constituents involved in the action of vasopressin.

## II. VASOPRESSIN ACTION

### A. The Role of Intramembrane Particle Aggregates

Since the initial descriptions by freeze-fracture electron microscopy of distinctive intramembrane particle aggregates in the apical membrane of the amphibian urinary bladder (7, 23, 25), a large number of studies from a number of independent laboratories have shown a remarkable correlation between the incidence of these structures and the water permeability response of the bladder to vasopressin (2, 9, 24, 26). Structures similar to aggregates have also been associated with hormone action in the toad skin (3) and mammalian collecting duct (17).

Although instances have been reported of quantitative discrepancies between the incidence of aggregates and measurements of water flow (9, 26, 30), these observations can be reasonably explained by the difficulty of obtaining accurate measurements of water permeability over short time periods when permeability is changing rapidly, and by the existence of an additional barrier to water flow in series with the apical membrane which can become limiting in unstretched vasopressin-treated bladders (28, 30). Recent observations indicating that cytochalasin B may have no effect on vasopressin's action in the absence of an osmotic gradient have been interpreted as indicating a dissociation between water permeability and the incidence of the aggregates (16). However, inhibition of aggregate appearance in the apical membrane by cytochalasin B has only been reported in the presence of an osmotic gradient (27), a condition in which all authors have found a striking inhibition of the water permeability response (4, 16, 27, 47).

The large number of studies correlating the incidence of aggregates with the water permeability response has led to the hypothesis that the aggregates may be the site of water channels. However, it should be kept in mind that the present evidence for this hypothesis rests on the correlation between the presence of aggregates and water permeability of the bladder. There is no direct evidence that the aggregates are the site of the water channels. Thus, it is possible to imagine that aggregates have some other role in the vasopressin response and that the water channels occur at another site. Antibody work might lead to direct evidence for the role of the aggregates in two ways: (1) if antibodies can be developed which specifically block water flow; and (2) if the protein(s) of the aggregates can be identified and affinity purified utilizing antibodies, reconstitution of the structure in phospholipid vesicles or planar bilayer membranes could provide the direct evidence for their function.

## B. Evidence for a Membrane Shuttle System

A major difficulty in early studies of the aggregates was a failure to recognize a possible functional rationale for the occurrence of these specific structures and the segregation of water channels into these limited domains of the apical membrane. In 1977, several laboratories independently recognized that the aggregates occur in the membrane of specific tubular vacuoles present in the cytoplasm of control bladders (21, 48). This observation led to the proposal that a specific "membrane shuttle" may be involved in the action of vasopressin (49). The physiological significance of this model is that a specialized cytoplasmic membrane with a high density of water channels would have no effect on epithelial permeability in the absence of hormone. However, if vasopressin leads to fusion of these "shuttle" membranes with the apical membrane, an insertion of a relatively small amount of membrane with aggregates could introduce enough water channels to lead to the dramatic increase in water permeability elicited by vasopressin.

In recent years a number of observations have been made that support the membrane shuttle hypothesis. First, vasopressin causes a significant reduction in the incidence of tubular vacuoles with aggregates in the cytoplasm as would be expected if the structures are transferred to the apical membrane (49). A second line of evidence for such a shuttle mechanism has been developed by Muller *et al.* (37). They have described deep invaginations which occur when cytoplasmic tubular vacuoles have fused with the apical membrane. In a number of conditions, there is a close correlation between the incidence of these structures, which Muller *et al.* have called "fusion events," and the incidence of aggregates in the apical membrane (37). Although these observations are support for a membrane shuttle mechanism, the exact significance of the fusion events is not certain. The finding of fusion events in control bladders (18) and the fact that they are as numerous 60 minutes after vasopressin addition as at 10 minutes (37) suggest that this structural feature may not be strictly associated with initiation of the water permeability change. Muller *et al.* have argued that fusion events are relatively static configurations, but the observation that tubular-shaped vacuoles become labeled by horseradish peroxidase or ferritin present in the luminal bath of vasopressin-treated bladders suggests that there may be some ongoing turnover of apical membrane material (15, 34, 51). Unfortunately, it is not technically possible to determine with current techniques if internalized peroxidase is located in vacuoles with aggregates. By localization of antibodies to the aggregates in vasopressin-treated tissue that has internalized peroxidase present in the bath, it should be possible to determine what fraction (if any) of the peroxidase-labeled vesicles contains aggregate material.

Similar studies could be useful in assessing the fate of aggregate material following removal of vasopressin from a stimulated bladder. While it is clear that the number of structures identifiable as aggregates decreases dramatically during reversal of the response (26, 40), it is not clear if aggregate material is dispersed in the apical membrane as has been suggested (26, 37) or internalized as intact aggregates via shuttle membranes. It is clear from peroxidase-labeling studies that a large amount of membrane is internalized during reversal (51), but we cannot be certain that this phenomenon necessarily involves aggregates.

Thus, though there is important structural evidence that a membrane shuttle mechanism is involved with the action of vasopressin, there is considerable ambiguity as to the precise characteristics of the mechanism. Antibodies could be extremely helpful in resolving these issues by providing a marker for the aggregates other than their structure as visualized by freeze-fracture electron microscopy.

There are also nonstructural observations that are consistent with the concept that addition of membrane to the apical surface has a role in the water permeability response. A number of laboratories have independently demonstrated an increase in capacitance with vasopressin (38, 45, 52). The most reasonable interpretation of these observations is that they reflect an increase in apical membrane area. We have shown that the increase in capacitance is blocked by methohexital (45), which is known to selectively inhibit the water permeability response and insertion of aggregates (24). However, since a 20–40% increase in capacitance is found, we cannot yet be certain that insertion of the aggregates, which represent only about 1% of the apical membrane (25), is the only factor responsible for the capacitance change. While it is possible that a large area of carrier membrane is inserted along with the aggregates, we must also keep in mind the possibility that membrane other than that associated with the aggregate system may fuse with the plasmalemma in response to vasopressin (15, 35, 46).

One basic prediction of the membrane shuttle hypothesis is that specific proteins are added to the apical membrane by vasopressin. While Scott and Slatin have reported a labeling of several apical membrane proteins associated with the hydroosmotic response (44), a similar study by Rodriguez and Edelman failed to detect any effect of vasopressin (41).

Taken together, a variety of evidence indicates that a membrane shuttle system may be involved in the toad bladder's response to vasopressin, but the evidence is far from compelling and important characteristics of the postulated mechanism are poorly defined.

## III. ANTIBODY ASSAY STRATEGY

### A. Overall Strategy

The membrane shuttle hypothesis suggests an approach to overcoming the fact that the constituents of the aggregates have not been identified or isolated. The hypothesis suggests that antibodies to apical membrane constituents inserted in response to vasopressin would not bind to control bladders from the luminal side but would bind to hormone-stimulated bladders. Antibodies to proteins not inserted in response to hormone can be expected to bind to the apical surface of both control and vasopressin-stimulated bladders. Even though the crude antigen preparations we have used also raise antibodies to nonapical membrane constituents, the large luminal surface of the intact bladder is used in our basic assay in such a way that only antibodies to this surface are evaluated. Thus, our basic screening approach is to examine binding of an antiserum to the apical surface of vasopressin-stimulated bladders compared to the binding of the same antiserum to control bladders. A major problem with this simple approach is that very immunogenic constituents present on the surface of control bladders result in such a high level of binding that vasopressin-specific antibodies are difficult to detect. Even if vasopressin-specific antibodies are detected, the presence of other antibodies severely limits the usefulness of such an antiserum. Fortunately, the fact that we have direct access to the apical surface of the toad urinary bladder means that control bladders can be used as an adsorbant to extract from antisera those antibodies that bind in the absence of vasopressin. When such adsorbed antisera are tested for binding to bladders, the capacity of our assay to detect antibodies to vasopressin-associated constituents is greatly increased.

## B. Preparation of Antigen

### 1. Isolated Toad Bladder Cells

Our isolation procedure basically follows that of Rodriguez *et al.* (42) except that toads are perfused with Ringer's via the ventricle to remove blood from the bladders and we mount the hemibladders as sacs on plastic tubes. Bladders are bathed in a standard aerated amphibian Ringer's solution (111 m*M* NaCl, 3.5 m*M* KCl, 2.5 m*M* $NaHCO_3$, 1.0 m*M* $CaCl_2$, pH 7.6–8.2) on both sides. After a 30-minute equilibration period, vasopressin (20 mU/ml) is added to the serosal side for 30 minutes. To maintain vasopressin-induced changes during cell isolation, bladders are lightly fixed by addition of glutaraldehyde (0.1% final concentration) to the mucosal surface for 1 minute. After washing glutaraldehyde from the mucosal side, bladders are scraped with a glass slide to remove the epithelium from the submucosa. The epithelial preparation is incubated for 10 minutes in Ringer's containing 0.1 m*M* CaCl and 0.15% collagenase. The resulting epithelial fragments are then resuspended in Ringer's containing 2 m*M* EDTA for 10 minutes. Cells and remaining fragments are then transferred to Ringer's containing 0.1 m*M* calcium to remove excess EDTA. Fragments are further incubated in Ringer's containing 0.1 m*M* calcium, 0.15% collagenase, and 133 U DNase/ml, and shearing forces are applied by drawing the suspension through 2.5-mm-diameter PE tubing into a 50-ml syringe. The supernatant of this solution contains mostly separated cells, although occasional groups of 2–6 cells also occur. The cells are fixed for an additional 4 minutes with 0.1% glutaraldehyde. Animals are immunized with $10^7$ cells in complete Freund's adjuvant.

### 2. Isolated Plasma Membrane

Our basic plasma membrane preparation is a combination (with slight modifications) of the approaches utilized by Hays *et al.* (19) and Chase and Al-Awqati (6). Bladders are blotted with filter paper to remove mucus, stretched on a Lucite sheet, and scraped with a glass slide. Scrapings are centrifuged and homogenized in a solution of 5.7% sucrose, 5 m*M* Tris, 1 m*M* EDTA, 1.0 m*M* $NaHCO_3$, and 30 μ*M* PMSF (phenylmethylsulfonyl fluoride). For every gram of tissue, 4 ml of homogenizing solution is added. A glass Dounce homogenizer with 11 strokes of a tight-fitting pestle is used for an initial homogenization. Since this homogenization does not break all cells, the pellet from a 1500 *g* spin (5 minutes in a Sorvall RC 5B) is rehomogenized. Following another spin at 1500 *g*, this pellet, which contains the very large membrane envelopes described by Hays *et al.* (19) and remaining unbroken cells, is spun at 170 *g* for 10 minutes. Most of these very large plasma membrane vesicles stay in the supernatant at this step and the unbroken cells are again in the pellet. Even greater yield can be obtained by additional homogenization of this pellet and repeating the centrifugation. The large membrane envelope fractions (i.e., material that pellets at 1500 *g* but not at 170 *g*) are combined, vortexed vigorously, and washed using spins at 2500 *g* (10 minutes) to remove mitochondria. This procedure provides the greatest yield of plasma membrane of any procedure we have examined. Our structural examination indicates that the plasma membrane recovered in this fraction is broken into large apical and basolateral membrane vesicles. The major contaminants of this fraction are nuclei and intracellular vesicles that remain trapped within the large envelopes.

Further purification and separation of apical from basolateral membrane is achieved by forcing the plasma membrane envelopes five times through a 22-gauge needle to produce smaller and more homogeneously sized vesicles and by running a sucrose step gradient, of 20, 40, and 60% (w/w) steps, at 65,000 *g* for 2 hours. An enriched preparation of apical membrane vesicles is recovered at the 40/60 interface as indicated both by studies with iodinated apical membranes (Chase and Al-Awqati, personal communication) and by our own structural examination of these fractions.

## C. Assay Procedures

### 1. Intact Bladder Assay

As in preparing cells and membranes, toads are first perfused with Ringer's via the ventricle to reduce contamination of the bladder with red cells (in this case it is necessary because an excess of red blood cells can provide a high background in our horseradish peroxidase-based assay). Hemibladders are mounted as sacs on tubes and bathed in Ringer's. Assays are always run with paired control and vasopressin-exposed (20 mU/ml) bladders that have been fixed for 5 minutes in 1% glutaraldehyde buffered by 0.1 *M* sodium cacodylate, pH 7.4. Washed bladders are removed from the tubes and stretched, luminal side up, across a Plexiglas sheet with pins in it to hold the bladder in place. Another Plexiglas sheet (0.8 cm thick) with 24 holes (7 mm diameter) is placed on top of the bladder to form the wells for the ELISA, with silicone sealing each well from those adjacent. Care is taken not to damage the bladder during mounting or to allow folds which increase nonspecific binding to the bladder. We have found that background is further decreased and specific binding increased by a Mucomyst treatment of the bladder surface (20% *N*-acetyl-L-cysteine in Ringer's for 4 hours). Wells are blocked with 3% nonimmune goat serum, 100 m*M* glycine in Ringer's for 30 minutes. Antisera and preimmune serum are diluted 1:500–1:10,000 with 1% goat serum in Ringer's and incubated in the wells (in duplicate) for 1 hour. Following five washes with Ringer's and an additional blocking with goat serum, biotinylated goat anti-rabbit serum (Vector Laboratories) diluted 1:1000 is added to the wells for 1 hour. After additional washing and blocking, ABC (avidin–biotin complex prepared from Avidin DH and biotinylated horseradish peroxidase H) diluted 1:10 is added to the wells for 1 hour. Following one more series of washes, the assay is completed by addition of substrate [20 mg *o*-phenylenediamine (OPD) in 50 ml citric acid–phosphate buffer, pH 5.0, with 20 μl $H_2O_2$ (30%) added immediately before use]. The reaction is stopped and color intensified by addition of 50 μl of 6 *N* HCl to each well. For quantitation, samples are read in a Gilford spectrophotometer with ELISA aspirating microcuvette at 492 nm.

### 2. Membrane Fractions

A more standard ELISA protocol utilizing microtiter plates is used to evaluate membrane fractions and solubilized proteins. Wells are coated with 5 μg of antigen overnight at 4°C. Wells are washed and the assay is carried out as above. For standardization of the assay we coat wells overnight with goat anti-rabbit antibody. Dilutions of affinity-purified antibody are used to prepare a standard curve.

3. ADSORPTION PROCEDURE

In order for polyclonal antisera to be useful in our studies, it is essential that antibodies to constituents expressed on the apical surface in non-vasopressin-treated bladders be removed from the antisera. Bladders are prepared as for the ELISA test described above except that larger wells (6 × 36 mm) are used to allow adsorption of 1 ml of serum. Antiserum is adsorbed at room temperature for 1-hour periods. We have found that adsorption in 9–12 wells effectively reduces antibody binding to the level of nonimmune serum when tested against control bladders.

# IV. CHARACTERIZATION OF ANTIBODIES

## A. Enzyme-Linked Immunosorbent Assay

As shown in Fig. 1, rabbits develop a good titer of anti-apical-membrane antibodies. After two boosting immunizations, good ELISA readings are obtained with 1:10,000 dilutions of antiserum. Although not yet as completely studied, we have also gotten good antibody responses in mice. The critical issue, however, is whether these animals have antibodies to constituents present in the

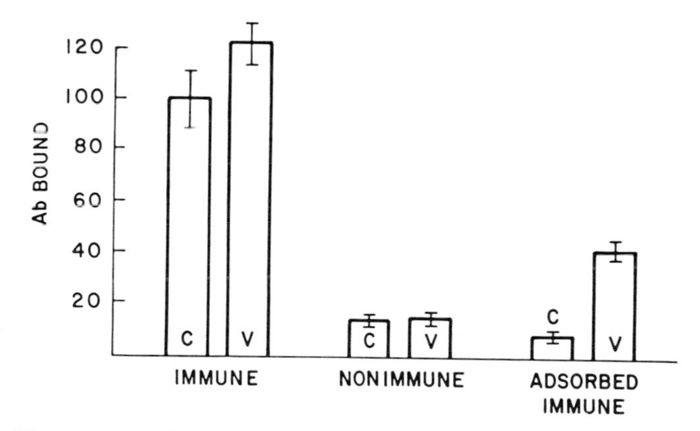


FIG. 1. Measurements of antibody binding to the apical surface of the toad urinary bladder by ELISA. Values from standard curves relating the optical density (OD) of the OPD reaction product at 492 nm to IgG bound have been normalized to 100% for the binding of the rabbit immune serum to control bladders. Antiserum is from a rabbit immunized with cells from vasopressin-treated bladders. For each assay, paired control (C) and vasopressin-stimulated (V) (20 mU/ml) bladders are evaluated for binding by the immune, nonimmune, and immune serum adsorbed 10 times to control bladders. Adsorption effectively reduces binding to the control bladder to background levels, but the same adsorbed antiserum is positive on the vasopressin-treated bladder.

apical membrane only after vasopressin stimulation. Even though more antibody is bound to vasopressin-stimulated bladders than to control bladders, it is clear that the high level of antibody binding to control constituents obscures any effect of vasopressin. We have found that repeated incubation of diluted serum in control wells reduces binding to control bladders down to background levels. When adsorbed antiserum is tested on vasopressin-treated bladders, a very positive ELISA is observed (Fig. 1). This result indicates that we have raised antibodies to contituents added to the apical membrane in the vasopressin response ($AVP^+$ antibodies). It is unlikely that this result could be due to a vasopressin-induced change in geometry (e.g., general increased membrane area or increased nonspecific trapping in the carrier membrane tubules) because (1) background binding by nonimmune serum is not altered by vasopressin, and (2) antisera adsorbed by vasopressin-stimulated bladders have similar binding to control and vasopressin-treated bladders. We cannot be certain that $AVP^+$ antigens are completely absent from the apical surface of control bladders, but their incidence (or conceivably the antibody affinity for them) must be much lower in control than in vasopressin-stimulated bladders.

These antisera have also been evaluated by the more traditional microtiter plate ELISA assay using isolated plasma membrane vesicles from the toad bladder. The antiserum raised against glutaraldehyde-fixed material is strongly positive even against unfixed vesicles. Thus while immunizing with fixed material may be desirable for structural localization studies, it does not preclude use of the antibodies for other purposes.

## B. Immunofluorescence Localization

The $AVP^+$ antibodies might be directed specifically against the aggregates, another component of the shuttle membrane, or possibly another component (e.g., sodium channels) that might be inserted into the membrane in response to vasopressin. If directed against the aggregates the antibody should localize in small patches on the surface of vasopressin-stimulated granular cells. Indeed, epifluorescence microscopy of adsorbed antibody visualized with FITC-conjugated avidin shows patches of label on granular cells associated with vasopressin treatment (Figs. 2 and 3). We are not certain that these patches represent labeling of individual aggregates. Since we have found that aggregates occur in groups, these patches of fluorescence may represent these groups. The apical surfaces of the mitochondria-rich cells are also labeled but this localization does not appear to be altered by hormone. Apparently, a component of mitochondria-rich cells is rather immunogenic, but this cell type is not present in sufficient numbers to remove these antibodies in the number of adsorption periods used.

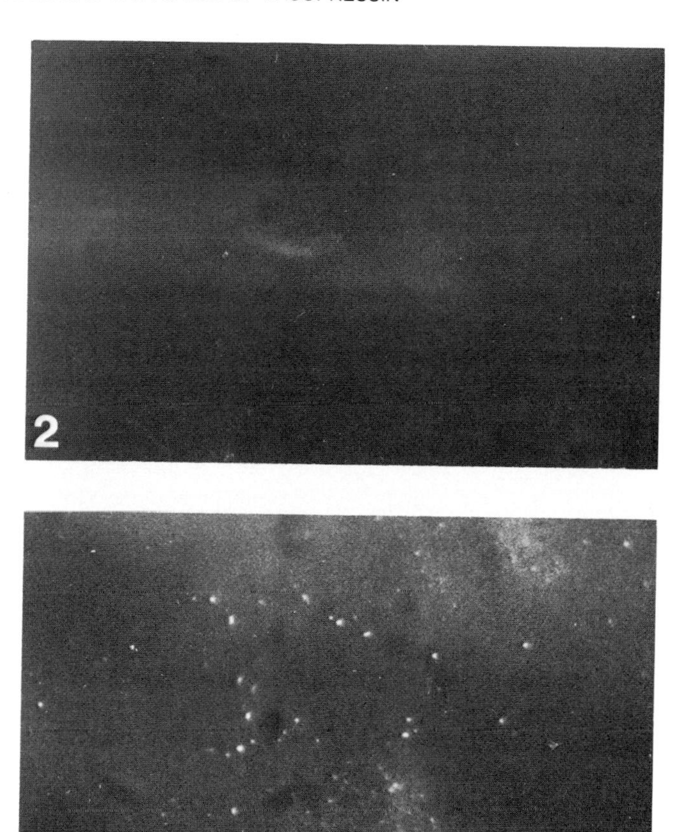


FIGS. 2 AND 3. Localization of adsorbed antiserum on the apical surface of the bladder by epifluorescence microscopy. Adsorbed antiserum and biotinylated anti-rabbit IgG were added as above and antibody was visualized by fluorescein–avidin conjugate. Very little label is visualized looking down on the surface of the control bladder (Fig. 2). Bright patches of fluorescence are observed on the surface of granular cells of the paired vasopressin-treated bladder (Fig. 3). ×1600.

## C. Electron Microscopic Localization

Localization of these antibodies by peroxidase-labeled avidin–biotin complex shows that unadsorbed antibody binds uniformly to the luminal membrane including the membrane associated with vasopressin-induced fusion events (arrow, Fig. 4). While labeling is lighter when adsorbed antibody is localized on vas-

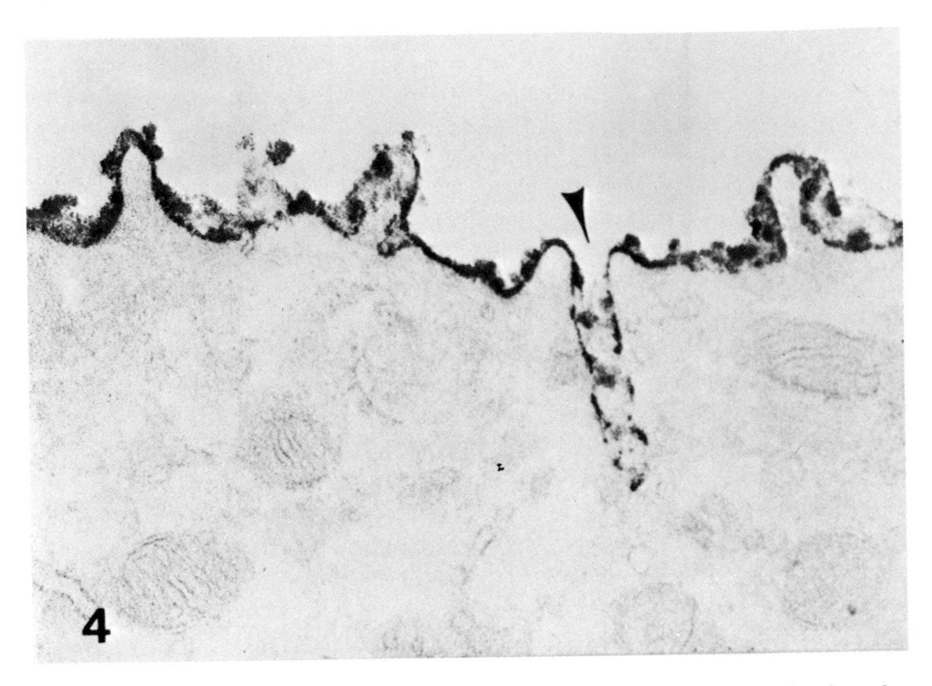


FIG. 4. Ultrastructural localization of unadsorbed antiserum added to the mucosal surface of a vasopressin-stimulated bladder. A 3,3′-diaminobenzidine (DAB) reaction was carried out before embedding. Label occurs uniformly along the apical surface, including the membrane of deep tubular membrane structures that have been called "fusion events" (arrow). ×58,000.

opressin-treated bladders, distinct patches approximately the area of aggregates (Fig. 5) and the membrane of fusion events (Fig. 6) are labeled. This labeling pattern is not found for control bladders, but in some experiments the tips of microvilli have label in both control and vasopressin-treated samples. This labeling of microvilli seems to be due to incomplete adsorption to control bladders. These preliminary observations indicate that the antisera may have antibodies directed against the aggregates, although it is also possible that they are directed against another membrane component associated with the vasopressin response.

While the approaches described above provide a means of characterizing those antibodies directed against apical surface constituents, they give no hint of the presence of additional antibodies to other bladder constituents. In addition, one would like to know if our putative antiaggregate antibodies label the cytoplasmic tubular vacuoles along with aggregates that are present in control bladders. To deal with those issues with the resolution required, we have evaluated procedures for localization of antibodies on thin sections using 5-nm colloidal gold coated with anti-rabbit IgG (10, 13, 14). For this work we have embedded bladders in the low-temperature, polar medium Lowicryl K 4M system which has recently been developed (1). Our experience is that this approach allows preservation of antigenic determinates with low nonspecific binding and good ultrastructural detail. We have found that addition of 0.1% Triton X-100 to antibody and wash solutions nearly eliminates nonspecific binding.

As might be expected when whole cells are used as antigen, these antisera label a host of cellular structures. Thus, such polyclonal antisera, while demonstrating that it is possible to raise antibodies to membrane constituents associated with the action of vasopressin, have limited usefulness.

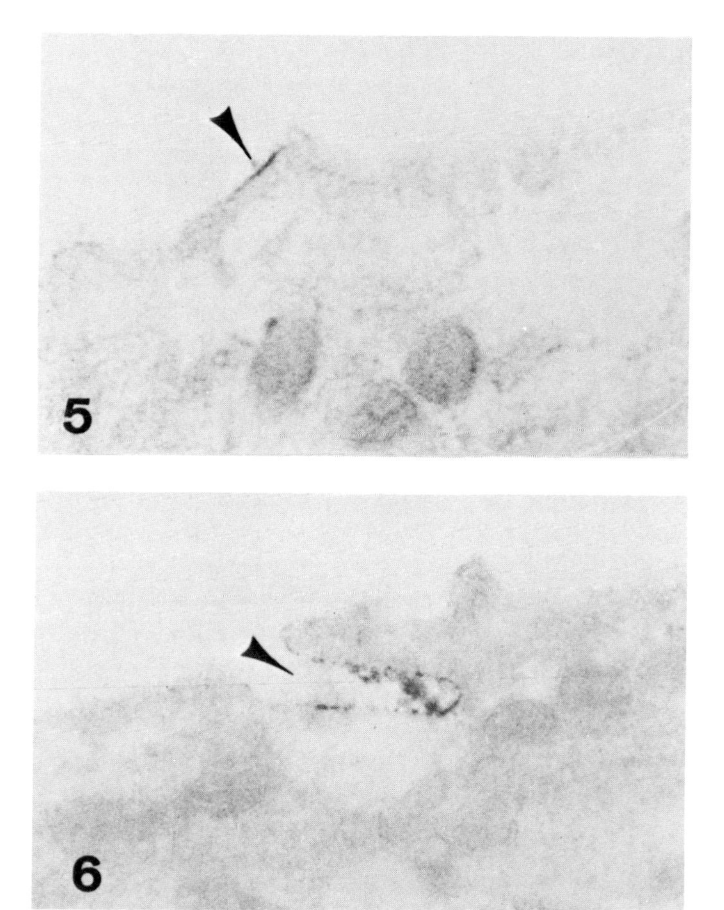


FIGS. 5 AND 6. Ultrastructural localization of adsorbed antiserum on the apical surface of a vasopressin-stimulated bladder. Label occurs in patches on the surface (arrow, Fig. 5) and associated with fusion events (arrow, Fig. 6). ×58,000.

## D. Antibodies to Isolated Apical Membrane

The value of our antibody for labeling apical and cytoplasmic membrane structures would clearly be greater if we could immunize with pure apical membrane containing aggregates. Wc have independently confirmed the finding of Hays *et al.* (19) that aggregates are retained in isolated membranes from vasopressin-treated bladders. Hays *et al.* used a bifunctional imidoester [dithiobis (succinimidyl) propionimidate, DTBP] with the idea of stabilizing the structure, but our observations indicate that DTBP may not be necessary, since aggregates can be identified in membranes isolated from vasopressin-treated bladders without such treatment. Although the possibility that DTBP and glutaraldehyde may increase the incidence of such structures has not been investigated, our observations indicate that the aggregates may be remarkably stable structures and that the distinctive features of the apical membrane are maintained following homogenization and sucrose density centrifugation.

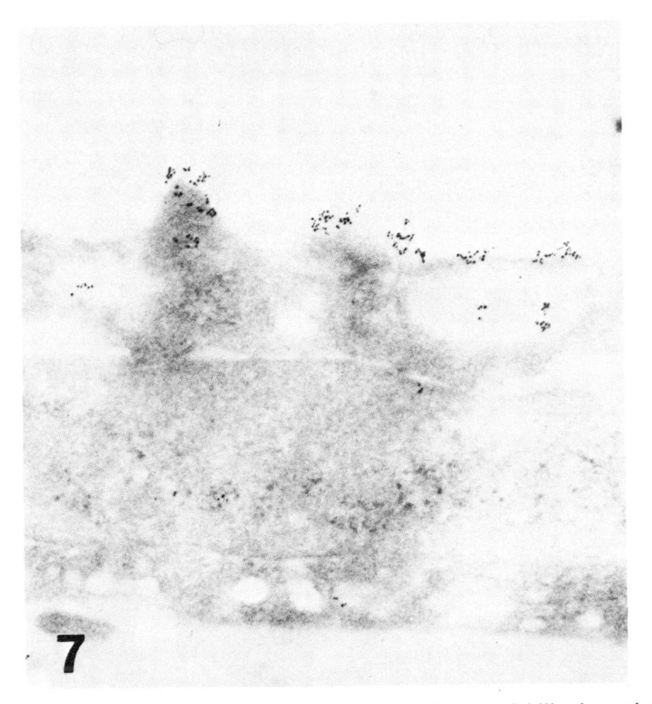


FIG. 7. Localization of rabbit antiserum raised to deoxycholate-solubilized membrane vesicles using colloidal gold. Although in this case antiserum is added to the section, antibody is predominantly localized on the apical membrane. This indicates that apical membrane components of the vesicle fraction used are an effective antigen following detergent solubilization. ×47,000.

We have immunized with detergent-solubilized membranes and obtained antibodies that localize largely to the apical surface in Lowicryl sections (Fig. 7). Not only does this result confirm the expected usefulness of membrane fractions for reducing the number of antibodies to contaminants, but it also shows that antibodies to unfixed membrane protein can be used for localization studies on glutaraldehyde-fixed cells.

## V. CONCLUDING REMARKS

There are a large number of important epithelial transport systems that have not been identified and purified. Current immunological techniques offer a powerful approach to these problems, provided a rapid and specific assay for specific transport systems can be developed. Utilizing recent insights into the action of vasopressin, we have developed antibodies to constituents added to the apical membrane of hormone-stimulated toad bladder. While useful in demonstrating the feasibility of our screening approach, immunizing with whole cells produces too many antibodies to non-apical-membrane constituents to be useful. Immunization with isolated membrane, however, produces antisera with greater specificity toward the apical membrane. The combination of immunological methods with biochemical and structural analysis offers a promising approach to understanding the physiological regulation of epithelial transport.

ACKNOWLEDGMENTS

We gratefully acknowledge the valuable assistance of Thomas Ardito in developing the colloidal gold localization and the extremely helpful discussions with G. Ojakian, M. Kashgarian, and D. Biemesderfer. This work was supported by NIH Grant AM 17433. I. Koeppen was supported by Grant Ko 784 from the Deutsche Forschungsgemeinschaft.

REFERENCES

1. Armbruster, B. L., Carlemalm, E., Chiovetti, R., Garavito, R. M., Hobot, J. A., Kellenberger, E., and Villiger, W. (1982). Specimen preparation using low temperature embedding resins. *J. Microsc.* **126,** 77–85.
2. Bourguet, J., Chevalier, J., and Hugon, J. S. (1976). Alterations in membrane-associated particle distribution during antidiuretic challenge in frog urinary bladder epithelium. *Biophys. J.* **16,** 627–639.
3. Brown, D., Grosso, A., and DeSousa, R. C. (1980). Isoproterenol-induced intramembrane particle aggregation and water flux in toad epidermis. *Biochim. Biophys. Acta* **596,** 158–164.
4. DeSousa, R. C., Grosso, A., and Rufener, C. (1974). Blockade of the hydroosmotic effect of vasopressin by cytochalasin B. *Experientia* **30,** 175–177.
5. Carlsen, J., Christiansen, K., and Bro, B. (1982). Purification of microvillus membrane vesicles from pig small intestine by immunoadsorbent chromatography. *Biochim. Biophys. Acta* **689,** 12–20.

6. Chase, H. S., Jr., and Al-Awqati, Q. (1981). Regulation of the sodium permeability of the luminal border of toad bladder by intracellular sodium and calcium. Role of sodium-calcium exchange in the basolateral membrane. *J. Gen. Physiol.* **77,** 693–712.
7. Chevalier, J., Bourguet, J., and Hugon, J. S. (1974). Membrane associated particles: Distribution in frog urinary bladder epithelium at rest and after oxytocin treatment. *Cell Tissue Res.* **152,** 129–140.
8. DeMey, J., Moeremans, M., Geuens, G., Nuydens, R., and DeBrabander, M. (1981). High resolution light and electron microscopic localization of tubulin with the IGS (Immuno Gold Staining) method. *Cell Biol. Int. Rep.* **5,** 889–898.
9. Dratwa, M., Tisher, C. C., Sommer, J. R., and Croker, B. P. (1979). Intramembrane particle aggregation in toad urinary bladder after vasopressin stimulation. *Lab. Invest.* **40,** 46–54.
10. Faulk, W. P., and Taylor, G. M. (1971). An immunocolloid method for the electron microscope. *Immunochemistry* **8,** 1081–1083.
11. Garcia-Perez, A., and Smith, W. L. (1983). Use of monoclonal antibodies to isolate cortical collecting tubule cells: AVP induces PGE release. *Am. J. Physiol.* **244:** (*Cell Physiol.* **13**), C211–C220.
12. Geering, K., Girardet, M., Bron, C., Kraehenbühl, J. P., and Rossier, B. C. (1982). Hormonal regulation of $(Na^+,K^+)$ATPase biosynthesis in the toad bladder. *J. Biol. Chem.* **257,** 10338–10343.
13. Geoghegan, W. D., and Ackerman, G. A. (1977). Adsorption of HRP, ovomucoid, and anti-immunoglobulin to colloidal gold for the indirect detection of Con A, WGA and goat anti-human immunoglobulin G on cell surfaces at the electron microscopic level: A new method, theory and application. *J. Histochem. Cytochem.* **25,** 1187–1200.
14. Geuze, H. J., Slot, J. W., van der Lay, P. A., and Scheffer, R. C. T. (1981). Use of colloidal gold particles in double-labelling immunoelectron microscopy of ultrathin frozen tissue sections. *J. Cell Biol.* **89,** 653–665.
15. Gronowicz, G., Masur, S. K., and Holtzman, E. (1980). Quantitative analysis of exocytosis and endocytosis in the hydroosmotic response of toad bladder. *J. Membr. Biol.* **52,** 221–235.
16. Hardy, M. A., and DiBona, D. R. (1982). Microfilaments and the hydroosmotic action of vasopressin in toad urinary bladder. *Am. J. Physiol.* **243:** (*Cell Physiol.* **12**), C200–C204.
17. Harmanci, M. C., Kachadorian, W. A., Valtin, H., and DiScala, V. A. (1978). Antidiuretic hormone-induced intramembranous alterations in mammalian collecting ducts. *Am. J. Physiol.* **235,** F440–F443.
18. Hays, R. M., Bourguet, J., and Chevalier, J. (1982). Membrane fusion in the action of antidiuretic hormone, determined with an ultrarapid freezing technique. *Kidney Int.* **21,** 276 (abstr.)
19. Hays, R. M., Bourguet, J., Satir, B. H., Franki, N., and Rapoport, J. (1982). Retention of antidiuretic hormone-induced particle aggregates by luminal membranes separated from toad bladder epithelial cells. *J. Cell Biol.* **92,** 237–241.
20. Herzlinger, D. A., Easton, T. G., and Ojakian, G. K. (1982). The MDCK epithelial cell line expresses a cell surface antigen of the kidney distal tubule. *J. Cell Biol.* **93,** 269–277.
21. Humbert, F., Montesano, R., Grosso, A., DeSousa, R. C., and Orci, L. (1977). Particle aggregates in plasma and intracellular membranes of toad bladder (granular cell). *Experientia* **33,** 1364–1367.
22. Kachadorian, W. A., and Levine, S. D. (1982). Effect of distension on ADH-induced osmotic water flow in toad urinary bladder. *J. Membr. Biol.* **64,** 181–186.
23. Kachadorian, W. A., Wade, J. B., and DiScala, V. A. (1975). Vasopressin: Induced structural change in toad bladder luminal membrane. *Science* **190,** 67–69.
24. Kachadorian, W. A., Levine, S. D., Wade, J. B. DiScala, V. A., and Hays, R. M. (1977). Relationship of aggregated intramembranous particles to water permeability in vasopressin-treated toad urinary bladder. *J. Clin. Invest.* **59,** 576–581.

25. Kachadorian, W. A., Wade, J. B., Uiterwyk, C. C., and DiScala, V. A. (1977). Membrane structural and functional responses to vasopressin in toad bladder. *J. Membr. Biol.* **30,** 381–401.
26. Kachadorian, W. A., Casey, C., and DiScala, V. A. (1978). Time course of ADH-induced intramembranous particle aggregation in toad urinary bladder. *Am. J. Physiol.* **234,** F461–F465.
27. Kachadorian, W. A., Ellis, S. A., and Muller, J. (1979). Possible roles for microtubules and microfilaments in ADH action on toad urinary bladder. *Am. J. Physiol.* **236,** F14–F20.
28. Kessler, S. W. (1976). Cell membrane antigen isolation with the staphylococcal protein A-antibody adsorbent. *J. Immunol.* **117,** 1482–1490.
29. Kohler, G., and Milstein, C. (1975). Continuous culture of fused cells secreting antibody of pre-defined specificity. *Nature (London)* **256,** 495–497.
30. Levine, S. D., and Kachadorian, W. A. (1981). Barriers to water flow in vasopressin-treated toad urinary bladder. *J. Membr. Biol.* **61,** 135–139.
31. Louvard, D., Reggio, H., and Warren, G. (1982). Antibodies to the golgi complex and the rough endoplasmic reticulum. *J. Cell Biol.* **92,** 92.-107.
32. Luzio, J. P., Newby, A. C., and Hales, C. W. (1976). A rapid immunological procedure for the isolation of hormonally sensitive rat fat-cell plasma membranes. *Biochem. J.* **154,** 11–21.
33. Mamelok, R. D., Liu, D., and Tse, S. (1982). Screening indicates that antibodies to the $Na^+$-dependent L-alanine transport system can be produced by hybridomas. *Kidney Int.* **21,** 281 (abstr).
34. Masur, S. K., Holtzman, E., Schwartz, I. L., and Walter, R. (1971). Correlation between pinocytosis and hydroosmosis induced by neurohypophyseal hormones and mediated by adenosine 3′,5′-cyclic monophosphate. *J. Cell Biol.* **49,** 582–594.
35. Masur, S. K., Holtzman, E., and Walter, R. (1972). Hormone-stimulated exocytosis in toad urinary bladder. *J. Cell Biol.* **52,** 211–219.
36. McDonough, A. A., Hiatt, A., and Edelman, I. S. (1982). Characteristics of antibodies to guinea pig ($Na^+ + K^+$)-adenosine triphosphatase and their use in cell-free synthesis studies. *J. Membr. Biol.* **69,** 13–22.
37. Muller, J., Kachadorian, W. A., and DiScala, V. A. (1980). Evidence that ADH-stimulated intramembrane particle aggregates are transferred from cytoplasmic to luminal membranes in toad bladder epithelial cells. *J. Cell Biol.* **85,** 83–95.
38. Palmer, L. G., and Lorenzen, M. (1983). Antidiuretic hormone-dependent membrane capacitance and water permeability in the toad urinary bladder. *Am. J. Physiol.* **244** (*Renal, Fluid Electrolyte Physiol.* **13**), F195–F204.
39. Papermaster, D. S., Lyman, D., Schneider, B. G., Labrienic, L., and Kraehenbühl, J. P. (1981). Immunocytochemical localization of the catalytic subunit of ($Na^+$, $K^+$)ATPase in *Bufo marinus* kidney, bladder and retina with biotinyl-antibody and streptavidin-gold complexes. *J. Cell Biol.* **91,** 273a (abstr).
40. Rapoport, J., Kachadorian, W. A., Muller, J., Franki, N., and Hays, R. M. (1981). Stabilization of vasopressin-induced membrane events by bifunctional imidoesters. *J. Cell Biol.* **89,** 261–266.
41. Rodriguez, H. J., and Edelman, I. S. (1979). Isolation of radioiodinated apical and basal-lateral plasma membranes of toad bladder epithelium. *J. Membr. Biol.* **45,** 215–232.
42. Rodriguez, H. J., Scholer, D. W., Purkerson, M. L., and Klahr, S. (1980). Isolation of epithelial cells from the toad bladder. *Am. J. Physiol.* **238,** F140–F149.
43. Roth, J., and Berger, E. G. (1982). Immunocytochemical localization of galactosyltransferase in HeLa cells: Codistribution with thiamine pyrophosphatase in trans-Golgi cisternae. *J. Cell Biol.* **93,** 223–229.
44. Scott, W. N., and Slatin, S. L. (1979). Alterations in lactoperoxidase catalyzed radio-iodination of membrane proteins associated with vasopressin-induced changes in tissue permeability to water. *Biochem. Biophys. Res. Commun.* **91,** 1038–1044.

45. Stetson, D. L., Lewis, S. A., Alles, W., and Wade, J. B. (1982). Evaluation by capacitance measurements of antidiuretic hormone induced membrane area changes in toad bladder. *Biochim. Biophys. Acta* **689,** 267–274.
46. Stetson, D. L., and Wade, J. B. (1983). Ultrastructural characterization of cholesterol distribution in toad bladder using filipin. *J. Membr. Biol.* **74,** 131–138.
47. Taylor, A., Mamelak, M., Reaven, E., and Maffly, R. (1973). Vasopressin: Possible role of microtubules and microfilaments in its action. *Science* **181,** 347–350.
48. Wade, J. B. (1978). Membrane structural specialization of the toad urinary bladder revealed by the freeze-fracture technique. III. Location, structure and vasopressin dependence of intramembrane particle arrays. *J. Membr. Biol.* **40,** (Special Issue), 281–296.
49. Wade, J. B. (1980). Hormonal modulation of epithelial structure. *Curr. Top. Membr. Transp.* **13,** 123–147.
50. Wade, J. B. (1981). Modulation of membrane structure in the toad urinary bladder by vasopressin. *Alfred Benzon Symp.* **15,** 422–430.
51. Wade, J. B., Stetson, D. L., and Lewis, S. A. (1981). ADH Action: Evidence for a membrane shuttle mechanism. *Ann. N. Y. Acad. Sci.* **372,** 106–117.
52. Warncke, J., and Lindemann, B. (1981). Effect of ADH on the capacitance of apical epithelial membranes. *Adv. Physiol. Sci., Proc. Int. Cong., 28th, 1980* **3,** 129–133.

# Chapter 13

# Molecular Modification of Renal Brush Border Maltase with Age: Monoclonal Antibody-Specific Forms of the Enzyme

*BERTRAM SACKTOR AND UZI REISS*

*Laboratory of Molecular Aging*
*National Institute on Aging*
*National Institutes of Health*
*Gerontology Research Center*
*Baltimore City Hospitals*
*Baltimore, Maryland*



## I. INTRODUCTION

Maltase (α-D-glucoside glucohydrolase, EC 3.2.1.20) is the major disaccharidase in rat renal brush border membranes (2, 11). Maltase-specific activities in renal cortex homogenates and brush border membranes prepared from senescent 24-month-old rats are decreased 30–40% relative to the activities of the enzyme from 6-month-old mature adults (7, 9). This decrement is not attributable to a change in the affinity of the enzyme for substrate, a difference in the stability of the enzyme in homogenates, or the presence of an activator in the young or an inhibitor in the aged (9). This prompted us to ask the question whether the decrease in the membrane-bound maltase activity with age results from an alteration in the enzyme protein per se or from a change in the enzyme's membrane environment, which is reflected secondarily as a loss in enzyme activity.



ISBN 0-12-153320-4

## II. PURIFICATION OF MALTASE

In order to answer this question, procedures for the solubilization of the enzyme in the absence of detergent and for the purification of maltase to homogeneity had to be developed. Solubilization was achieved by the partial hydrolysis of the brush border membrane with papain, which selectively cleaved maltase and a few other proteins from the membrane (1, 6). Purification to homogeneity was facilitated by incorporating in the procedure a new affinity chromatography step in which tris(hydroxymethyl)aminomethane (Tris), a competitive inhibitor of the enzyme, was used as the affinity ligand (8). Utilizing these innovations, an isolation technique was established which could be completed in 2 days with the recovery of ~40% of the total maltase activity in the homogenate (8). When the enzyme preparations were examined by polyacrylamide gel electrophoresis, in the presence and absence of SDS, and by immunodiffusion, they were found to be homogeneous (8).

The purification of maltase from young (6-month-old) and aged (25-month-old) rats is compared in Table I. The specific activity of membrane-bound enzyme in the renal cortex homogenate from senescent animals was decreased ~30% relative to the specific activity of the enzyme from the young adult animals. When the enzymes from rats of the two ages were solubilized and purified to homogeneity the same decrement with age was seen. This finding suggested that the decrease in maltase activity with age resulted from an alteration in the enzyme per se, rather than from a change in the enzyme's membrane environment. Because, as shown in Table I, the percentage of recovery of total activity and the degree of purification were essentially the same for the enzymes from the 6- and 25-month-old animals, the possibility that the age-related dif-

TABLE I
PURIFICATION OF RENAL MALTASE FROM YOUNG AND OLD RATS[a]

| | Specific activity (units/mg) | | Recovery of activity (%) | | Yield of protein (%) | | Purification (fold) | |
|---|---|---|---|---|---|---|---|---|
| Fraction | Young | Old | Young | Old | Young | Old | Young | Old |
| Crude homogenate | 0.040 (0.035–0.044) | 0.027 (0.024–0.031) | 100 | 100 | 100 | 100 | 1 | 1 |
| Pure enzyme | 46.10 (42.1–52.1) | 32.50 (29.0–35.3) | 38 | 44 | 0.038 | 0.040 | 1152 | 1204 |

[a] Results are the average of three different preparations. Values in parentheses indicate the range. For each enzyme preparation from young and old animals, kidneys from eight rats were used. The total soluble protein in the crude homogenates varied from 2.7 to 2.8 g. These initial values are designated as 100% yield of protein. Data are from Reiss and Sacktor (9).

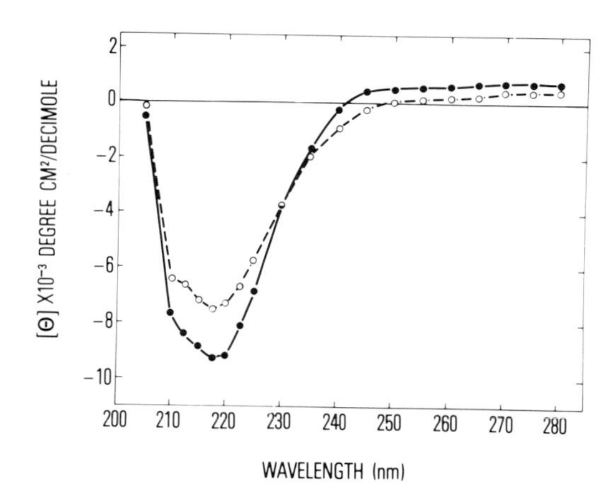


FIG. 1. Typical circular dichroism spectra of maltase from young (○) and old (●) rat kidneys. Data are from Reiss and Sacktor (9).

ferences in specific activities of the pure enzymes were due to the selective purification of an altered form of the enzyme was largely precluded.

The pure "young" and "old" enzymes did not differ measurably in molecular weight (1,300,000, comprised of tetramers of identical subunits, each having a molecular weight of 335,000), electrophoretic mobility, affinity for maltose, and amino acid composition (9). Young and old maltase did differ in their circular dichroism spectra. Representative spectra, illustrated in Fig. 1, show that both the young and old enzymes presented typical β structure patterns. However, old maltase contained more helical structure than did the young enzyme, suggesting a conformational alteration with age (9). Immunotitration of pure maltase from young and old rats using antiserum produced against pure young maltase also revealed a change in the old enzyme. In a reaction in which the initial activities of both enzymes were adjusted to the same level, more antiserum was needed to precipitate 50% of the activity of the old enzyme than that of the young enzyme (9).

## III. PRODUCTION OF ANTIBODIES TO "YOUNG" AND "OLD" ENZYME

Because of the differences in the circular dichroism spectra of the young and old enzyme, perhaps suggesting a conformational change, and the differences in immunotitration, perhaps suggesting an antigenic alteration, we next asked whether monoclonal antibodies specific to young and old forms of the enzyme could be produced to support a hypothesis of a molecular modification of maltase with age. If this be the case, then we might resolve the important question: is the decrement in specific activity of maltase in the old animal due to (1) a partial decrease in the catalytic activity of all enzyme molecules in the aged rat or (2) the presence of catalytically inactive enzyme molecules whose proportion increases relative to the active species with age?

Monoclonal antibodies to maltase from young and old rats were prepared by the hybridoma technique (5), described in detail elsewhere (10). In brief, pure young and old maltase (150 μg of enzyme protein of each), together with complete Freund's adjuvant, was injected intraperitoneally into BALB/c female mice. Two additional injections, without adjuvant, were given at 3-week intervals. Four days after the last injection, spleen cells were harvested and fused with myeloma cells (x63-Ag 8.6.5.3.). The cells were grown in hypoxanthine– aminopterin–thymine (HAT) medium (3). The surviving hybrids were tested for production of specific anti-maltase antibodies by enzyme-linked immunosorbent assay (ELISA) (4) in which the anti-maltase IgG bound to maltase was monitored by rabbit anti-mouse IgG conjugated with horseradish peroxidase. Positive cultures were tested for reactivity with young and old maltase separately and were then cloned by limiting dilutions. The resulting monoclonal antibody-generating cells were injected into BALB/c mice to obtain increased production of antibodies in the ascites fluid.

## IV. CHARACTERIZATION OF ANTIBODIES

Four cultures were established that were stable producers of anti-maltase antibodies. They were of the $IgG_1$ subclass, as determined by immunodiffusion using rabbit anti-mouse IgG antibodies. Immunofluorescent studies indicated that the monoclonal antibodies to maltase bound specifically to proximal tubular cell brush border membranes, the locus of renal maltase (Fig. 2).

Figure 3 shows titration curves for the binding of two monoclonal antibodies, designated 8B1G6 and 7G10H3, to pure maltase from mature and aged rats, when either equal maltase activity or equal enzyme protein was applied to ELISA plates and the amount of IgG bound estimated by a coupled reaction with horseradish peroxidase-conjugated rabbit anti-mouse IgG. Monoclone 8B1G6 in-

FIG. 2. Demonstration by immunofluorescence that the monoclonal antibody to maltase bound specifically to the renal proximal tubule brush border membrane, the locus of maltase activity in the kidney. (A) illustrates a low-power view of the cortex showing that fluorescence was confined to proximal tubules and (B) shows in a higher magnification that the antibody bound only to the brush border of the tubule. Antibodies that recognized maltase from 6-month-old rats and kidneys from rats of the same age were used. The anti-mouse IgG was produced in the goat and was conjugated with fluorescein isothiocyanate. (We acknowledge the collaboration of Ms. Ruby Slusser, Department of Pathology, Baltimore City Hospitals in this experiment.)

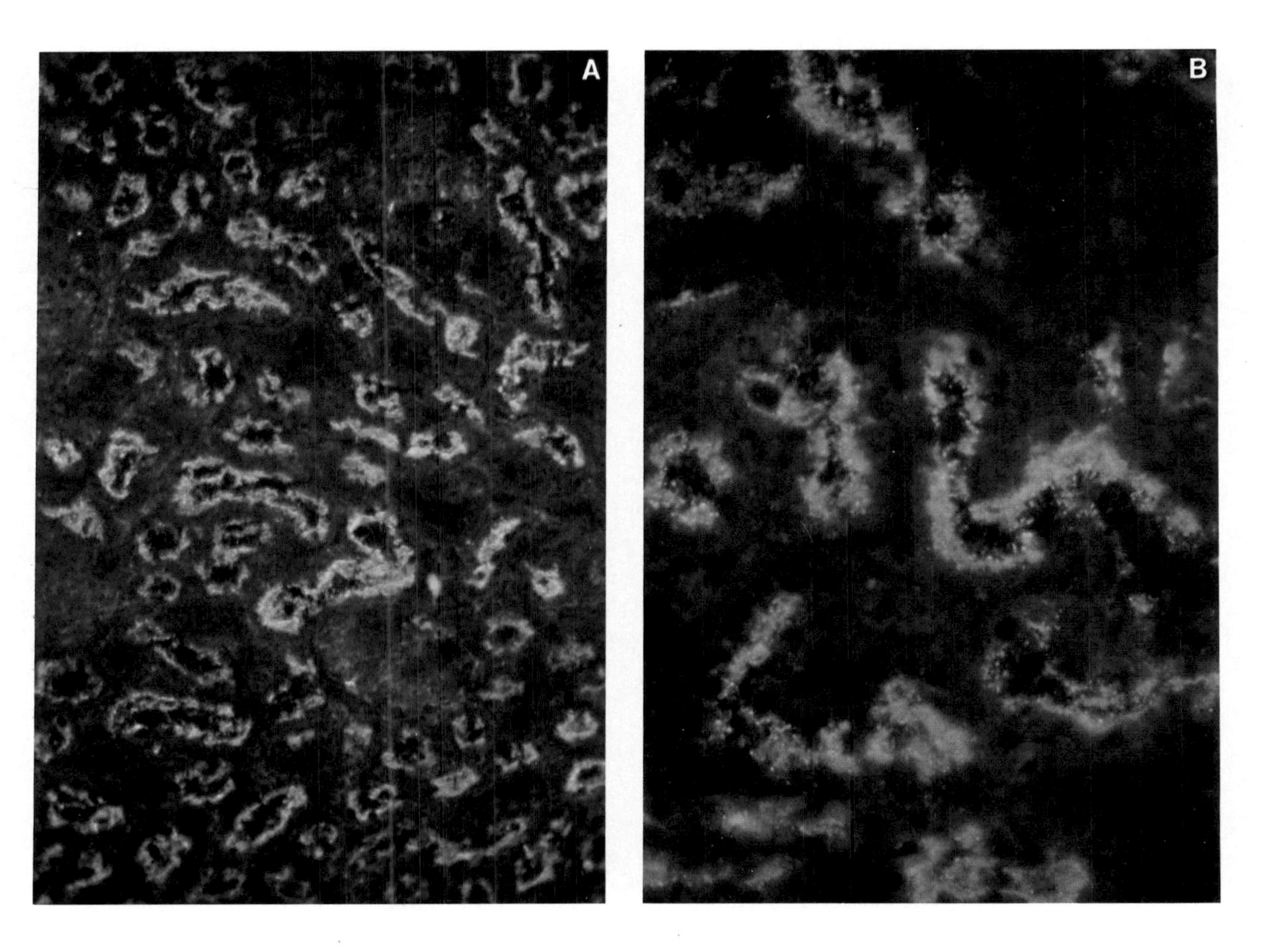
A
B

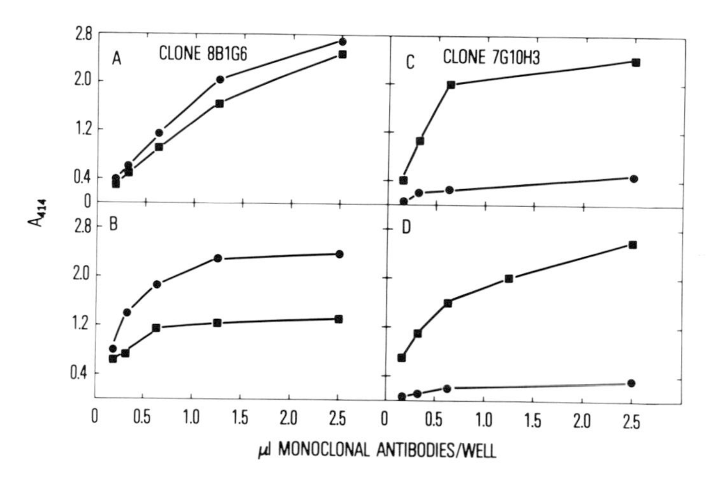


FIG. 3. The binding of monoclonal antibodies 8B1G6 (A,B) and 7G10H3 (C,D) to pure maltase from 6- (●) and 24- (■) month-old rats, determined by the ELISA assay. In (A) and (C) equal maltase activities (0.025 units) and in (B) and (D) equal enzyme proteins (1 µg) were applied to the wells. Data are from Reiss and Sacktor (10).

teracted with maltase prepared from both young and old animals. When equal units of enzyme activity were applied, the titration curves were similar (Fig. 3A). However, when equal amounts of enzyme protein were applied (i.e., with old maltase, less active enzyme was used because of its lower specific activity), there was decreased binding of antibody 8B1G6 to the old enzymes (Fig. 3B). In contrast, monoclonal antibody 7G10H3 bound predominantly to maltase from senescent rats. There was little recognition of the young enzyme, when either the activities (Fig. 3C) or the protein contents (Fig. 3D) of the two maltase preparations were held constant.

The two types of monoclonal antibodies were also distinct in their ability to immunoprecipitate maltase activity. Figure 4 shows that monoclonal antibody 1F12E1, which behaved like monoclonal antibody 8B1G6 in binding to young and old maltase, immunoprecipitated maltase activity in both the young and old enzyme preparations. In contradistinction, monoclonal antibody 7G10H3, although it was capable of binding the maltase protein from aged animals (Fig. 3), did not immunoprecipitate activity when incubated with enzyme preparations from senescent or old rats.

These results suggested that monoclonal antibodies 8B1G6 and 1F12E1 recognized the active form of maltase, found predominantly in the enzyme preparation from young animals but present, although less prevalent, in the enzyme isolated from aged animals. On the other hand, monoclonal antibodies 7G10H3 (and 2E1C10) were directed against an inactive form of maltase, found mainly in the enzyme preparation from senescent rats. Indeed, antibody 7G10H3 could be

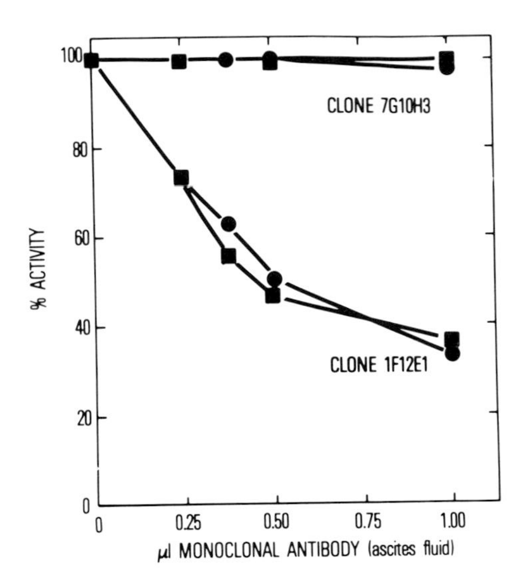


FIG. 4. The precipitation of maltase activity with monoclonal antibodies 1F12E1 and 7G10H3. Maltase preparations from young (●) and aged (■) rats (0.12 units of activity) were incubated with the indicated amounts of antibody. Data are from Reiss and Sacktor (10).

used to estimate the relative amounts of the inactive enzyme moiety in various mixtures of maltase preparations from 6- and 24-month-old animals. When the mixture was 0% old enzyme and 100% young enzyme, the antibody bound was only 15% of that bound when the mixture represented 100% old and 0% young enzyme preparations. As the percentage of old maltase in a fixed total protein mixture was increased, antibody bound increased linearly.

TABLE II

AFFINITY BINDING OF $^{125}$I-LABELED MALTASE TO IgG–PROTEIN A–SEPHAROSE 4B COLUMNS[a]

| | Monoclone | | | | | |
|---|---|---|---|---|---|---|
| | 1F12E1 | | | 7G10H3 | | |
| Maltase | cpm | μg enzyme | % active | cpm | μg enzyme | % inactive |
| Young | 2830 (2100–4130) | 2.83 | 85 | 500 (300–760) | 0.50 | 15 |
| Old | 1040 (910–1240) | 1.04 | 41 | 1510 (1100–2210) | 1.51 | 59 |

[a] $^{125}$I-labeled young and old maltase preparations were placed on protein A–Sepharose 4B columns to which monoclones 1F12E1 and 7G10H3 were attached. After the columns were washed, 0.1 *M* acetic acid was added to remove the IgG–enzyme complex. The eluants were counted for radioactivity. Values in parentheses indicate the range. Data are from Reiss and Sacktor (10).

Because of the findings that the monoclonal antibodies would bind to specific forms of maltase, immunoaffinity columns were developed, using monoclonal antibodies 1F12E1 and 7G10H3 as the ligands, to separate the active from the inactive forms of the enzyme and to determine quantitatively the relative proportions of active and inactive enzyme that was in the maltase preparations from 6- and 24-month-old animals. Table II shows that 85% of the total maltase from 6-month-old rats bound to monoclone 1F12E1 and, thus, was in the active form. On the other hand, 15% of the total maltase from the 6-month-old animal bound to antibody 7G10H3 and, therefore, was in the inactive form. In contrast, with the maltase preparations from aged animals, only about 40% of the total enzyme was in the active form whereas 60% was inactive.

## V. CONCLUSIONS

1. The antigenicity of renal brush border membrane maltase is altered during the aging process. These findings indicate that young and old maltase have different epitopes and, thus, the data are consistent with and may reflect the age-dependent conformational change of the enzyme, suggested by the alteration in its circular dichroism spectrum.
2. Active and inactive forms of maltase are present and these can be separated using specific monoclonal antibodies as ligands.
3. The increased prevalence of the inactive form of the enzyme in the aged rat can largely account for the decreased maltase-specific activity found in the senescent animal.
4. The presence of fully active and inactive enzyme molecules negates the possibility that the difference in specific activity between the enzyme from young and old animals is due exclusively to a partial decrease in the catalytic activity of all maltase molecules. The possibility that transitional forms exist which may be partially active has not been precluded, however.
5. The relative proportions of active and inactive enzyme in the kidney of an animal of any age can be determined. Thus, a biological assessment of the aging process for the kidney, in addition to chronological age, may be envisioned.

The isolation of two maltase species, one predominantly in young rats and the other mostly in the senescent animal, now permits for the first time the direct examination of critical questions on the mechanism of aging, e.g.: Are new gene products expressed during the aging process? Do molecular changes occur in preformed proteins and, if so, what are the natures of these chemical alterations? These questions can now be approached by a logical extension of the technique and findings reported in this chapter.

## REFERENCES

1. Benson, R. L., Sacktor, B., and Greenawalt, J. W. (1971). Studies on the ultrastructural localization of intestinal disaccharidases. *J. Cell Biol.* **48,** 711–716.
2. Berger, S. J., and Sacktor, B. (1970). Isolation and biochemical characterization of brush borders from rabbit kidney. *J. Cell Biol.* **47,** 637–645.
3. de St. Groth, S. F., and Scheidegger, D. (1980). Production of monoclonal antibodies: Strategy and tactics. *J. Immunol. Methods* **35,** 1–21.
4. Engvall, E., and Perlman, P. (1971). Enzyme-linked immunosorbent assay (ELISA). Quantitative assay of immunoglobulin G. *Immunochemistry* **8,** 871–874.
5. Kohler, G., and Milstein, C. (1975). Continuous cultures of fused cells secreting antibody of predefined specificity. *Nature (London)* **256,** 495–497.
6. Noronha-Blob, L. (1979). Effects of papain on enzymic and transport functions of isolated rabbit renal brush border membrane vesicles. *Fed. Proc. Fed. Am. Soc. Exp. Biol.* **38,** 838.
7. O'Bryan, D., and Lowenstein, L. M. (1974). Effect of aging on renal membrane-bound enzyme activities. *Biochim. Biophys. Acta* **339,** 1–9.
8. Reiss, U., and Sacktor, B. (1981). Kidney brush border membrane maltase: Purification and properties. *Arch. Biochem. Biophys.* **209,** 342–348.
9. Reiss, U., and Sacktor, B. (1982). Alteration of kidney brush border membrane maltase in aging rats. *Biochim. Biophys. Acta* **704,** 422–426.
10. Reiss, U., and Sacktor, B. (1983). Monoclonal antibodies to renal brush border membrane maltase: Age-associated antigenic alterations. *Proc. Natl. Acad. Sci. U.S.A.* **80,** 3255–3259.
11. Sacktor, B. (1968). Trehalase and the transport of glucose in the mammalian kidney and intestine. *Proc. Natl. Acad. Sci. U.S.A.* **60,** 1007–1014.

# Part III

# Biochemical Characterization of Transport Proteins

# Chapter 14

# Sodium–D-Glucose Cotransport System: Biochemical Analysis of Active Sites

*R. KINNE,[1] M. E. M. DA CRUZ,* AND J. T. LIN*

*Department of Physiology and Biophysics*
*Albert Einstein College of Medicine*
*Bronx, New York*
*and*
**Laboratório Nacional de Engenheria*
*e Technologia Industrial*
*Lisbon, Portugal*



## I. INTRODUCTION

Renal and intestinal transport of glucose is characterized by the following common features: it is highly specific, excluding sugars modified at the C-2 position, it is sodium dependent, and it can be inhibited by phlorizin (for review see Ullrich, 1979). The elucidation of the molecular mechanisms involved in this transport was greatly facilitated by using isolated brush border membrane vesi-

[1]Present address: Max-Planck-Institut für Systemphysiologie, Dortmund, Federal Republic of Germany.




ISBN 0-12-153320-4

cles, in which the composition of the fluid bathing the external and cytoplasmic faces of the transport system can be modified at will and in a well-defined fashion (Murer and Kinne, 1980). The information gained at this level using either D-glucose or sodium transport as indicator for the transport system or phlorizin as probe for the first step in the translocation, the binding of D-glucose to the transport system, can be summarized as follows. The transport system shows a stoichiometry of one sodium ion cotransported with one D-glucose molecule in intestine and renal cortex (Hopfer and Groseclose, 1980; Turner and Moran, 1982), but higher stoichiometries have been observed in kidney outer medulla and small intestine (R. J. Turner, private communication; Kaunitz *et al.*, 1982). Each transporter seems to bind one phlorizin molecule (Turner and Moran, 1982). The binding of phlorizin is defined by the nature of the sugar moiety and by the nature of the aglycon (for review see Kinne, 1976). The affinity of the phlorizin binding site is increased in the presence of sodium, and the same is found for the affinity of the transport system for D-glucose. The sodium–D-glucose cotransport is electrogenic (Murer and Hopfer, 1974; Sacktor, 1980) and is increased when the inside of the membrane vesicles is negative compared to the incubation medium.

The transport molecule is an intrinsic membrane protein which can be removed from the membrane only when the integrity of the lipid bilayer is destroyed (Kinne and Faust, 1977; Crane *et al.*, 1976, 1979). It is inserted asymmetrically in the membrane, as shown for example by a higher sensitivity of the cytoplasmic face to proteases and SH-group reagents (Malathi *et al.*, 1980; Klip *et al.*, 1980).

In order to elucidate the molecular mechanism of the cotransport, essentially two different approaches have been used during recent years, the kinetic approach and the biochemical approach. In kinetic analyses, the application of models derived from enzymatic reactions involving two substrates provided evidence for a random bi–bi mechanism or an isoordered bi–bi mechanism with glide symmetry for the translocation of sodium and glucose (Hopfer and Groseclose, 1980; Turner and Silverman, 1981). These kinetic considerations were extended recently to propose a similarity of the cotransport system with a gated channel (Toggenburger *et al.*, 1982). The equivalent gate is supposed to be negatively charged in the unloaded form.

The biochemical approach followed mainly three avenues which will be reviewed separately below: (1) attempts to isolate the transport protein and to determine its molecular weight in the isolated or in the membrane-bound form, (2) the characterization of amino acid residues related to the active sites of the molecule by chemical modification of the transport system in intact membranes, and (3) the use of reconstitution experiments to get information about the interaction between the membrane lipids and the transport system.

## II. THE POSSIBLE POLYMERIC STRUCTURE OF THE SODIUM–D-GLUCOSE COTRANSPORT SYSTEM

Attempts to isolate the sodium–D-glucose cotransport system have involved negative extraction procedures (Klip *et al.*, 1979a), in which by successive use of proteases and detergents all membrane transport proteins except the one related to phlorizin binding or glucose transport are removed from the membrane, as well as positive extraction of the transport system from the membranes followed by enrichment through various chromatographic procedures (Crane *et al.*, 1979). An example of the latter is work performed in our laboratory on the sodium–D-glucose cotransport system from calf kidney cortex (Lin *et al.*, 1981). The first step in this isolation procedure is the preparation of large quantities of brush border membrane. The membrane vesicles are then exposed to Triton X-100 or octylglucoside, which according to uptake studies under zero gradient isotope exchange conditions extracts the sodium-dependent D-glucose transport system from the membranes. The transport activity can be recovered in the membrane extract in reconstitution experiments. For reconstitution the membrane proteins are incorporated into liposomes formed with a large excess of natural or artificial lipids. The transporter can be purified by affinity chromatography of the membrane extract on a phlorizin polymer that possesses the features essential for a high-affinity interaction between phlorizin and the glucose transporter, namely a mobile D-glucose residue and a hydrophobic aglycon residue (Lin and Kinne, 1980). In the presence of sodium, the transport system is retained by the affinity column and can be eluted by high concentrations of D-glucose. The enrichment of the fraction is then followed by reconstitution of the various protein fractions into liposomes. The highest published figure on enrichment, as determined by comparison with the sodium-dependent D-glucose uptake in intact membranes, is about 30-fold. In SDS gel electrophoresis the purified fraction shows an enrichment of specific polypeptide chains to which an apparent molecular weight of 60,000–70,000 can be assigned. For the transport system purified from rabbit kidney a molecular weight of 160,000 (Malathi *et al.*, 1980) has been found; however, the molecular weight was determined under conditions in which aggregation of proteins or incomplete dissociation cannot be excluded.

Another approach to determining the molecular weight is to label the transport protein specifically, using either affinity labeling with a pseudosubstrate such as azidophlorizin (Hosang *et al.*, 1981) or substrate protection conditions (Thomas, 1973; Smith *et al.*; 1975; Lemaire and Maestracci, 1978). In those experiments, molecular weights of ~30,000 and ~50,000, respectively, have been assigned to the transport system (see Table I). Since all of the above-mentioned experiments were performed after disintegration of the membrane (and probably also of the transport system into single polypeptide chains), the apparent molecular

TABLE I

APPARENT MOLECULAR WEIGHT OF THE SODIUM–D-GLUCOSE COTRANSPORT SYSTEM AS DETERMINED BY SUBSTRATE PROTECTION, PROTEIN PURIFICATION, OR RADIATION INACTIVATION

| Reference | Apparent molecular weight (×1000) | Remarks |
|---|---|---|
| Thomas (1973) | 30<br>60 | Substrate protection experiments using phlorizin and *N*-ethylmaleimide in rat kidney brush border membranes |
| Smith *et al.* (1975) | 31.5 | Substrate protection experiments using D-glucose or phlorizin and *p*-chloromercuribenzoate in pig intestinal brush border membranes |
| Lemaire and Maestracci (1978) | 51<br>42 | Substrate protection experiments using phlorizin and *p*-chloromercuribenzoate or *N*-ethylmaleimide in rabbit intestinal brush border membranes |
| Koepsell *et al.* (1983) | 52 | Purification from pig kidney brush border membrane extracts, monitoring sodium-dependent D-glucose transport activity after reconstitution |
| Klip *et al.* (1979b) | 60–70 | Negative purification method, monitoring phlorizin binding to rabbit intestinal brush border membranes |
| Lin *et al.* (1981) | 60–70 | Affinity chromatography of pig kidney brush border membrane extracts, monitoring sodium-dependent D-glucose transport activity after reconstitution |
| Malathi *et al.* (1980) | 160 | Purification from rabbit kidney brush border membrane extracts, monitoring sodium-dependent D-glucose transport activity after reconstitution |
| Turner and Kempner (1982) | 110 | Radiation inactivation of the phlorizin binding site in rabbit kidney brush border membranes |
| Lin *et al.* (unpublished observation) | 230 | Radiation inactivation of the phlorizin binding site in calf kidney brush border membranes |
| Lin *et al.* (unpublished observation) | 345 | Radiation inactivation of sodium-dependent D-glucose tracer exchange in calf kidney brush border membrane vesicles |

weights do not necessarily reflect the molecular size of the transport system in the intact membrane. Radiation inactivation offers the advantage of determining the molecular size of the transport system without membrane disruption and protein extraction (Harmon *et al.*, 1982; Lo *et al.*, 1982). Studies performed recently in Dr. Turner's laboratory (Turner and Kempner, 1982) show that in brush border membranes isolated from rabbit kidney the component involved in phlorizin binding has a target size corresponding to a molecular weight of 110,000. By contrast, we have recently found that the number of the sodium-dependent phlorizin binding sites in the brush border membrane isolated from calf kidney decreases drastically by irradiation at low energy ($<$10 Mrad). From these studies a molecular target size of 230,000 has been estimated. In parallel studies, sodium-dependent D-glucose transport determined under zero gradient tracer exchange conditions in calf kidney brush border membrane vesicles was more sensitive to radiation inactivation than phlorizin binding to the same membranes. Under a situation in which diffusional uptake of D-glucose in the presence of potassium as well as the intravesicular space remained almost unaltered, a target size for the sodium–D-glucose cotransport system of $\sim$345,000 was found. These data can be interpreted to indicate that the sodium–D-glucose cotransport system is probably a multimeric system composed of a phlorizin binding unit, which also contains the D-glucose binding site of the transport system, and of the polypeptide subunits essential for the translocation of sugar and sodium across the membranes. The latter polypeptide chains might be involved in the gating process for sodium and glucose, in the interaction of the transport system with the membrane lipids, or in both.

## III. CHARACTERIZATION OF THE FUNCTIONALLY ACTIVE SITES

From the statements made above three functionally essential or active sites of the transport system can be defined: the sugar binding site that also binds phlorizin, the gating site that interacts with sodium, and the portions of the protein that interact with the lipid phase of the membrane.

### A. Sugar Binding Site

Studies on the amino acid residues linked to the sugar and phlorizin binding site have shown that organic and inorganic mercury compounds as well as *N*-ethylmaleimide (NEM) inhibit the binding of phlorizin to isolated brush border membranes (Bode *et al.*, 1970; Thomas *et al.*, 1972; Kinne, 1976). Under certain conditions *N*-ethylmaleimide blocks only the high-affinity, sodium-de-

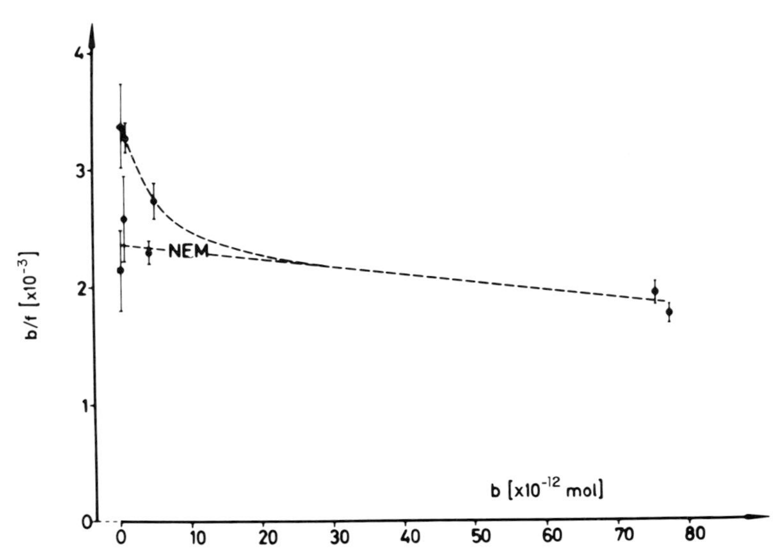


FIG. 1. Inhibition of phlorizin binding to brush border fragments by *N*-ethylmaleimide (NEM). The membranes were preincubated for 30 minutes at 37°C in the presence of NEM, and the binding of phlorizin was then determined. The amount of phlorizin, b, is expressed per unit of alkaline phosphatase in the membrane; f represents the free concentration of phlorizin. A similar inhibition of binding is observed when, in the absence of NEM, sodium is removed from the incubation medium or when high concentrations of D-glucose are used. The data are taken from Bode *et al.* (1970) with kind permission of the publisher.

pendent, D-glucose-inhibitable phlorizin binding sites related to the sodium–D-glucose cotransport system (Fig. 1). The interaction of NEM with the protein is inhibited by low concentrations of phlorizin, and about three NEM-sensitive sites are protected by one molecule of phlorizin (see Table II). This so-called substrate protection effect indicates that the reactivity of certain SH groups is altered in the presence of phlorizin. This protection is either due to steric hindrance, i.e., the

TABLE II

PROTECTION OF *N*-ETHYLMALEIMIDE (NEM) BINDING SITES BY PHLORIZIN AND D-GLUCOSE[a]

| | Number of NEM binding sites protected ($10^{-10}$ mol/mg protein) | Number of phlorizin binding sites blocked ($10^{-10}$ mol/mg protein) |
|---|---|---|
| Phlorizin | 2.4 | 0.80 |
| D-Glucose followed by phlorizin | 2.4 | — |
| D-Glucose | 12.6 | — |
| D-Glucose and phlorizin | 12.1 | — |

[a] Data are taken from Thomas *et al.* (1972).

inhibitor and the substrate bind at the same site of the protein molecule, or due to a substrate-induced change in protein conformation which alters the reactivity of an amino acid residue. For the sake of simplicity we are assuming that the former is the case for the sodium–D-glucose cotransport system. The same SH groups were protected when an incubation with D-glucose and NEM was followed by an incubation with phlorizin and NEM. These results indicate that these groups are very closely involved in the binding of D-glucose to the external surface of the transport molecule (for further discussion, see Kinne, 1976). When only D-glucose was used as a protective reagent, additional NEM molecules did not react with the transport molecule (Table II). In these experiments, the incubation conditions were such that D-glucose had sufficient time to enter the membrane vesicles and thus was probably protecting sites located at the external face as well as at the cytoplasmic face of the transport molecule. It can be hypothesized therefore that the majority of the SH groups protected by D-glucose alone are located at the cytoplasmic face of the D-glucose transporter. This finding is in agreement with the observation that the transport system is much more sensitive to inhibition by SH-group reagents when exposed to them from the cytoplasmic side than it is from the external side (Klip *et al.*, 1979b). Assuming that under the reaction conditions used NEM reacted only with SH groups, one therefore can conclude that the D-glucose binding site of the protein is rich in SH groups and that the binding sites at the cytoplasmic face and the external face differ in their properties. The latter point is also supported by the result that L-glucose, which is secreted by the renal proximal tubule (Baumann and Huang, 1969) using the sodium–glucose cotransport system, stimulates phlorizin binding to the brush border membranes (Frasch *et al.*, 1970).

## B. Sodium Binding Site

Following observations made recently in squid axon (Brodwick and Eaton, 1978), $Na^+,K^+$-ATPase (Cantley *et al.*, 1978), and the amiloride-sensitive sodium channel in tight epithelia (Park and Fanestil, 1980), we investigated a possible role of tyrosine in the interaction of the sodium–D-glucose cotransport system with sodium. Primarily one compond [7-chloro-4-nitrobenzo-2-oxa-1,3-diazole (NBD-Cl)], shown to be relatively specific for tyrosine residues, was investigated. This compound forms a condensation product with tyrosine residues and SH groups, and under proper experimental conditions the former reaction prevails. In addition, the reaction product with tyrosyl residues is more stable against reduction by thiols than is the reaction product with SH groups (Ferguson *et al.*, 1975; Cantley *et al.*, 1978). It could be demonstrated that sodium-dependent D-glucose transport was inhibited after treatment of brush border vesicles with NBD-Cl both under sodium gradient conditions and under

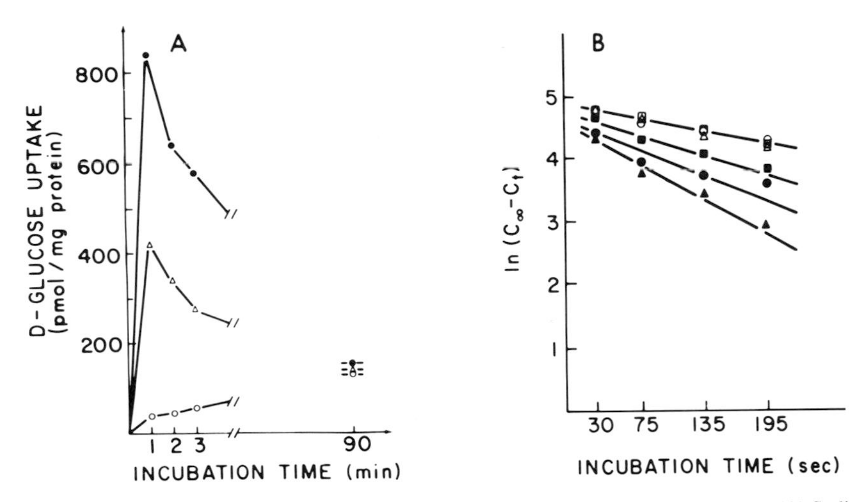


FIG. 2. Effect of NBD-Cl on the D-glucose uptake by renal brush border membranes. (A) Sodium gradient conditions: membranes were preincubated with 0.167 m*M* NBD-Cl for 60 minutes at pH 7.4 and 25°C. D-Glucose uptake was determined under gradient conditions. (●) Control membranes in 100 m*M* NaSCN, (△) NBD-Cl-treated membranes in 100 m*M* NaSCN. (B) Tracer exchange conditions: vesicles were preincubated with 0.075 or 0.15 m*M* NBD-Cl, washed, and loaded with 0.1 m*M* D-glucose, 100 m*M* KSCN or NaSCN. Uptake by control membranes without NBD-Cl in NaSCN (▲) or KSCN (△). Uptake in NaSCN (●) or KSCN (○) by membranes treated with 0.075 m*M* NBD-Cl. Uptake in NaSCN (■) or KSCN (□) by membranes treated with 0.15 m*M* NBD-Cl. $C_\infty$ represents the amount of tracer present in the vesicles at equilibrium; $C_t$ is the amount of tracer present in the vesicle after a certain time of incubation. Data were redrawn from Lin *et al.* (1982).

zero gradient tracer exchange conditions (see Fig. 2). Kinetic analysis revealed that the maximal velocity of the transport with respect to its interaction with sodium was decreased whereas the apparent affinity of the transport system for sodium was not changed (Lin *et al.*, 1982). This result suggested that NBD-Cl blocked the interaction of the transport system with sodium in an all-or-none fashion, and thus probably directly interfered with the binding of sodium to the cation binding site of the transport molecule. When substrate protection experiments were performed, this assumption was confirmed. As shown in Fig. 3, the inactivation of the transport system by NBD-Cl was significantly lower in the presence of high concentrations of sodium compared to potassium. D-Glucose, at a concentration where the sodium–D-glucose cotransporter should be saturated with glucose, did not protect the transport molecule from inactivation. From these results it can be hypothesized that a tyrosine group (or groups) is involved in the interaction of sodium with the sodium–D-glucose cotransport system. Whether this group or, as postulated by Toggenburger *et al.* (1982), a carboxyl group is functioning as the equivalent gate of a gated channel remains to be elucidated.

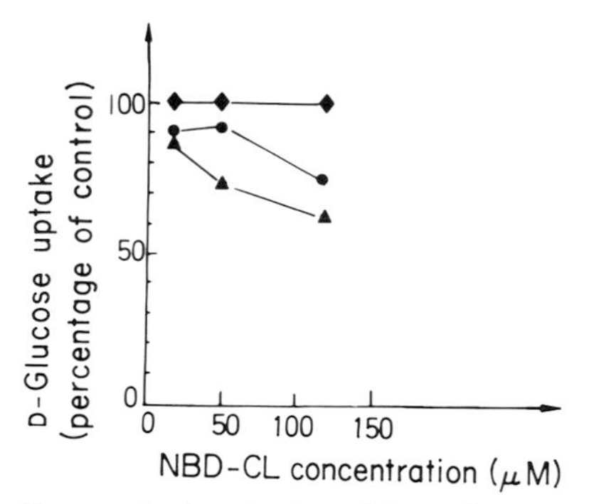


FIG. 3. Effect of sodium on the inactivation of the sodium–D-glucose cotransport system by NBD-Cl. Membrane vesicles were preincubated with 50, 100, or 150 $\mu M$ NBD-Cl in the presence of 500 m$M$ NaSCN or 500 m$M$ KSCN for 15 minutes. After intensive washing, D-glucose uptake was determined in the presence of a 100 m$M$ NaSCN gradient. As control, membrane vesicles were incubated under identical condition, i.e., 500 m$M$ NaSCN, but without NBD-Cl (♦). Membrane with NBD-Cl and 500 m$M$ NaSCN (●), membrane with NBD-Cl and 500 m$M$ KSCN (▲). The data are taken from Lin *et al.* (1982) with kind permission of the publisher.

## C. Interaction with Membrane Lipids

A third functionally important portion of the sodium–D-glucose cotransport system is the region of the system interacting with the membrane lipids. Since the natural membrane is composed of a variety of lipids, the question arises whether there is a specific lipid requirement of the transport which would indicate a strong interaction between one membrane lipid species with the hydrophobic area of the transport system. We approached this question by studying the temperature dependence of sodium-dependent D-glucose transport in natural membranes and in proteoliposomes reconstituted from brush border membrane extracts and phospholipids whose transition temperature is well defined. The underlying hypothesis was that if there were an annulus of tightly bound specific lipids around the transport system, these lipids should protect the transport system from the changes occurring in the bulk phase of the phospholipids. In following this line of reasoning, experiments with intact brush border membranes were first performed. Sodium-dependent D-glucose uptake under zero gradient tracer exchange conditions showed a well-defined transition temperature as evident from the breakpoint found in the Arrhenius plot (Fig. 4). The apparent breakpoint occurs at 15°C. Sodium-dependent phosphate transport and the intrinsic membrane protein alkaline phosphatase show break points at similar temperatures (DeSmedt and Kinne, 1981).

A phospholipid mixture used frequently in reconstitution studies is asolectin, a mixture of phospholipids with a transition temperature of 0°C. When the sodium–D-glucose cotransport system is incorporated into asolectin, no difference

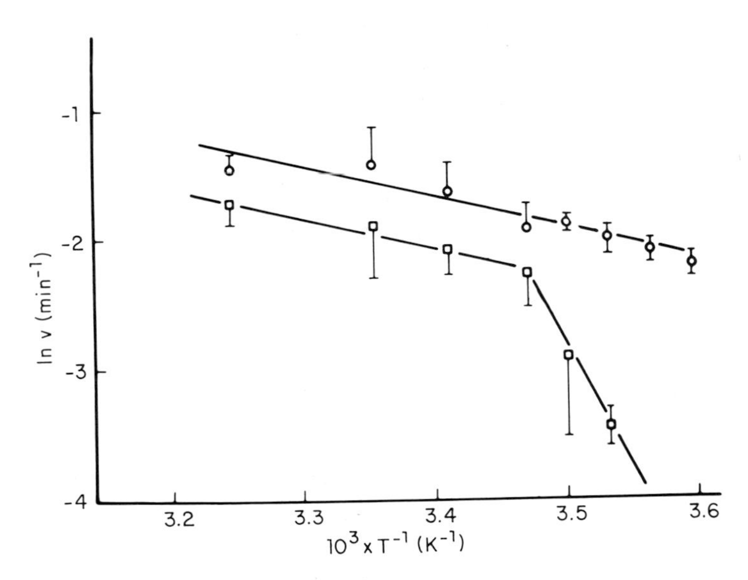


FIG. 4. Arrhenius plot for the D-glucose tracer exchange in renal brush border membranes. D-Glucose uptake by hog kidney brush border membranes was determined under tracer exchange conditions in a medium containing 100 m*M* NaSCN, 20 m*M* Tris–HEPES (pH 7.4), 15 m*M* KSCN, 100 m*M* mannitol, and 0.1 m*M* D-glucose, in the presence (○) and in the absence (□) of 0.5 m*M* phlorizin. The apparent exchange rate of D-glucose ($v$) was determined at various temperatures as indicated in the figure. Data are taken from DeSmedt and Kinne (1981) with kind permission of the publisher.

in sodium-dependent glucose uptake into asolectin proteoliposomes is observed in the range between 25 and 4°C, quite in contrast to the findings in the intact brush border membrane (Da Cruz, 1982). This result alone indicates that the phospholipids surrounding the sodium–D-glucose cotransport system can be exchanged quite easily and that for proper functioning of the transport system no cholesterol or specific phospholipid composition is required. This assumption was tested further by choosing a phospholipid that has a higher transition temperature than the brush border phospholipids. Dimyristoylphosphatidylcholine (DMPC) was chosen for this purpose. The sodium–D-glucose cotransport system incorporated in DMPC vesicles shows all the essential properties of the transport system in the intact brush border membrane. The proteoliposomes formed after incorporation of the transport system exhibit a distinct phase transition between 23 and 24°C (Da Cruz *et al.*, 1983). When the sodium-dependent D-glucose uptake into these proteoliposomes is determined under zero gradient tracer exchange conditions, a break point in the Arrhenius plot is observed at the same temperature (Fig. 5). Thus the transition temperature observed for the transport reflects the transition temperature of the bulk lipids, making the existence of an appreciable layer of boundary lipids around the transport systems very unlikely.

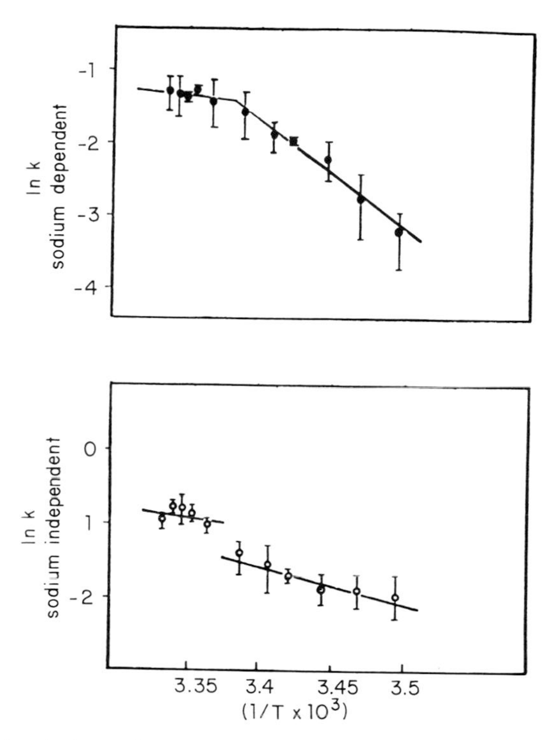


FIG. 5. Arrhenius plot for D-glucose tracer exchange in proteoliposomes. Proteoliposomes reconstituted from solubilized hog kidney brush border membranes and DMPC were preequilibrated in a buffer containing 0.1 m*M* D-glucose, 150 m*M* KCl (○) or 50 m*M* KCl and 100 m*M* NaSCN (●), and 20 m*M* Tris–HEPES (pH 7.4) at 37°C. D-Glucose uptake at various temperatures was measured in the same incubation medium after addition of 50 μCi of D-[$^3$H]glucose. The natural logarithms of the apparent D-glucose exchange rates, *k*, estimated at different temperatures were plotted against the temperature. Data are taken from Da Cruz (1982) and Da Cruz *et al.* (1983).

In the current state of knowledge we cannot, however, decide whether there are subtle differences in the function of the transport system depending on the headgroups of the phospholipids or the length and the nature of the acyl chains of the phospholipids. For such investigations the number of transport proteins incorporated and their orientation in the proteoliposomes have to be known in order to allow the determination of the turnover number.

## IV. SUMMARY AND CONCLUSIONS

The data presented above indicate that we are slowly beginning to understand the biochemical reactions underlying sodium–D-glucose cotransport. The molecular target size of the sodium–D-glucose cotransport system has been determined to be 345,000. Investigation of the inactivation of the sodium-dependent phlorizin binding with the same method as used in the study of the D-glucose transport (radiation inactivation) reveals that the functional phlorizin binding site in the calf brush border membrane has a molecular weight of 230,000, which is lower than that needed for the D-glucose transport and which suggests that there are one or two additional polypeptides linked, not to phlorizin binding or glucose binding, but to sugar and sodium translocation across the membrane. In addition, evidence has been accumulated that the phlorizin binding unit is composed of monomer subunits. It remains to be elucidated whether these subunits are identical or not. Using substrate protection experiments, the role of SH groups in sugar and phlorizin binding to the external and cytoplasmic faces of the transporter could be demonstrated. Similar experiments suggest a role of tyrosine in the interaction of the transport system with sodium. In addition, reconstitution studies demonstrated that the areas of the transport protein that interact with the lipid bilayer seem to have a rather low specificity for the lipids of the membrane. This low specificity makes the presence of a closely bound boundary lipid very unlikely. Attempts to isolate the proteins involved in the transport continue, and in the isolated system the as yet hypothetical assignment of different functions to different polypeptides should be possible. It could be imagined, for example, that the glucose binding site and the sodium binding site are located on different polypeptides and that a varying ratio of α units (an α unit equals a sodium binding site) and β units (a β unit equals a substrate recognition site) might be the way nature establishes varying stoichiometries in different sodium cotransport systems. Thus, the higher stoichiometry of sodium–sugar interaction in the medullary part of the kidney and in the intestine might be explained by a 2 α:1 β or a 3 α:1 β ratio. However, such speculations are far from being proven.

## ACKNOWLEDGMENTS

The work presented above was partly supported by NIH Grant GM 27859, start-up funds from the Albert Einstein College of Medicine (BRSG 613-4610), and the Max-Planck Society. Some of the data in this chapter are from a thesis to be submitted by M. E. M. Da Cruz to the Johann Wolfgang Goethe University, Frankfurt, Federal Republic of Germany, in partial fulfillment of the requirements for her Ph.D. degree.

## REFERENCES

Baumann, K., and Huang, K. D. (1969). Micropuncture and microperfusion study of L-glucose secretion in rat kidney. *Pfluegers Arch.* **305,** 155–166.

Bode, F., Baumann, K., Frasch, W., and Kinne, R. (1970). The binding of phlorizin to the brush border fraction of rat kidney. *Pfluegers Arch.* **315,** 53–65.

Brodwick, M. S., and Eaton, D. C. (1978). Sodium channel inactivation in squid axon is removed by high internal pH or tyrosine-specific reagents. *Science* **200,** 1494–1496.

Cantley, L. C., Gelles, J., and Josephson, L. (1978). Reaction of (Na-K) ATPase with 7-chloro-4-

nitrobenzo-2-oxa-1,3-diazole: Evidence for an essential tyrosine at the active site. *Biochemistry* **17,** 418–415.

Cleland, W. W. (1963). The kinetics of enzyme-catalyzed reactions with two or more substrates or products, I., II. & III. *Biochim. Biophys. Acta* **67,** 104–137, 173–187, 188–196.

Crane, R. K., Malathi, P., and Preiser, H. (1976). Reconstitution of specific $Na^+$-dependent D-glucose transport in liposomes by triton-x-100 extracted proteins from purified brush border membranes of rabbit kidney cortex. *FEBS Lett.* **67,** 214–216.

Crane, R. K., Malathi, K. P., and Fairclough, P. (1979). Some characteristics of kidney $Na^+$-dependent glucose carrier reconstituted into sonicated liposomes. *Am. J. Physiol.* **234,** E1–E5.

Da Cruz, M. E. M. (1983). The use of proteoliposomes to study the sodium-D-glucose cotransport system from hog kidney. Ph.D. Thesis, J. W. Goethe Univ., Frankfurt, Federal Republic of Germany.

Da Cruz, M. E. M., Kinne, R., and Lin, J. T. (1983). Temperature dependence of D-glucose transport in reconstituted liposomes. *Biochim. Biophys. Acta* **732,** 691–698.

DeSmedt, H., and Kinne, R. (1981). Temperature dependence of solute transport and enzyme activities in hog renal brush border membrane vesicles. *Biochim. Biophys. Acta* **648,** 207–253.

Ferguson, S. J., Lloyd, W. J., Lyons, M. H., and Radda, G. K. (1975). Specific modification of mitochondrial ATPase. *Eur. J. Biochem.* **54,** 117–126.

Frasch, W., Frohnert, P. P., Bode, F., Baumann, K., and Kinne, R. (1970). Competitive inhibition of phlorizin binding by D-glucose and influence of sodium: A study on isolated brush border membrane of rat kidney. *Pfluegers Arch.* **320,** 265–284.

Harmon, J. T., Kahn, G. R., Kempner, E. S., and Johnson, M. L. (1982). Examination of the insulin receptor in the membrane by radiation inactivation studies. *Proc. Serono Symp.* **41,** 37–44.

Hopfer, U., and Groseclose, R. (1980). The mechanism of $Na^+$-dependent D-glucose transport. *J. Biol. Chem.* **255,** 4453–4462.

Hosang, M., Gibbs, E. M. Diedrich, D. F., and Semenza, G. (1981). Photoaffinity labeling and identification of (a component of) the small intestinal $Na^+$,D-glucose transporter using 4-azidophlorizin. *FEBS Lett.* **130,** 244–248.

Kaunitz, J. D., Gunther, R., and Wright, E. (1982). Involvement of multiple sodium ions in intestinal D-glucose transport. *Proc. Natl. Acad. Sci. U.S.A.* **79,** 2315–2318.

Kinne, R. (1976). Properties of the glucose transport system in the renal brush border membrane. *Curr. Top. Membr. Transp.* **8,** 209–267.

Kinne, R., and Faust, R. G. (1977). Incorporation of D-glucose, L-alanine and phosphate transport systems from rat renal brush border membranes into liposomes. *Biochem. J.* **168,** 311–314.

Klip, A., Grinstein, S., and Semenza, G. (1979a). Partial purification of sugar carrier or intestinal binding component by selective extractions. *J. Membr. Biol.* **51,** 47–73.

Klip, A., Grinstein, S., and Semenza, G. (1979b). Transmembrane disposition of the phlorizin binding protein of intestinal brush-border. *FEBS Lett.* **99,** 91–96.

Klip, A. Grinstein, S., and Semenza, G. (1980). The small intestinal sodium D-glucose cotransporter is inserted in the brush border membrane asymmetrically. *Ann. N.Y. Acad. Sci.* **358,** 374–377.

Koepsell, H., Menuhr, H., Ducis, I., and Wissmüller, Th. F. (1983). Partial purification and reconstitution of the $Na^+$–D-glucose cotransport protein from pig renal proximal tubules. *J. Biol. Chem.* **258,** 1888–1894.

Lo, M. M. S., Barhard, E. A., and Dolby, J. O. (1982). Size of acetylcholine receptors in the membrane. An improved version of the radioactive inactivation method. *Biochemistry* **21,** 2210–2217.

Lemaire, J., and Maestracci, D. (1978). Labeling of a glucose binding protein in the rabbit intestinal brush border membrane. *Can. J. Physiol. Pharmacol.* **56,** 760–770.

Lin, J. T., and Kinne, R. (1980). A phlorizin polymer as affinity gel for sugar binding proteins. *Angew. Chem., Int. Ed. Engl.* **19,** 540–541.

Lin, J. T., Da Cruz, M. E. M., Riedel, S., and Kinne, R. (1981). Partial purification of hog kidney sodium-D-glucose cotransport system by affinity chromatography on a phlorizin polymer. *Biochim. Biophys. Acta* **640,** 43–54.

Lin, J. T., Stroh, A. M., and Kinne, R. (1982). Renal sodium-D-glucose cotransport system: Involvement of tyrosyl-residues in sodium-transporter interaction. *Biochim. Biophys. Acta* **692,** 210–217.

Malathi, P., Preiser, H., and Crane, R. K. (1980). Protease-resistant integral brush border membrane proteins and their relationship to sodium dependent transport of D-glucose and L-alanine. *Ann. N.Y. Acad. Sci.* **358,** 253–266.

Murer, H., and Hopfer, U. (1974). Demonstration of electrogenic $Na^+$-dependent D-glucose transport in intestinal brush border membranes. *Proc. Natl. Acad. Sci. U.S.A.* **72,** 484–488.

Murer, H., and Kinne, R. (1980). The use of isolated membrane vesicles to study epithelial transport processes. *J. Membr. Biol.* **55,** 89–95.

Park, C. S., and Fanestil, D. D. (1980). Covalent modification and inhibition of an epithelial sodium channel by tyrosine-reactive reagents. *Am. J. Physiol.* **8,** F299–F306.

Sacktor, B. (1980). Electrogenic and electroneutral $Na^+$-gradient-dependent transport system in the brush border membrane vesicles. *Curr. Top. Membr. Transp.* **13,** 291–299.

Smith, M. W., Ferguson, D. S., and Burton, K. A. (1975). Glucose- and phlorizin-protected thiol groups in pig intestinal brush border membranes. *Biochem. J.* **147,** 617–619.

Thomas, L. (1973). Isolation of *N*-ethylmaleimide-labeled phlorizin-sensitive D-glucose binding protein of brush border membrane from rat kidney cortex. *Biochim. Biophys. Acta* **291,** 454–464.

Thomas, L., Kinne, R., and Frohnert, P. P. (1972). *N*-Ethylmaleimide labelling of a phlorizin-sensitive D-glucose binding site of brush border membrane from the rat kidney. *Biochim. Biophys. Acta* **290,** 125–133.

Toggenburger, G., Kessler, M., and Semenza, G. (1982). Phlorizin as a probe of the small intestinal $Na^+$,D-glucose cotransporter. A model. *Biochim. Biophys. Acta* **688,** 557–571.

Turner, R. J., and Kempner, E. S. (1982). Radiation inactivation studies on the renal brush-border membrane phlorizin-binding protein. *J. Biol. Chem.* **257,** 10794–10797.

Turner, R. J., and Moran A. (1982). Heterogenity of sodium-dependent D-glucose transport sites along the proximal tubule: Evidence from the vesicle studies. *Am. J. Physiol.* **242,** F406–F424.

Turner, R. J., and Silverman, M. (1981). Interaction of phlorizin and sodium with the renal brush-border membrane D-glucose transporter: Stoichiometry and order of binding. *J. Membr. Biol.* **58,** 43–55.

Ullrich, K. J. (1979). Renal transport of organic solutes. *In* "Membrane Transport in Biology" (G. Giebisch, D. C. Tosteson, and H. H. Ussing, eds.), pp. 413–448. Springer-Verlag, Berlin and New York.

# Chapter 15

# Probing Molecular Characteristics of Ion Transport Proteins

*DARRELL D. FANESTIL,*
*RALPH J. KESSLER, AND*
*CHUN SIK PARK*

*Division of Nephrology*
*Department of Medicine*
*University of California, San Diego*
*La Jolla, California*



## I. INTRODUCTION

In this chapter two strategies for the study of ion transport proteins in apical plasma membranes of epithelial cells are considered. The first strategy takes advantage of the hydrophobic (i.e., lipophilic) properties of membrane proteins. These lipophilic properties enable proteins to span the hydrophobic interior of the lipid bilayer (Capaldi, 1982). We present evidence for the existence in renal brush borders of a hydrophobic protein with the ability to bind phosphate with high affinity and selectivity. The second strategy capitalizes on ligand-specific protection against inactivation by chemicals that react with selectivity for specific chemical groups (e.g., amino acid residues) on proteins. We describe, in the urinary bladder of the toad, protection of the $Na^+$ channel by amiloride.




ISBN 0-12-153320-4

## II. A PHOSPHATE-BINDING PROTEOLIPID

Some proteins are so lipophilic that they are soluble in the organic solvents used to extract lipids away from the bulk of proteins in tissues. In his work on extraction of lipids, Folch (Folch and Lees, 1951) coined the term ''proteolipid'' to describe these proteins that are soluble in mixtures of chloroform and methanol. Their unique solubility appears to be a consequence of a coating of lipids that may be either tightly absorbed or, in some cases, covalently attached to the protein (Schlesinger, 1981). Such proteolipids have been extracted from a variety of cellular membranes. One of the proteolipids in mitochondrial membranes is a putative phosphate carrier protein (Guerin and Napias, 1978). Similarly, Kessler and colleagues (Kessler *et al.*, 1982b) in our laboratories have found evidence for a phosphate-binding proteolipid in brush borders of rodent kidneys. Brush borders, prepared by a method involving precipitation of other cellular elements by divalent cations (Booth and Kenny, 1974), were extracted according to the protocol shown in Fig. 1. The assay for binding of phosphate by the proteolipid takes advantage of the ability of the proteolipid to partition the bound

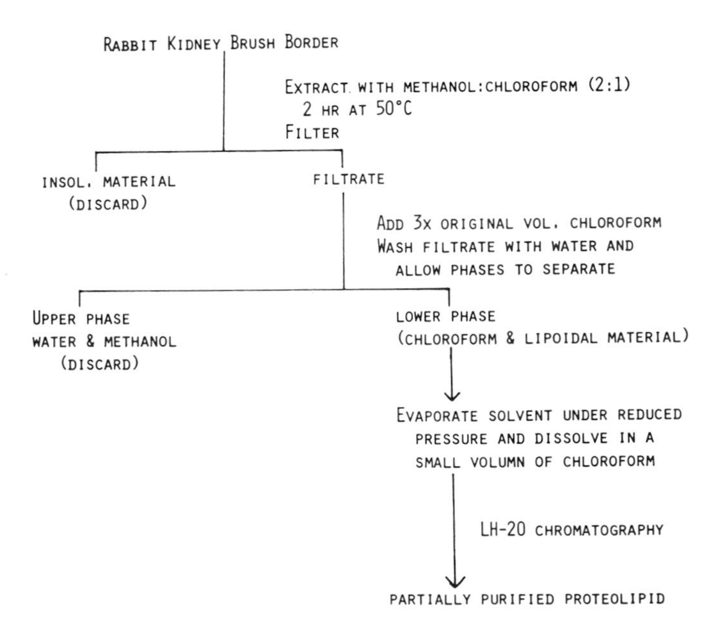


FIG. 1. Extraction and partial purification of proteolipid.

STEP 1. ORGANIC PHASE: CONTAINS 0.5 ML OF 1-BUTANOL/CHLOROFORM/METHANOL (150:50:25) ± THE PROTEOLIPID.

STEP 2. BINDING: [$^{32}$P]PHOSPHATE AND ANY HYDROPHILIC ADDITIONS ARE ADDED IN 10 µl OF WATER TO THE ORGANIC PHASE AND BRIEFLY MIXED TO FORM A SINGLE PHASE. THE MIXTURE IS ALLOWED TO INCUBATE FOR 10 MIN AT ROOM TEMPERATURE.

STEP 3. PHASE SEPARATION: ADDITION OF 0.6 ML OF 2M SUCROSE IS FOLLOWED BY VIGOROUS MIXING AND CENTRIFUGATION.

STEP 4. BOUND PHOSPHATE IS QUANTITATED FROM THE AMOUNT OF RADIOACTIVITY IN AN ALIQUOT OF THE ORGANIC PHASE.

FIG. 2. Phosphate binding assay.

phosphate into an organic phase, leaving unbound phosphate in the aqueous phase. The assay is described further in Fig. 2. Binding of phosphate increases in a saturating fashion as phosphate concentration increases (Fig. 3), but the relationship produces a sigmoidal curve, suggestive of cooperative interactions. Indeed, a Hill plot of the data shown in Fig. 3 yields a straight line with a slope of 1.92, a finding consistent with positive cooperativity in the binding of phosphate. The concentration of phosphate at which the binding is half-saturated ($K_{0.5}$), as derived from the Hill plot, is 8 $\mu M$.

Other features of the binding of phosphate by the proteolipids are briefly described here. First, the binding is not inhibited by sulfhydryl-reactive reagents, including the lipophilic reagent, phenylmercuric acetate. This clearly distinguishes our brush border phosphate-binding proteolipid from that in mito-

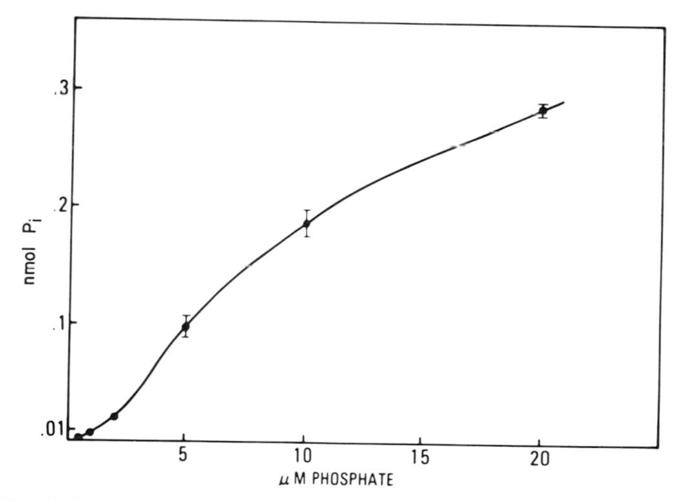


FIG. 3. The binding of phosphate by the renal brush border proteolipid is shown as a function of the concentration of inorganic phosphate. Data from Kessler *et al.* (1982b).

chondria, for the latter is readily inhibited by such reagents (Kadenback and Hadvary, 1973). Second, there is little dependence of the binding on the pH of the aqueous phase of the assay in the range of 6.5–8.0. Third, the binding of phosphate is selective for inorganic phosphate in that the organic phosphates, ATP and cyclic AMP, are not bound. Fourth, the binding is selective for the phosphate anion, since a 250-fold molar excess of such anions as lactate, sulfate, chloride, nitrate, or acetate inhibits binding of phosphate by $<50\%$. Fifth and in marked contrast, arsenate is a potent inhibitor of phosphate binding. When phosphate concentration is varied between 5 and 100 $\mu M$ in the absence or presence of two concentrations of arsenate, Lineweaver–Burke analysis indicates arsenate is a competitive inhibitor with an apparent $K_I$ of 27 $\mu M$. Sixth, chelators of divalent metals (EDTA, citrate, oxalate) inhibit binding of phosphate by the proteolipid. Seventh, amino-reactive chemicals (e.g., disulfonic stilbenes or dinitrofluorobenzene) inhibit binding of phosphate by the proteolipid. And eighth, the binding activity can be partially purified either by chromatography on Sephadex LH-20 or by precipitation with diethyl ether. Both purification maneuvers concentrate protein that migrates on highly cross-linked SDS gels (under reducing conditions) with an apparent molecular weight of ~3000.

Future studies must test more vigorously our postulate that this phosphate-binding proteolipid is the phosphate recognition site (or phosphate carrier moiety) of the $Na^+$-dependent phosphate transport system in brush borders. Thus far, we have developed several lines of circumstantial evidence that are consistent with this postulate. First, the $K_{0.5}$ (8 $\mu M$) for binding of phosphate is about one-tenth the apparent $K_m$ (80 $\mu M$) in brush borders, a finding predicted by the Gouy–Chapman theory (McLaughlin, 1977). This difference in apparent $K_{0.5}$ is a result of the net negative charge present on the surface of the brush border but absent in the assay where the solubilized proteolipid binds phosphate. Second, the $K_m$ for phosphate of the transport system in brush borders is not dependent upon sodium (Cheng and Sacktor, 1981); neither is the binding of phosphate by the proteolipid. Third, the $K_m$ for phosphate of the transport system in brush borders is relatively insensitive to pH in the range of 6.5–8.0 (Burckhardt *et al.*, 1981); so is the binding of phosphate by the proteolipid. Fourth, binding of phosphate by the proteolipid is inhibited by the amino-reactive reagent, dinitrofluorobenzene; so is the transport of phosphate by brush borders (Kessler *et al.*, 1982b). Fifth, binding of phosphate by the proteolipid is inhibited by EDTA; so is the transport of phosphate by brush borders (Kessler *et al.*, 1982a). And sixth, $Na^+$-dependent transport of phosphate by brush borders is competitively inhibited by arsenate; so is the binding of phosphate by the proteolipid (Kessler *et al.*, 1982b).

Based on these studies, we have advanced the name *phosphorin* for this proteolipid that binds phosphate. We propose that phosphorin is the likely phosphate-binding component of a multisubunit $Na^+$-dependent phosphate transport system in renal brush borders.

## III. FUNCTIONAL RESIDUES OF THE $NA^+$ CHANNEL

We now direct attention to a second strategy, pursued by Park and colleagues in our laboratories, for defining characteristics of ion transport proteins. We chose to study transepithelial transport of sodium by the urinary bladder of the toad, *Bufo marinus*. This strategy required an initial study of potential functional residues (or groups) on the amiloride-sensitive $Na^+$ channel at the apical or mucosal surface of this epithelium. [We use the term "channel" in preference to "carrier" or "permease" because of the accumulating evidence in favor of the "channel" nature of this entry step for sodium in tight epithelia (Lindemann and VanDriessche, 1977; Palmer *et al.*, 1982).] We gained initial information about functional groups on the $Na^+$ channel by examining the effect of mucosal pH upon amiloride-sensitive short-circuit current (SCC) (Fig. 4). We analyzed this curve by assuming that the residues titrating between pH 3.0 and 5.5 and between 5.5 and 9.0 were fully protonated at the lower of the two pH values and fully deprotonated at the higher of the two. Hill plots of the fractional change in SCC as a function of pH yielded apparent $pK_a$ values of 4.2 and 6.7, respectively. The Hill coefficients (0.94 and 0.93, respectively) were compatible with the interpretation that titration of only one residue was responsible for the increase in SCC between pH 3.0 and 5.5 and that titration of one other residue was responsible for the decrease in SCC between pH 5.5 and 9.0. The most likely residue that can titrate with a $pK_a$ of 4.2 is a carboxyl. The possible nature of the residue with a $pK_a$ of 6.7 is considered later. In the following paragraphs, further evidence is examined for the involvement of a carboxyl residue in the functioning of amiloride-sensitive $Na^+$ channel.

We tested a variety of reagents that react with selectivity for carboxyls. These included five carbodiimides (selected to provide compounds that varied in nature from very hydrophilic to very hydrophobic; Hoare and Koshland, 1967; Abrams and Barren, 1970) and EEDQ (*N*-ethoxycarbonyl-2-ethoxy-1,2-dihydroquinoline), a reagent of intermediate hydrophobicity (Belleau *et al.*, 1969; Herz and Packer, 1981). Hydrophilic carbodiimides produced no inhibition of SCC until concentration of the reagents exceeded $10^{-2}$ *M*. The most potent reagents for inhibiting SCC were the two amphipathic reagents, DPC (*N,N'*-diisopropylcarbodiimide) and EEDQ. From these findings we suggest that carboxyl residues are involved in the transport of sodium and, further, that the important carboxyls are in an environment of intermediate lipophilicity. A wide spectrum of experiments was conducted using EEDQ. Based on a variety of direct and indirect parameters we conclude that EEDQ, when added from the

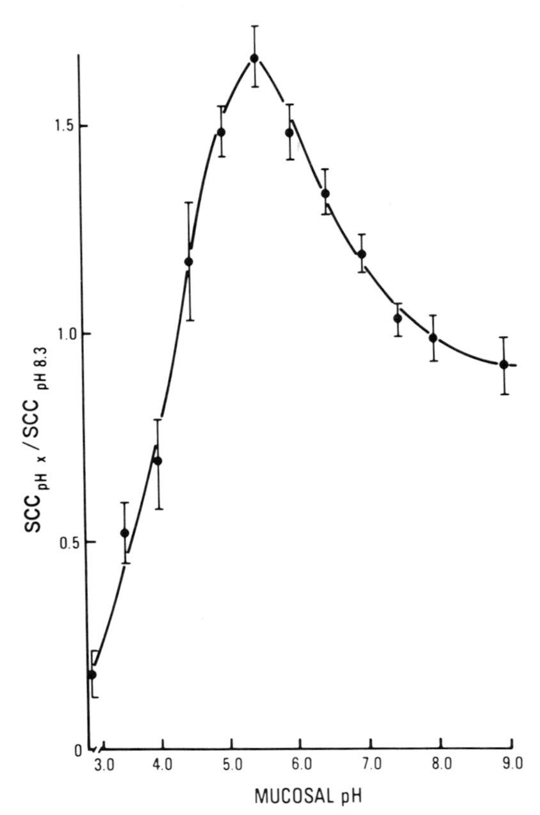


FIG. 4. Dependence of SCC on the mucosal pH. Octa-bladders from one toad were stabilized at mucosal pH 8.3. When SCC was stabilized, the pH of the mucosal Ringer's of each octa-bladder was varied from 3.0 to 5.5 in 0.5–1.0 pH intervals in one series of experiments and from 5.5 to 9.0 in another series of experiments. The stabilized SCC at each mucosal pH is expressed as the ratio of SCC to that at pH 8.3. Each point is mean ± SE from eight octa-bladders.

mucosal side, inhibits the $Na^+$ entry step across the apical membrane without effect on (1) cellular energy metabolism, (2) $Na^+$ pump activity, (3) ADH-stimulated water flow, (4) the permissive action of aldosterone on ADH-stimulated water flow, and (5) transepithelial proton transport (Fanestil and Park, 1981). One important evidence supporting our conclusion is shown in Table I, which demonstrates that permeabilization of the apical barrier to monovalent cations with amphotericin B (Lichtenstein and Leaf, 1965) enables the tissues previously inhibited by mucosal EEDQ to increase their SCC to the same absolute level as found in control tissues not treated with EEDQ.

Figure 5 shows a scheme for the reaction of EEDQ with carboxyls. Note that the activated carboxyl subsequently reacts with a nucleophile (the latter is depicted as $R{-}NH_2$ in the figure). The nucleophile shown in the reaction could be

TABLE I
REVERSAL OF THE EEDQ-INHIBITED SCC BY MUCOSAL AMPHOTERICIN B[a]

| | Pre-EEDQ | EEDQ | Amphotericin B |
|---|---|---|---|
| Control | 24.6 ± 3.7 | 23.3 ± 3.2 | 37.3 ± 3.3 |
| Experimental | 25.5 ± 3.8 | 3.7 ± 0.6 | 36.3 ± 3.1 |
| *p* | >0.1 | <0.001 | >0.1 |

[a] Values are means ± SE of SCC (μA). When SCC was stabilized (pre-EEDQ), EEDQ ($5 \times 10^{-4}$ *M*) was added to the mucosal Ringer's of the experimental bladders. Sixty minutes later EEDQ was washed out with EEDQ-free fresh Ringer's and SCC was allowed to stabilize (EEDQ). Amphotericin B (13 μg/ml) was added to the mucosal Ringer's of both control and experimental bladders. The maximal SCCs after amphotericin B were recorded (amphotericin B). *p* values are from paired comparison of SCCs of the control and experimental bladders ($N = 8$).

either exogenous or endogenous. Moreover, an endogenous nucleophile could be a nucleophilic group on the same peptide as the activated carboxyl or on an adjacent peptide. To test for reactivity with an exogenous nucleophile, we provided glycine methyl ester (GME) in the mucosal solution. The addition of EEDQ in the presence of GME could have resulted in two possible effects. First, the GME might react with the activated carboxyl to produce an inactive $Na^+$ channel. Thus, GME might accelerate the inhibition of SCC produced by EEDQ. Second, the reaction with GME might occur, but the $Na^+$ channel might not be inactivated. In this latter case, the carboxyls that had reacted with GME (the exogenous nucleophile) would not be available for reaction with the endogenous

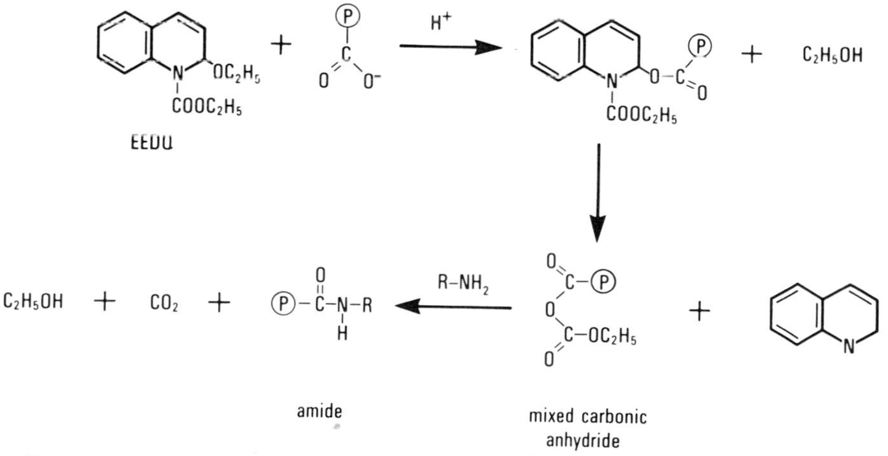

FIG. 5. Chemical reaction scheme for carboxyl modification by *N*-ethoxycarbonyl-2-ethoxy-1,2-dihydroquinoline (EEDQ) (Belleau *et al.*, 1969; Herz and Packer, 1981). R–$NH_2$ represents either endogenous or exogenous nucleophiles. Ⓟ represents a protein or polypeptide chain.

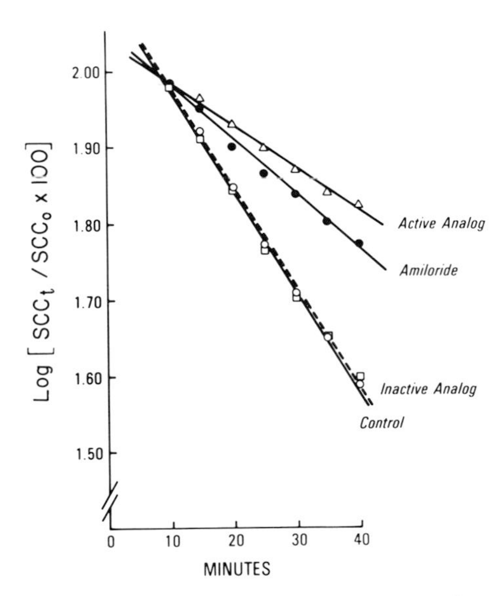


FIG. 6. Rate of inhibition of SCC by mucosal NBD-Cl ($2.5 \times 10^{-5}$ *M*) in the absence (control) and presence of amiloride and its two analogs (at the same concentrations given in the text. NBD-Cl was added to the mucosal Ringer's of four matched quarter-bladders, and SCC was recorded at 10-minute intervals. The rate of inhibition of SCC by NBD-Cl was significantly decreased by amiloride and an active analog (benzamil) but not by an inactive analog ($N = 9$). Data from Park and Fanestil (1980).

nucleophile. We found that EEDQ decreased SCC less when $10^{-2}$ *M* GME was present (55% inhibition without and 30% inhibition with GME). From this result we infer that the EEDQ-mediated reaction of the carboxyl on the $Na^+$ channel with an endogenous nucleophile results in inactivation or closure of the $Na^+$ channel, whereas reaction with an exogenous nucleophile does not. An implication of this conclusion is that the carboxyl, itself, could be reacted with or bound by an agent or drug without necessarily producing inhibition of sodium transport. The importance of this implication will become apparent after the effect of analogs of amiloride is discussed.

The most impressive evidence that EEDQ is reacting with a functional residue on the $Na^+$ channel comes as a result of use of the strategy of site-specific protection. We capitalize on the high specificity of amiloride for the $Na^+$ channel to produce such site-specific protection. We employed not only amiloride, but also benzamil (which has 10-fold higher activity than does amiloride) and an inactive analog of amiloride with 5-amino group substitution (Cuthbert and Fanelli, 1978). For comparison, we mention our previous experiments with the tyrosine-specific reagent, NBD-Cl (7-chloro-4-nitrobenzo-2-oxa-1,3-diazole). NBD-Cl inhibited $Na^+$ transport in the urinary bladder of toads apparently by blocking the $Na^+$ entry step across the apical membrane (Park and Fanestil, 1980). This is similar to the action of EEDQ, as described above. Amiloride ($2 \times 10^{-7}$ *M*) and benzamil ($2 \times 10^{-8}$ *M*), but not the inactive analog ($2 \times 10^{-7}$ *M*), protected against the inhibition of $Na^+$ transport by NBD-Cl, as shown in Fig. 6. In marked contrast, amiloride and its active as well as *inactive* analogs protected against the inhibition of $Na^+$ transport by EEDQ. For example, the half-time for inhibition of SCC by mucosal EEDQ was 70 minutes in the absence of amiloride, 127 minutes with amiloride ($2 \times 10^{-7}$ *M*), 136 minutes with benzamil ($2 \times 10^{-8}$ *M*), and 117 minutes with the analog ($2 \times 10^{-7}$ *M*) that did not inhibit SCC. The implications of this difference in the protective ability of the inactive analog with EEDQ vs NBD-Cl are discussed below.

What is the chemical nature of the endogenous nucleophile that, according to the scheme we are developing, protonates with a $pK_a$ of 6.7? Could t be the same nucleophile that reacts with EEDQ-activated carboxyls to produce inactivation of $Na^+$ channels? Could it be the tyrosine that is reactive with NBD-Cl? Could it be a sulfhydryl? These questions remain to be explored more fully.

## IV. HOW DOES AMILORIDE INHIBIT THE $Na^+$ CHANNEL?

We now describe a working model of functional residues on the $Na^+$ channel and advance an explanation for the mechanism whereby the model can account for inhibition of the $Na^+$ channel by amiloride. This model is based upon the data presented in the prior section. Until more information is available, we propose the simplest model consistent with our findings: the $Na^+$ channel has two functional residues, a carboxyl and a nucleophile, at the mucosal surface of the channel. (In this simplest of models, we assume that the endogenous nucleophile that reacts with the EEDQ-activated carboxyl, the group that titrates with $pK_a$ 6.7, and the tyrosyl that reacts with NBD-Cl are one and the same. Reality may be much more complex.) As mucosal pH is lowered from 9.0 to 5.5, the nucleophile titrates and results in greater conductance of the $Na^+$ entry step. As pH decreases below 5.5, titration of the carboxyl results in inactivation of $Na^+$ channels. At neutral pH, the cationic guanidium of amiloride (Fig. 7) binds electrostatically with the carboxyl on the $Na^+$ channel. We postulate that this one-point binding does not inactivate the channel. Thus, the inactive analog of amiloride (Fig. 7), which has the normal guanidinium structure of amiloride, can bind to the carboxyl without inhibiting sodium transport. This explains how analogs of amiloride that are either active or inactive on SCC can decrease the rate at which EEDQ activates the carboxyl to inhibit SCC. Moreover, the high affinity of the $Na^+$ channel for benzamil can also be explained by our knowledge, based on the fact that amphipathic carboxyl-reactive reagents are most

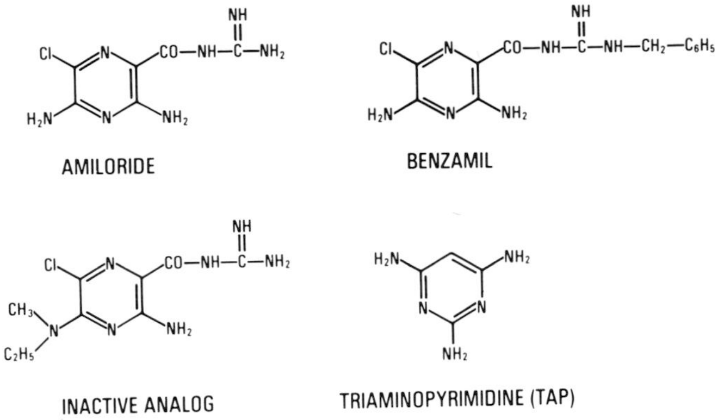

AMILORIDE

BENZAMIL

INACTIVE ANALOG

TRIAMINOPYRIMIDINE (TAP)

FIG. 7. Chemical structures of amiloride and its analogs.

inhibitory, that the carboxyl on the $Na^+$ channel is in a partially lipophilic environment. Benzamil, because it has a benzene at the charged guanidinium (Fig. 7), is more lipophilic than amiloride and therefore, we postulate, has more ready access to the amphipathic carboxyl. According to this model, the guanidinium portion of amiloride and its analogs "addresses" the molecule to the correct location on the $Na^+$ channel, but this one-site binding of the molecule to the carboxyl produces no inhibition of SCC. The inhibition of SCC (the true "message" carried by amiloride) is "delivered" by the pyrazine ring to a second site on the $Na^+$ channel. In our simplest of models, this second site is the nucleophile on the $Na^+$ channel. Substitutions on the pyrazine ring, such as the ethylmethyl substitution for $-NH_2$ [the inactive analog we used (Fig. 7)], interfere with delivery of the message or, in other words, this substitution on the pyrazine ring prevents the inactive analog from interacting with the nucleophile. In our simple model, this nucleophile is identical with the tyrosine that is reactive with NBD-Cl. Therefore, the nucleophile–tyrosine is available for reaction with NBD-Cl fully as well in the presence as in the absence of the pharmacologically inactive analog of amiloride. In contrast, the pharmacologically active analogs of amiloride do deliver the message by interacting with the nucleophile–tyrosine on the $Na^+$ channel and, thereby, decrease the reactivity of the tyrosine with NBD-Cl (Fig. 6). From this model of how amiloride inhibits the $Na^+$ channel in two steps, we can also advance an explanation for the ability, albeit with low affinity, of 2,4,6-triaminopyrimidium (TAP) to inhibit $Na^+$ channels (Fanestil and Vaughn, 1979; Balaban *et al.*, 1979). TAP (Fig. 7) lacks the guanidinium of amiloride and, therefore, lacks the high affinity moiety that addresses the molecule to the carboxyl on the $Na^+$ channel. However, we postulate that the message carried by the ring structure of TAP is sufficiently close to that of the pyrazine ring of amiloride that, when TAP is present at $\sim 10^{-3} M$, it can interact with the nucleophile that is essential for the function of the $Na^+$ channel.

We advance this simple, two-site model of the amiloride-sensitive $Na^+$ channel not because we believe it to be complete or necessarily fully correct, but because we believe such models provide the basis for the design of further experimental techniques and protocols that will eventually enable the construction of more reliable models of the amiloride-sensitive $Na^+$ channel.

## V. SUMMARY

A variety of strategies are being developed for the identification and characterization of membrane proteins involved in the transit of ions and other solutes across the apical plasma membrane of epithelial cells. In this chapter we have described briefly our work with two of these strategies. In the first strategy, we took advantage of the hydrophobic properties of membrane proteins to find a proteolipid (lipophilic protein) in renal brush borders that binds inorganic phosphate with high affinity and selectivity. We propose this protein, phosphorin, as a likely candidate for the phosphate-binding component of $Na^+$-dependent phosphate transport in renal brush borders. In the second strategy, we took advantage of the specificity of amiloride for the $Na^+$ channel in a high resistance epithelium to help define the involvement of a carboxyl residue and a nucleophile in the normal function of the sodium channel and in the mechanism with which amiloride locates and inhibits the $Na^+$ channel with selectivity.

## REFERENCES

Abrams, A., and Baron, C. (1970). Inhibitory action of carbodiimides on bacterial membrane ATPase. *Biochem. Biophys. Res. Commun.* **41,** 858–863.

Balaban, R. S., Mandel, L. J., and Benos, D. J. (1979). On the cross-reactivity of amiloride and 2,4,6 triaminopyrimidine (TAP) for the cellular entry and tight junctional cation permeation pathways in epithelia. *J. Membr. Biol.* **49,** 363–390.

Belleau, B., Ditullio, V., and Godin, D. (1969). The mechanism of irreversible adrenergic blockade by *N*-carbethoxydihydroquinolines—Model studies with typical serine hydrolases. *Biochem. Pharmacol.* **18,** 1039–1044.

Booth, A. G., and Kenny, A. J. (1974). A rapid method for the preparation of microvilli from rabbit kidney. *Biochem. J.* **142,** 575–581.

Burckhardt, C. T., Stern, H., and Murer, H. (1981). The influence of pH on phosphate transport into rat renal brush border membrane vesicles. *Pfluegers Arch.* **390,** 191–197.

Capaldi, R. A. (1982). Structure of intrinsic membrane proteins. *Trends Biochem. Sci.* **7,** 292–295.

Cheng, L., and Sacktor, B. (1981). Sodium gradient-dependent phosphate transport in renal brush border membrane vesicles. *J. Biol. Chem.* **256,** 1556–1564.

Cuthbert, A. W., and Fanelli, G. M. (1978). Effects of some pyrazinecarboxamides on sodium transport in frog skin. *Br. J. Pharmacol.* **63,** 139–149.

Fanestil, D. D., and Park, C. S. (1981). Effects of inhibitors of $F_0 + F_1$ proton-translocating ATPase on urinary acidification. *Am. J. Physiol.: Cell Physiol.* **240,** C201–C206.

Fanestil, D. D., and Vaughn, D. A. (1979). Inhibition of short-circuit current by triaminopyrimidine in isolated toad urinary bladder. *Am. J. Physiol.: Cell Physiol.* **236,** C221–C224.

Folch, J., and Lees, M. (1951). Proteolipids, a new type of tissue lipoproteins. Their isolation from brain. *J. Biol. Chem.* **191,** 807–817.

Guerin, M., and Napias, C. (1978). Phosphate transport in yeast mitochondria: Purification and characterization of a mitoribosomal synthesis dependent proteolipid showing a high affinity for phosphate. *Biochemistry* **17,** 2510–2516.

Herz, J. M., and Packer, L. (1981). Structural involvement of carboxyl residues in the photocycle of bacteriorhodopsin. *FEBS Lett.* **131,** 158–164.

Hoare, D. G., and Koshland, D. E. (1967). A method for the quantitative modification and estimation of carboxylic acid groups in protein. *J. Biol. Chem.* **242,** 2447–2453.

Kadenback, B., and Hadvary, P. (1973). Specific binding of phosphate by a chloroform-soluble protein from rat-liver mitochondria. *Eur. J. Biochem.* **39,** 21–26.

Kessler, R. J., Vaughn, D. A., and Fanestil, D. D. (1982a). Inhibition of the $Na^+$ dependent phosphate uptake in kidney brush border by the combination of a divalent ionophore and EDTA. *Fed. Proc. Fed. Am. Soc. Exp. Biol.* **41,** 1365.

Kessler, R. J., Vaughn, D. A., and Fanestil, D. D. (1982b). Phosphate-binding proteolipid from brush border. *J. Biol. Chem.* **257,** 14311–14317.

Lichtenstein, N. S., and Leaf, A. (1965). Effect of amphotericin B on the permeability of the toad bladder. *J. Clin. Invest.* **44,** 1328–1342.

Lindemann, B., and VanDriessche, W. (1977). Sodium-specific membrane channels of frog skin are pores: Current fluctuations reveal high turnover. *Science* **195,** 292–294.

McLaughlin, S. (1977). Electrostatic potentials at membrane-solution interfaces. *Curr. Top. Membr. Transp.* **9,** 71–144.

Palmer, L. G., Li, J. H.-Y., Lindemann, B., and Edelman, I. S. (1982). Aldosterone control of the density of sodium channels in the toad urinary bladder. *J. Membr. Biol.* **64,** 91–102.

Park, C. S., and Fanestil, D. D. (1980). Covalent modification and inhibition of an epithelial sodium channel by tyrosine-reactive reagents. *Am. J. Physiol. Renal Fluid Electrolyte Physiol.* **239,** F299–F306.

Schlesinger, M. J. (1981). Proteolipids. *Annu. Rev. Biochem.* **50,** 193–206.

# Chapter 16

# Aldosterone-Induced Proteins in Renal Epithelia

*MALCOLM COX AND MICHAEL GEHEB*[1]

*Renal Electrolyte Section, Medical Service*
*Philadelphia Veterans Administration Medical Center*
*and*
*Department of Medicine*
*University of Pennsylvania School of Medicine*
*Philadelphia, Pennsylvania*



[1]Present address: Department of Medicine, Wayne State University, Harper-Grace Hospital, Detroit, Michigan.




ISBN 0-12-153320-4

# I. INTRODUCTION

Since the early 1960s it has been known that aldosterone-stimulated $Na^+$ transport in renal epithelia is dependent on genetic derepression and new protein synthesis (17). The natriferic effect of the hormone has been examined in a variety of species, but its cellular mechanism of action has been studied most extensively in the toad urinary bladder. In this classic *in vitro* model of the mammalian distal nephron, aldosterone (and other adrenal corticosteroids) bind to specific cytosolic receptors and to two classes of nuclear receptors; occupation of the high-affinity, low-capacity (Type I) nuclear receptor correlates with the natriferic activity of the steroid (19, 45).

Aldosterone stimulates the incorporation of $[^3H]$uridine into 9–12 S non-methylated RNA (i.e., RNA species with characteristics of mRNA) with a time course concordant with that of aldosterone-stimulated $Na^+$ transport (66). The synthesis of this putative mRNA is blocked by spironolactone (53), and non-natriferic concentrations of glucocorticoids are devoid of stimulatory activity (66). Aldosterone-stimulated $Na^+$ transport appears to be dependent on the production of both polyadenylated and nonpolyadenylated mRNA (64, 84, 85). Aldosterone also stimulates rRNA synthesis in the toad urinary bladder, but the physiological implications of this are unclear. Although 3′-deoxycytidine (a "specific" inhibitor of rRNA synthesis) does not inhibit the early phase of aldosterone-stimulated $Na^+$ transport, its effects on the late phase of $Na^+$ transport (when new ribosome synthesis may be important) have not been evaluated (68, 86).

Most of the older evidence that aldosterone-stimulated $Na^+$ transport is mediated through new protein synthesis rests on inhibitor studies (16,77). More recently, several investigators have identified proteins that may be related to aldosterone's effect on $Na^+$ transport (see Section III). Before discussing these developments in detail we consider what is known about the subcellular site of action of aldosterone. Such information could prove helpful in focusing the search for the putative aldosterone-induced proteins (AIPs).

# II. SUBCELLULAR SITE OF ACTION OF ALDOSTERONE

## A. The Ussing–Skou Two-Barrier Model

Analysis of the nature and subcellular site of action of the proteins induced by aldosterone has been largely based on the well-known "two-barrier" Ussing–Skou model. In this model, $Na^+$ entry into the cell across the apical cell membrane is a passive process (the magnitude of which is determined by the electrochemical gradient across this portion of the plasma membrane), and the extrusion of $Na^+$ from the cell across the basolateral cell membrane is an active process (performed by the ATP-driven $Na^+$ pump, the enzymatic equivalent of which is the ouabain-sensitive $Na^+$,$K^+$-ATPase).

Within this framework, AIPs have been assigned three distinct, but not necessarily independent, regulatory roles (17). In the "$Na^+$ permease" hypothesis, the AIP is presumed to facilitate entry of $Na^+$ into the cell (by forming part of, or in some way activating, the apical $Na^+$ channel). In the "metabolic" hypothesis, the AIP is presumed to increase the provision of energy, in the form of ATP, to the $Na^+$ pump (by stimulating mitochondrial oxidative phosphorylation). In this hypothesis, the AIP could be a rate-limiting enzyme in the tricarboxylic acid cycle or could, in some other way, influence the activity of this cycle. In the "pump" hypothesis, the AIP is presumed to facilitate extrusion of $Na^+$ from the cell (by forming part of, or in some way activating, the basolateral $Na^+$ pump).

## B. Apical $Na^+$ Channel

Evidence consistent with an effect of the hormone on the apical permeability barrier includes an increase in apical $Na^+$ conductance (measured or estimated by a variety of different techniques) in aldosterone-stimulated anuran and mammalian epithelia (9, 13, 14, 22, 49, 54, 58, 70, 78, 81), and an increase in the presumptive $Na^+$ transport pool in toad urinary bladder epithelial cells treated with aldosterone (10, 12, 33, 48, 58).

In probing potential effects of aldosterone on apical membrane permeability, considerable use has been made of the epithelial $Na^+$ channel blocker, amiloride. This pyrazine diuretic is believed to bind either to a component of the $Na^+$ channel itself or to a channel-regulatory moiety in close proximity to the channel (4). Exposure of toad urinary bladders to aldosterone increased the number of high-affinity binding sites for [$^{14}$C]amiloride but had no effects on amiloride-binding affinity (14). Since actinomycin D and cycloheximide abolished aldosterone's effects on [$^{14}$C]amiloride binding, these results are consistent with the induction of new $Na^+$ channels. Park and Edelman (quoted in 18) and Crabbe (11) reported a decrease in the sensitivity of amphibian epithelial $Na^+$ transport to amiloride following treatment with aldosterone, an effect that is also compatible with the induction of new $Na^+$ channels.

Using shot noise analysis, Palmer *et al.* (58) have found qualitatively similar results: aldosterone increased the number of $Na^+$ channels but did not change the single-channel $Na^+$ current or the on- and off-rate constants for $Na^+$ channel blockade by amiloride. While these observations are consistent with the "permease" hypothesis, many important issues (dose dependence, time course, mineralocorticoid specificity, etc.) have yet to be addressed. The development of amiloride analogs that bind irreversibly to $Na^+$ channel proteins would represent a major step forward in our ability to examine whether such proteins are induced by aldosterone.

## C. Oxidative Phosphorylation

Although there is little doubt that aldosterone increases apical membrane $Na^+$ permeability, it has become equally clear that this is by no means the sole effect of the hormone. For example, microelectrode studies in toad skin have shown that aldosterone increases the conductance of the basolateral, as well as the apical, cell membrane (13, 54). In addition, aldosterone-stimulated $Na^+$ transport has been shown to be dependent on the metabolic state of the tissue (20, 81–83). Indeed, even the short-term maintenance of apical $Na^+$ permeability in the toad urinary bladder appears to require unimpeded energy metabolism (57, 58). Moreover, aldosterone has been demonstrated to increase the metabolic driving force (thermodynamic affinity) for epithelial $Na^+$ transport (2, 71, 80).

There is considerable information regarding the biochemical basis of aldosterone's effects on energy metabolism. Aldosterone augments the activity of several mitochondrial enzymes, most notably that of citrate synthase, in the toad urinary bladder (38, 42, 43), the mammalian kidney (40, 41, 47), and the rabbit cortical collecting tubule (52). These responses correlate with the dose dependence and time course of aldosterone-stimulated $Na^+$ transport and appear to be mineralocorticoid specific. In addition, the increase in citrate synthase activity appears to be independent of the entry of $Na^+$ into the cell (42, 51). However, in only one instance (47) has the increase in enzyme activity been shown to result from enzyme induction. Moreover, in several cultured renal epithelial cell lines (derived from the toad kidney and toad urinary bladder), aldosterone increases $Na^+$ transport without an associated increase in citrate synthase activity (38).

## D. $Na^+$,$K^+$-ATPase

Regardless of whether the action of aldosterone involves an increase in apical $Na^+$ entry or in the provision of energy (or both), the Ussing–Skou model requires that the final common pathway must involve an increase in the activity of the $Na^+$ pump in the basolateral cell membrane. Most attempts to demonstrate an aldosterone-dependent increase in $Na^+$,$K^+$-ATPase activity in the mammalian kidney have been flawed by uncertainties as to whether the observed effects correlated with the dose dependence and time course of the hormone's physiologic effects, whether the observed effects were mineralocorticoid specific, or whether cellular heterogeneity may have obscured a response in a small subpopulation of cells (8, 23, 35, 39, 44, 62, 72). However, recent studies in isolated rabbit cortical collecting tubules have shown that physiological amounts of aldosterone stimulate $Na^+,K^+$-ATPase with a time course similar to that of hormone-stimulated $Na^+$ transport and that this effect is spirolactone sensitive and, presumably therefore, mineralocorticoid specific (37, 60). However, for reasons that are unclear, Doucet and Katz (15) were unable to reproduce these results in either mouse or rabbit cortical collecting tubules.

The studies of Petty *et al.* (60) are of particular interest since the aldosterone-related increase in $Na^+,K^+$-ATPase activity was blocked by pretreatment with amiloride. This suggests that $Na^+,K^+$-ATPase activation is secondary to increased entry of $Na^+$ into the epithelial cell rather than being a "primary" effect of the hormone. Studies with cultured toad kidney epithelial cells (the A6 cell line) have also demonstrated an amiloride-sensitive increase in $Na^+,K^+$-ATPase activity following incubation with aldosterone (34). Indeed, the activation of $Na^+,K^+$-ATPase subsequent to increased uptake of $Na^+$ has been observed in a variety of different cells (6, 53, 69, 79).

Most attempts to demonstrate effects of aldosterone on the $Na^+$ pump in toad urinary bladder epithelial cells have been unsuccessful (18, 36, 78). Recently, however, monospecific antibodies to the $\alpha$ and $\beta$ subunits of the $Na^+$ pump have been employed to show that aldosterone produces a dose-dependent, parallel increase in the synthesis of both subunits after incubation with the hormone for 3 hours (25). This effect appears to be mineralocorticoid specific in that it was antagonized by spironolactone, and to be independent of changes in apical membrane permeability in that it was amiloride insensitive. As a result, it has been postulated that the "late" response to aldosterone in the toad urinary bladder is dependent on the synthesis of new $Na^+$ pumps. However, it should be recognized that the appearance of new pump subunits in whole cell extracts does not address the question of whether these subunits are assembled and inserted into the basolateral cell membrane (i.e., whether they are able to influence epithelial $Na^+$ transport). Indeed, insertion of new pumps into the cell membrane (and their subsequent expression as measured by an increase in $Na^+,K^+$-ATPase activity) may be dependent on entry of $Na^+$ into the cell, a fact that would explain the amiloride sensitivity of $Na^+,K^+$-ATPase activity previously demonstrated in the isolated rabbit cortical collecting tubule (60) and the A6 cell line (34). Clearly, more information concerning the postsynthesis processing and subcellular localization of newly induced pump subunits and the amiloride sensitivity of this process will be required before induction of the $Na^+$ pump can be considered to be a proven determinant of aldosterone-stimulated $Na^+$ transport.

In summary, it has become increasingly clear that it may not be possible to localize the "primary" effect of aldosterone to one specific cellular process. Rather, available evidence suggests that aldosterone influences several mutually interdependent processes which, collectively, determine the hormone's natriferic effect. It is clear, therefore, that until *all the proteins* induced by the hormone have been identified and characterized, we are unlikely to be able to describe the cellular mechanisms by which aldosterone modulates epithelial $Na^+$ transport in a comprehensive fashion. The development of reliable tools for the detailed analysis of epithelial cell protein synthesis and the ability to precisely relate biochemical changes to their electrophysiological counterparts are prerequisites for such an analysis. Recent advances toward this goal, and some potential future directions, are described in the remainder of this chapter.

## III. ALDOSTERONE-INDUCED PROTEINS: DETECTION

### A. Model Systems

Because of its relative simplicity of organization, the ease with which $Na^+$ transport can be monitored (using short-circuit current), and its apparent electrophysiological similarity to the mammalian distal nephron, the toad urinary bladder has been extensively employed in studies of aldosterone-stimulated $Na^+$ transport in general and of aldosterone-induced protein synthesis in particular. Indeed, with a single exception (46, 47) in which rat renal medullary slices were employed, all presently available studies of AIP synthesis have utilized anuran epithelia (3, 5, 25, 26–28, 34, 61, 73–75). Although it provides a larger amount of starting material, the renal medulla is extremely heterogeneous and does not lend itself to detailed physiological and biochemical correlative studies.

The use of the toad urinary bladder is not without problems of its own. Despite its relative simplicity, it is composed of several epithelial cell types, a basement membrane, and an underlying stroma of connective tissue and smooth muscle (50). The epithelium is rather simply separated from the underlying submucosa, but one is still left with a heterogeneous population of epithelial cells. Despite numerous efforts (32, 65, 76) there appears to be no satisfactory way of completely separating the different epithelial cell types and, therefore, of localizing AIP synthesis to one particular subpopulation of cells with certainty.

Scott and co-workers have performed extensive studies to identify AIPs in the toad urinary bladder. For the most part, these investigators have used disaggregated epithelia (separated into mitochondria-rich and granular cell fractions) as starting material (73–75). More recently, because of concerns regarding the viability of these separated cells (32), new protein synthesis in the intact urinary bladder epithelium has also been examined (61). In both cases aldosterone stimulated the synthesis of several proteins in mitochondria-rich cells, but similar proteins were consistently absent in granular cells. Since there is excellent evidence that it is the granular cell that is involved in $Na^+$ transport (50), it is difficult to integrate these observations into present concepts of anuran epithelial $Na^+$ transport.

Another problem with the toad urinary bladder is that unless large numbers of bladders are pooled, the amount of starting material is small. Unfortunately, pooling tissue from different animals, as was done in the protocol employed by Scott and co-workers (61, 73–75), makes correlative studies of the hormone's electrophysiological and biochemical efforts difficult, if not impossible. Because of this limitation, the assay used should be sufficiently sensitive to reliably detect new protein synthesis in *single* bladders. In such circumstances, correlative studies become feasible and AIP synthesis can be related to $Na^+$ transport (see Sections IV,B and C).

In recent years a number of renal epithelial cell lines have been established in culture. Of most relevance to the present discussion are the A6 cell line (derived from the kidney of the toad, *Xenopus laevis*) and the TB-6c and TB-M cell lines (derived from the urinary bladder of the toad, *Bufo marinus*). Under appropriate conditions, these cells form high-resistance epithelia with many of the characteristics of their parent tissues. The advantages of these cell lines for studies of epithelial transport phenomena have been extensively reviewed (31). For biochemical purposes at least three possible advantages should be considered: (1) the availability of relatively large amounts of potentially homogeneous starting material (unfortunately, while the A6 cell line is composed of a single cell type, this is not the case for the currently available bladder cell lines), (2) ready access to the entire circumference of the plasma membrane and the ability to selectively label apical and basolateral plasma membranes, and (3) the potential for the development of mutant cell lines that manifest discrete alterations in one or more components of the cellular machinery underlying $Na^+$ transport. The latter possibility is most exciting but, to our knowledge, has yet to be exploited. However, cultured cells have been used to examine the biochemical concomitants of aldosterone-induced $Na^+$ transport in a number of recent studies (5, 34, 38).

## B. Assays

Three rather distinct types of assays have been traditionally employed to detect new protein synthesis: enzyme kinetics, antibodies, and radiolabeled amino acids. The use of enzyme kinetics is based on the fact that the rate of a reaction ($V_{max}$) is related to the amount of active enzyme present (provided there is no change in the affinity of the enzyme for its substrate). The technique presupposes that (1) the protein of interest is an enzyme, (2) one knows in advance what enzyme to examine, and (3) the increase in $V_{max}$ is due to the synthesis of new enzyme rather than to activation of preformed inactive enzyme molecules. Obviously, therefore, its use is rather limited. Nonetheless, in the appropriate circumstances, it can be a powerful tool; the best application of this approach is found in studies of the induction of citrate synthase in the rat kidney (47).

The use of antibodies to detect AIP synthesis in renal epithelia has been rather limited to date. The studies of Geering *et al.* (25) with regard to the possible induction of $Na^+,K^+$-ATPase (see Section II,D) and the studies of Law and Edelman on the induction of rat renal medullary citrate synthase (47) are notable exceptions. However, with the development of hybridoma technology and with the increasing use of antibodies as probes of cellular function, we are likely to see increasing use of immunological techniques to identify AIPs. Antibodies may also prove invaluable in the identification of aldosterone-related alterations in plasma membrane structure and function. For example, it is possible that they could be used to determine whether new $Na^+$ pump subunits induced by aldosterone (25) are inserted into the basolateral cell membrane. One wonders, also, whether they could be used to examine the apical $Na^+$ channel. A word of caution: the mere appearance of a protein that is titrated with a specific antibody does not necessarily indicate that the protein has been newly synthesized. The preexistence of the protein in a form that is not recognized by, or is inaccessible to, the antibody has to be considered.

The incorporation of radiolabeled amino acid precursors has been the basis for the vast majority of studies of aldosterone-induced proteins reported to date. The development, in recent years, of amino acids of very high specific activity (e.g., [$^{35}$S]methionine and [$^{35}$S]cysteine, 1000 Ci/mmol; [*methyl*-$^{3}$H]methionine, 100 Ci/mmol) has enabled the labeling and subsequent detection of very small amounts of protein. The difficulty is not so much in the labeling of the protein of interest, but in its subsequent separation from other labeled proteins.

Most studies (3, 25, 27, 34, 46, 61, 73–75) have used one-dimensional polyacrylamide gel electrophoresis (1D-PAGE) for this purpose, but the resolving power of this technique (especially with regard to membrane proteins) is poor. Thus, although in initial studies using this technique we were able to identify AIPs in both membrane-rich and cytosolic fractions of toad bladder epithelial cells (27), the sensitivity of the technique was so limited that appropriate electrophysiological and biochemical correlative studies could not be performed.

## C. Two-Dimensional Polyacrylamide Gel Electrophoresis

We next turned to the use of two-dimensional polyacrylamide gel electrophoresis (2D-PAGE), a technique with much greater resolving power—one that separates proteins on the basis of differences in both isoelectric point and molecular weight (55). Whereas 1D-PAGE is only capable of resolving a few dozen proteins in a particular subcellular fraction, 2D-PAGE is able to distinguish several hundred. Consequently, the ability to examine new protein synthesis is increased by several orders of magnitude. The technique is reasonably simple to perform (28,55) and remarkably reproducible.

Nonetheless, there are several problems with 2D-PAGE that need to be considered. First, only relatively small amounts of protein (usually less than 100 μg) can be subjected to electrophoresis. The technique is analytical not preparative, a problem if purification of the protein of interest is desired. Second, not all cell proteins are equally well discriminated; basic proteins and low-molecular-weight proteins are particular problems in this regard. Fortunately, modifications of the original technique are available that alleviate some of these problems (30,56). Third (and this is a problem common to all protein electrophoretic techniques), one can never assume that all the proteins in a particular fraction, especially membrane-rich fractions, have been solubilized. Thus, even if proteins are identified under a particular set of conditions, it should not be assumed that these are the only ones induced by the hormone. Fourth, because of its extreme sensitivity, 2D-PAGE is apt to detect even minor amounts of proteolytically altered derivatives of a single induced protein. Whenever multiple induced proteins are detected, their relationship to one another should be considered and the question of proteolysis during tissue processing must be addressed.

Finally, it is important to realize that 2D-PAGE was not developed as a quantitative technique. Unlike 1D-PAGE, in which it is not uncommon to excise particular portions of the gel, elute the radiolabeled proteins, and determine their activity by direct counting (usually in a liquid scintillation counter) (3, 34, 61, 73–75), 2D-PAGE has relied almost exclusively on autoradiography. While the latter is extremely sensitive it does not lend itself easily to quantitation. However, some success has been obtained by applying computer techniques to the analysis of the two-dimensional protein maps (1, 7, 24).

## IV. ALDOSTERONE-INDUCED PROTEINS: CHARACTERIZATION BY TWO-DIMENSIONAL POLYACRYLAMIDE GEL ELECTROPHORESIS

### A. Physical Characteristics

Using two-dimensional polyacrylamide gel electrophoresis and autoradiography of [$^{35}$S]methionine-labeled proteins, we have been able to identify several AIPs in subcellular fractions of epithelial cells derived from *single* urinary bladders of the toad, *B. marinus* (28). Maximal natriferic concentrations of aldosterone (~$10^{-7}$ *M*) increase the synthesis of a characteristic "cluster" of four to six membrane-associated proteins (p*I* ~5.5–6.0; $M_r$ ~70,000–80,000) (Fig. 1) and a single cytosolic protein (p*I* ~5.7; $M_r$ ~ 70,000) (Fig. 2). More recent studies have shown that approximately half-maximal natriferic concentrations of aldosterone ($10^{-8}$ *M*) have similar effects, although induction of the cytosolic protein may be less consistent under these circumstances (26). The multiple

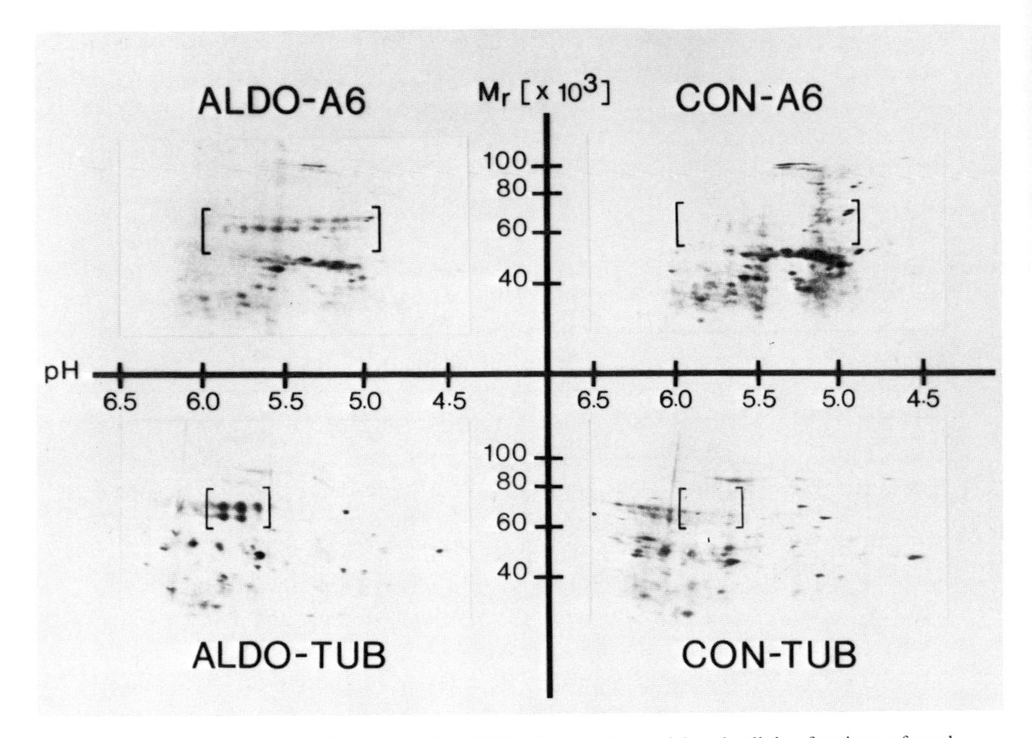


FIG. 1. Aldosterone-induced proteins (AIPs) in membrane-rich subcellular fractions of toad kidney epithelial cells (A6 cells) and toad urinary bladder epithelial cells (TUB). The brackets in the upper left autoradiograph delineate the AIPs observed in the membrane-rich fraction derived from A6 cells; the brackets in the lower left autoradiograph delineate the AIPs observed in an analogous fraction derived from toad urinary bladder epithelial cells. Autoradiographs from non-aldosterone-treated controls are shown on the right. pH and molecular weight ($M_r$) scales are provided to facilitate comparision of the proteins in the two types of cells. ALDO, aldosterone ($1.4 \times 10^{-7}$ *M*); CON, control. Reprinted with permission from Blazer-Yost *et al.* (5).

membrane-associated AIPs do not appear to be proteolytically derived components of a single induced protein, since the inclusion of protease inhibitors during tissue processing does not alter their appearance.

It is of interest to compare the approximate molecular weights of these proteins with those of $Na^+,K^+$-ATPase and citrate synthase. The $\alpha$ and $\beta$ subunits of toad kidney $Na^+,K^+$-ATPase have apparent molecular weights (determined by SDS–gel electrophoresis) of ~96,000 and ~60,000, respectively (29). The value for the $\alpha$ subunit agrees well with that determined (by a variety of different techniques, including sedimentation equilibrium) for $\alpha$ subunits derived from other tissues (21, 59). The true molecular weight of the $\beta$ subunit, which is heavily glycosylated, has been more difficult to determine. Using sedimentation equilibrium, the glycosylated (native) and nonglycosylated $\beta$ subunits have mo-

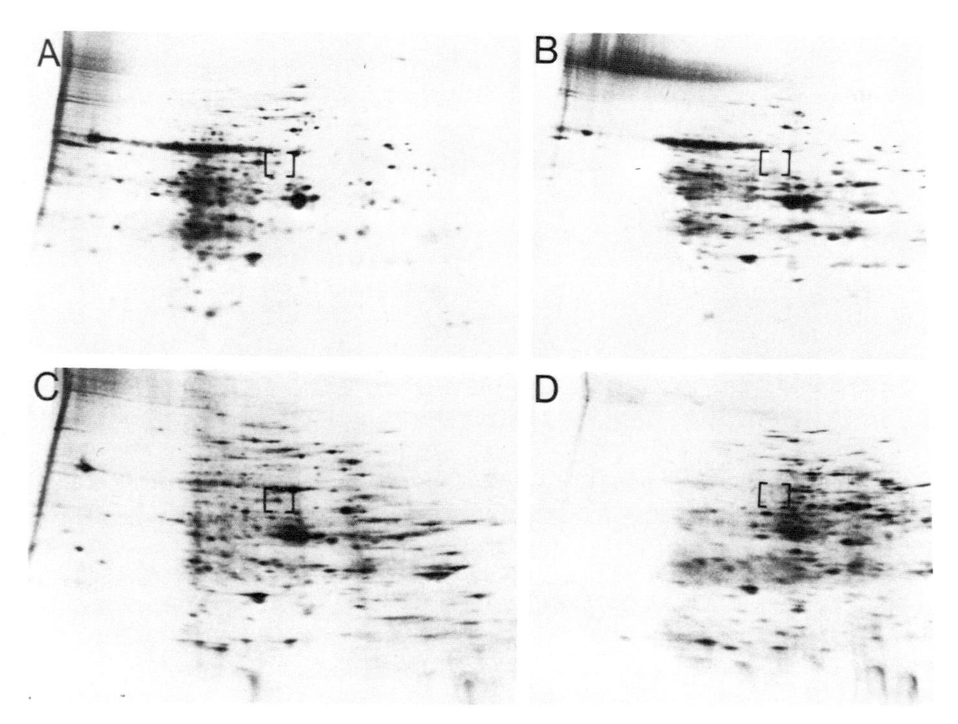


FIG. 2. Cytosolic aldosterone-induced proteins (AIP) in toad urinary bladder epithelial cells. The upper autoradiographs represent matched hemibladders incubated with (A) or without (B) aldosterone ($1.4 \times 10^{-7}$ *M*). The lower autoradiographs represent paired hemibladders incubated with aldosterone alone (C) or with both aldosterone and spironolactone (D); the antagonist-to-agonist ratio was 2000:1. The cytosolic AIP is present in the hemibladders treated with aldosterone alone (brackets in A and C) but is absent in the control hemibladder (B) and in the hemibladder treated with both aldosterone and spironolactone (D).

lecular weights of ~41,000 and ~32,000, respectively (21). In contrast, using SDS–gel electrophoresis and/or gel filtration in guanidinium chloride, the apparent molecular weight of the β subunit (isolated from a variety of tissues) has been found to range from ~36,000 to ~65,000. Although the molecular weight of the α subunit can be accurately assessed by SDS–gel electrophoresis, such is not the case for the β subunit. (59). It would appear, therefore, that the membrane-associated proteins we have identified are probably not the α subunit of $Na^+,K^+$-ATPase. However, given the uncertainty regarding the molecular weight of the β subunit and its "anomalous" electrophoretic behavior, the possibility that these proteins represent the β subunit may warrant further investigation.

The native form of citrate synthase (isolated from several different mammalian tissues) has a molecular weight of ~100,000, but in SDS it may dissociate into

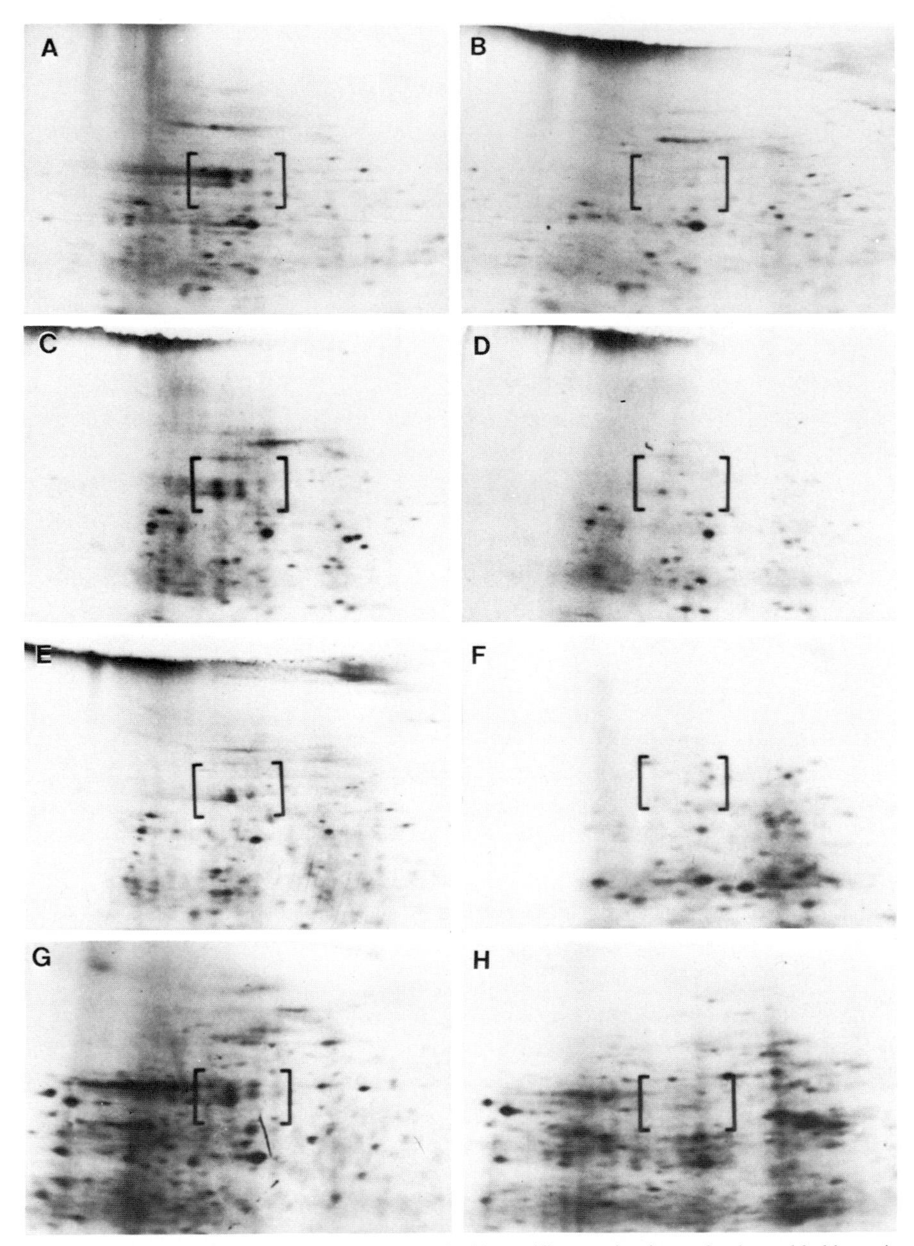


FIG. 3. Membrane-associated mineralocorticoid-specific proteins in toad urinary bladder epithelial cells. Each horizontal pair of autoradiographs represents matched hemibladders. The brackets delineate that region of the gel in which the proteins induced by aldosterone would normally be located. The steroids employed were (A) aldosterone ($1.4 \times 10^{-8}$ *M*); (B) aldosterone plus spironolactone ($1.4 \times 10^{-8}$ *M* and $2.8 \times 10^{-5}$ *M*, respectively); (C) dexamethasone ($1.4 \times 10^{-7}$ *M*); (D) dexamethasone plus spironolactone ($1.4 \times 10^{-7}$ *M* and $2.8 \times 10^{-4}$ *M*, respectively); (E) corticosterone ($1.4 \times 10^{-7}$ *M*); (F) corticosterone plus spironolactone ($1.4 \times 10^{-7}$ *M* and $2.8 \times$

two similar subunits with molecular weights of ~53,000 (87). Scott and Skipski (unpublished data quoted in 61) have apparently isolated citrate synthase, with a molecular weight of ~48,000, from toad myocardium; if this represents the subunit, it is in reasonable agreement with the mammalian data. It would appear, therefore, that the proteins we have identified do not represent citrate synthase.

Proteins with similar characteristics have been identified in membrane-rich (but not in cytosolic) fractions of cells of the A6 line derived from the kidney of the toad, *X. laevis* (5) (Fig. 1). The molecular weights of the AIPs are very similar in the two types of epithelial cells; proteins belonging to two discrete molecular weight classes (in the general range of 70,000–80,000) are present in both epithelia. However, with regard to the isoelectric points of the proteins, differences exist between the two types of cells. AIPs similar to those present in the toad urinary bladder appear to be present in the A6 cells as well, but the latter also contain several additional AIPs with more acidic isoelectric points (Fig. 1). The reason for the existence of these more acidic proteins in A6 cells is unknown, but differences may exist in the posttranslational processing (e.g., glycosylation, phosphorylation) of AIPs in the two epithelia.

Whatever the case, the similarity of the AIPs in these two different renal epithelia is consistent with an important role for these proteins in aldosterone-stimulated $Na^+$ transport. However, the apparent absence of a cytosolic AIP in A6 cells remains to be explained. Interestingly, in the toad urinary bladder, the cytosolic AIP has essentially identical characteristics (size, isoelectric point) to one of the membrane-associated AIPs (in Fig. 1, the more acidic of the two lower molecular weight AIPs). The relationship of the cytosolic AIP to the membrane-associated AIPs, and their respective roles in the cellular response to aldosterone, have yet to be defined.

## B. Relationship to Aldosterone-Stimulated $Na^+$ Transport

Our studies of aldosterone-induced protein synthesis in toad urinary bladders were predicated on the development of biochemical techniques of sufficient sensitivity that the biochemical and electrophysiological responses to the hormone could be monitored in the same tissue (a single hemibladder or quarter-bladder). Under such circumstances, detailed correlative studies become feasible and the induced proteins can be related to aldosterone-stimulated $Na^+$ transport.

---

$10^{-4}$ *M*, respectively); (G) cortisol ($7.0 \times 10^{-6}$ *M*); (H) cortisol ($5 \times 10^{-8}$ *M*). In each case, natriferic concentrations of the corticosteroids induced proteins with very similar characteristics (A, C, E, and G). Spironolactone, in concentrations that completely blocked steroid-induced $Na^+$ transport, inhibited the synthesis of these proteins (B, D, and F). Nonnatriferic concentrations of cortisol did not induce the synthesis of these proteins (H).

The induction of both the cytosolic and the membrane-associated AIPs appears to be necessary for expression of the hormone's natriferic effect. Actinomycin D (at concentrations that block aldosterone-stimulated $Na^+$ transport and new mRNA synthesis, but have no effect on basal $Na^+$ transport) abolishes their synthesis (28). In addition, the induction of these proteins is not secondary to enhanced epithelial $Na^+$ transport. Vasopressin-stimulated $Na^+$ transport in the toad urinary bladder is not associated with the synthesis of proteins similar to the AIPs, and AIP synthesis in both the toad urinary bladder and in A6 cells is unaffected by preincubation with amiloride (5, 28).

The synthesis of these proteins is not a nonspecific steroid effect. For example, spironolactone, a specific mineralocorticoid antagonist in renal epithelia (63, 67), inhibits their synthesis (26) (Figs. 2 and 3). In addition, while non-natriferic concentrations of cortisol ($5 \times 10^{-8}$ *M*) do not induce similar proteins (28), natriferic concentrations of this steroid ($7 \times 10^{-6}$ *M*) induce proteins with identical characteristics to the AIPs (Fig. 3). Natriferic concentrations of other corticosteroids (e.g., $10^{-7}$ *M* corticosterone, $10^{-7}$ *M* dexamethasone) also induce proteins with identical characteristics to the AIPs; both the biochemical and the electrophysiological (natriferic) effects of these steroids are inhibited by spironolactone (Fig. 3). The induction of these proteins by natriferic concentrations of glucocorticoids should not be surprising, since glucocorticoids can bind to the mineralocorticoid receptor (19, 45). Thus, it appears that whenever a corticosteroid stimulates $Na^+$ transport the characteristic proteins are induced.

## C. Quantitation of Aldosterone-Induced Protein Synthesis

Taken together, the above studies strongly suggest that the proteins we have identified are involved in aldosterone-stimulated $Na^+$ transport. However, they in no way prove that the synthesis of these proteins is an absolute prerequisite for the natriferic effect of the hormone. This question would be best addressed by examining the magnitude and time course of AIP synthesis and relating these parameters to the concentration dependence and time course of aldosterone-stimulated $Na^+$ transport. The toad urinary bladder would be ideal for such experiments, provided that a quantitative measure of AIP synthesis was available.

Recently, using double-isotope labeling and two-dimensional polyacrylamide gel electrophoresis, we have developed an "assay" of AIP synthesis that appears to meet this specification. In a typical experiment, [$^{35}$S]methionine is incubated with the aldosterone-treated tissue and [*methyl*-$^{3}$H]methionine is incubated with the control tissue. Isotope reversal experiments have shown that the results are not due to differences in the cellular processing of the two labeled species.

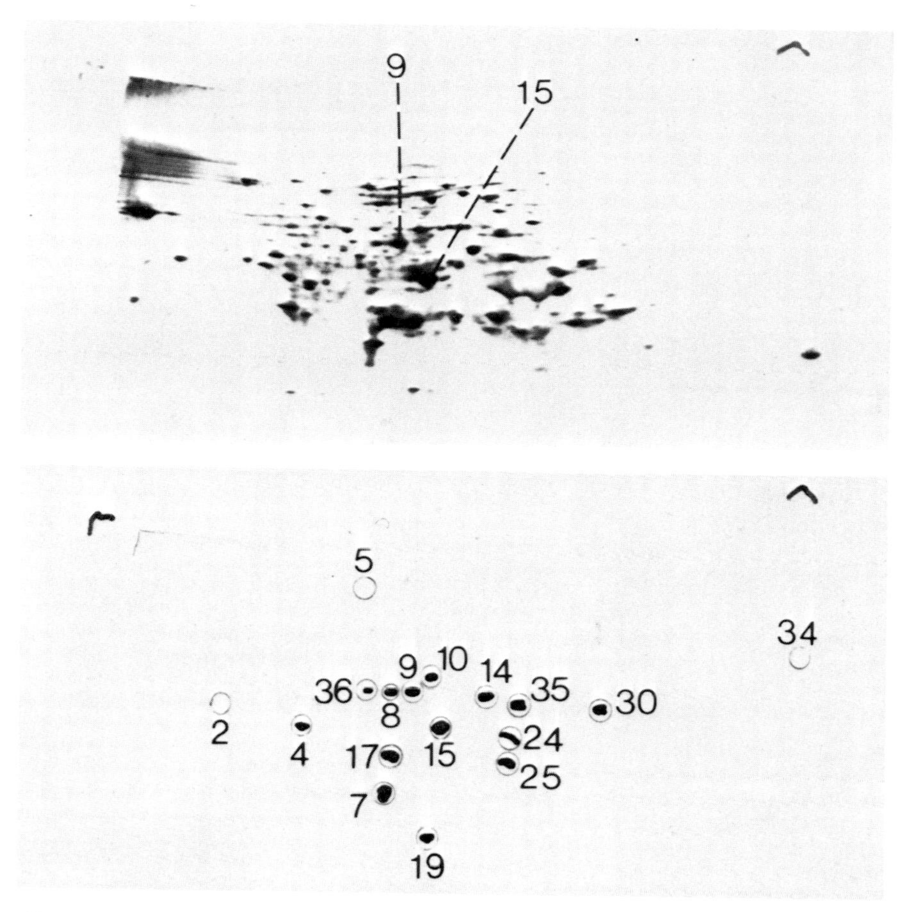


FIG. 4. Autoradiographic "template" of toad urinary bladder epithelial cell cytosolic proteins. The actual autoradiograph is shown in the upper panel. A schematic with the numbering system employed in Tables I and II is shown in the lower panel. The dried gel, from which the autoradiograph is derived, is overlaid with the autoradiograph, and radioactively labeled proteins, corresponding to specific autoradiographic "dots," are excised from the gel with a steel punch (see text).

Following the incubations, the epithelial cells from the two pieces of tissue are combined and then processed jointly in the usual fashion (28).

The autoradiographs prepared from the dried gels are used as templates (Fig. 4) for the excision of specific portions of the gel with a steel punch. In this fashion, "cores" of dried gel, measuring as little as 2–3 mm in diameter and containing specific proteins of interest, can be easily obtained. Appropriate measures are taken to ensure that the proteins are completely released from the gel cores prior to liquid scintillation counting. In this regard, we have found it important to reswell the dry cores prior to their extraction with soluene 350. In

TABLE I
DOUBLE-LABEL INCORPORATION[a]

| Gel location | Incorporation (dpm) $^3H$ | $^{35}S$ | Ratio ($^{35}S/^3H$) | Normalized ratio |
|---|---|---|---|---|
| 2 | 3,337 | 2,958 | 0.89 | 0.86 |
| 4 | 5,630 | 7,179 | 1.28 | 1.24 |
| 5 | 282 | 284 | 1.01 | 0.98 |
| 7 | 11,642 | 9,756 | 0.84 | 0.82 |
| 8 | 2,938 | 3,066 | 1.04 | 1.01 |
| 9 | 3,994 | 9,538 | 2.39 | 2.32 |
| 10 | 8,531 | 11,280 | 1.32 | 1.28 |
| 14 | 8,007 | 7,852 | 0.98 | 0.95 |
| 15 | 25,918 | 32,917 | 1.27 | 1.23 |
| 17 | 6,603 | 6,303 | 0.95 | 0.92 |
| 19 | 3,489 | 4,024 | 1.15 | 1.12 |
| 24 | 7,369 | 6,895 | 0.94 | 0.91 |
| 25 | 12,488 | 12,190 | 0.98 | 0.95 |
| 30 | 5,260 | 5,427 | 1.03 | 1.00 |
| 34 | 144 | 148 | 1.02 | 0.99 |
| 35 | 7,133 | 6,643 | 0.93 | 0.90 |
| 36 | 3,267 | 2,623 | 0.80 | 0.78 |

[a] The locations of the cytosolic AIP (protein at location 9) and of selected proteins not induced by aldosterone are shown in Fig. 4. The mean (±1 SD) $^{35}S/^3H$ incorporation ratio in this experiment is $1.11 \pm 0.36$ ($n = 17$); the only protein with a ratio greater than 2 SD from this mean is at location 9. If location 9 is excluded, the mean (±1 SD) $^{35}S/^3H$ incorporation ratio is $1.03 \pm 0.15$ ($n = 16$). The normalized ratios were obtained by dividing each individual incorporation ratio by the mean incorporation ratio (calculated with 9 excluded). The mean (±1 SD) normalized incorporation ratio was $1.00 \pm 0.15$ and the only protein with a ratio greater than 2 SD from this mean is at location 9.

addition, we have found that dimilume 30 completely eliminates the chemiluminescence that is a very real problem when counting low levels of $^3H$. Standard double-label quench curves are employed to correct for $^{35}S$ spillover and the results are expressed as absolute $^{35}S$ and $^3H$ counts and as isotope ratios (Table I).

In any given experiment it is also useful to excise a number of proteins whose synthesis is not altered by aldosterone. The mean isotope incorporation ratio of these noninduced proteins can then be employed to "normalize" the incorporation ratios of induced proteins, thus allowing for comparisons between different experiments (Table II).

We initially utilized this approach to confirm the existence of the cytosolic and membrane-associated AIPs previously detected autoradiographically (Figs. 1 and 2) and are now in the process of using it to examine the dose dependence of AIP

TABLE II
DOUBLE-LABEL ANALYSIS OF CYTOSOLIC PROTEIN SYNTHESIS[a]

| | Ratio | | Normalized ratio | |
|---|---|---|---|---|
| Experiment | Location 9 | Mean ±1 SD | Location 9 | 2 SD |
| A | 3.24 | 2.03 ± 0.44 | 1.66 | 0.33 |
| B | 1.36 | 0.98 ± 0.13 | 1.42 | 0.20 |
| C | 3.58 | 1.72 ± 0.56 | 2.23 | 0.36 |
| D | 2.39 | 1.11 ± 0.36 | 2.32 | 0.30 |

[a] Four separate experiments were performed. In each case, one hemibladder of a matched pair was treated with aldosterone ($1.4 \times 10^{-7}$ *M*) for 10 hours; the other hemibladder served as a control. The aldosterone-treated hemibladder received [$^{35}S$]methionine, and the control hemibladder [*methyl*-$^{3}H$]methionine, for the last 4 hours of the incubation period. Cytosolic and membrane-rich fractions were obtained as previously described (28) and new protein synthesis in each fraction was analyzed as described in the text. In all four experiments, the only cytosolic protein with a ratio greater than 2 SD from the mean incorporation ratio was located at position 9. (The analysis of Experiment D is shown in detail in Table I.) The characteristic AIPs (Fig. 1) in the membrane-rich fraction were, likewise, identified (data not shown).

synthesis. Preliminary indications are that it can discriminate between the relative amounts of protein induced by different concentrations of aldosterone (in the range $10^{-9}$–$10^{-7}$ *M*). Experiments relating new protein synthesis to the time course of aldosterone's natriferic response are also planned. These data should provide a definitive answer to whether the proteins we have identified have an integral role in aldosterone-stimulated $Na^+$ transport.

## V. SUMMARY

There is evidence that aldosterone influences all three of the major components of the $Na^+$ transport pathway in high-resistance epithelia: the apical membrane $Na^+$ permease, the generation of ATP, and the $Na^+$,$K^+$-ATPase pump. Which of these represents the primary effect of the hormone is unclear. Indeed, a single, primary effect may not exist. Rather, the hormone may modulate all of these processes in a coordinated and interdependent fashion, the net result of which is enhanced epithelial $Na^+$ transport. Under such circumstances it is unlikely that the induction of one, and only one, protein will completely account for the natriferic effect of the hormone. Instead, aldosterone may increase $Na^+$ transport by inducing the synthesis of several proteins, each of which regulates a particular component of the cell's $Na^+$ transport machinery.

If aldosterone does act in a multifactorial fashion, it will be important to identify and characterize all the induced proteins. Only then will a comprehensive picture of the cellular events underlying hormone action emerge. In this regard one should keep in mind that the natriferic response to aldosterone has a relatively long time course and that it may be heterogeneous in nature (81). Indeed, it is possible that different proteins may be induced at different times during the response. Consequently, it will be important not only to conduct a comprehensive search for all induced proteins but also to examine their temporal interrelationships.

The availability of sensitive, reliable, and quantitative assays of AIP synthesis is a prerequisite for such precise correlative studies. The studies of Geering *et al.* (25) are an example of how antibodies can be used as the basis for such an assay, and it is likely that immunochemical techniques will see much greater use in the identification of AIPs in the future. The development of the double-label 2D-PAGE technique may also prove extremely helpful. However, its successful application will likely be dependent on at least partial purification of the proteins we have identified.

In this regard, the determination of the precise subcellular location of the membrane-associated AIPs will be an important first step. Consequently, the development of a reliable method for the subcellular fractionation of toad bladder epithelial cells is one of our next goals. At a minimum, such a method should be able to obtain plasma membranes essentially free of mitochondrial contamination and be able to separate apical from basolateral plasma membranes. Not only would this improve our assay capability but it would also provide considerable insight into the function of the membrane-associated AIPs (apical membrane permease versus basolateral membrane pump versus mitochondrial enzyme). Other purification techniques are more speculative and would be dependent on the chemical characterization of the induced proteins. For example, we have preliminary evidence that the AIPs we have identified are glycoproteins. If this is the case, lectin chromatography may prove useful in their purification. Of course, the development of specific antibodies to these proteins would also be immensely helpful. However, this implies the prior ability to at least partially purify the proteins of interest.

In summary, despite intensive investigation, the delineation of the cellular functions of the proteins induced by aldosterone is still at a relatively early stage. Success will ultimately be dependent on (1) the development of reliable and quantitative means for the identification of all proteins induced by the hormone, (2) the precise correlation of the hormone's electrophysiologic and biochemical effects (including the determination of potential interrelationships between the various proteins), (3) the subcellular localization of the induced proteins, and (4) the purification and chemical characterization of these proteins. Anything less than such a comprehensive approach is unlikely to yield a meaningful picture of how aldosterone modulates epithelial $Na^+$ transport.

## ACKNOWLEDGMENTS

Much of our work would have been impossible without the skilled technical assistance of Eileen Hercker and Gary Huber. We also express our appreciation to Drs. Irwin Singer and Joseph Handler and to Ms. Bonnie Blazer-Yost for many helpful discussions. We would also like to thank Dr. Bernard Rossier for providing us with unpublished data. Ms. Renee Pannell provided able secretarial assistance. Our research has been supported by the Philadelphia Veterans Administration Medical Center (Research Fund 103.14M), the National Institutes of Health (Grant 1-R01-AM-21454), and the American Heart Association (Pennsylvania Affiliate). M. C. is an Established Investigator of the American Heart Association and M. G. is a Research and Education Associate of the Veterans Administration.

## REFERENCES

1. Alexander, A., Cullen, B., Emigholz, K., Norgard, M. V., and Monahan, J. J. (1980). A computer program for displaying two-dimensional gel electrophoresis data. *Anal. Biochem.* **103,** 176–193.
2. Beauwens, R., Beaujean, V., and Crabbe, J. (1982). The significance of changes in thermodynamic affinity induced by aldosterone in sodium-transporting epithelia. *J. Membr. Biol.* **68,** 11–18.
3. Benjamin, W. B., and Singer, I. (1974). Aldosterone-induced protein in toad urinary bladder. *Science* **186,** 269–272.
4. Benos, D. J. (1982). Amiloride: A molecular probe of sodium transport in tissues and cells. *Am. J. Physiol.* **242,** C131–C145.
5. Blazer-Yost, B., Geheb, M., Preston, A., Handler, J., and Cox, M. (1982). Aldosterone-induced proteins in renal epithelia. *Biochim. Biophys. Acta* **719,** 158–161.
6. Boardman, L., Huett, M., Lamb, J. F., Newton, J. P., and Polson, J. M. (1974). Evidence for the genetic control of the sodium pump density in HeLa cells. *J. Physiol. (London)* **241,** 771–794.
7. Bossinger, J., Miller, M. J., Kiem-Phong, Vo, Geiduschek, E. P., and Nguyen-Huu Xuong (1979). Quantitative analysis of two-dimensional electrophoretograms. *J. Biol. Chem.* **254,** 7986–7998.
8. Charney, A. N., Silva, P., Besarab, A., and Epstein, F. H. (1974). Separate effects of aldosterone, DOCA, and methylprednisolone on renal Na-K-ATPase. *Am. J. Physiol.* **227,** 345–350.
9. Civan, M. M., and Hoffman, R. E. (1971). Effect of aldosterone on electrical resistance of toad bladder. *Am. J. Physiol.* **220,** 324–328.
10. Crabbe, J. (1974). Influence of aldosterone on active sodium transport by toad bladder: A kinetic approach. *Acta Physiol. Scand.* **90,** 417–426.
11. Crabbe, J. (1980). Decreased sensitivity to amiloride of amphibian epithelia treated with aldosterone. *Pfluegers Arch.* **383,** 151–158.
12. Crabbe, J., and deWeer, P. (1969). Relevance of sodium transport pool measurements in toad bladder tissue for the elucidation of the mechanism whereby hormones stimulate active sodium transport. *Pfluegers Arch.* **313,** 197–221.
13. Crabbe, J., and Nagel, W. (1982). Analysis of cellular reaction to glucose of toad skin treated with aldosterone. *Pfluegers Arch.* **393,** 130–132.
14. Cuthbert, A. W., and Shum, W. K. (1975). Effects of vasopressin and aldosterone on amiloride binding in toad bladder epithelial cells. *Proc. R. Soc. London* **189,** 543–575.

15. Doucet, A., and Katz, A. I. (1981). Short-term effect of aldosterone on Na-K-ATPase in single nephron segments. *Am. J. Physiol.* **241,** F273–F278.
16. Edelman, I. S. (1968). Aldosterone and sodium transport. *In* "Functions of the Adrenal Cortex" (K. McKerns, ed.), pp. 79–133. Appleton, New York.
17. Edelman, I. S. (1978). Candidate mediators in the action of aldosterone on $Na^+$ transport. *In* "Membrane Transport Processes" (J. F. Hoffman, ed.), Vol. 1, pp. 125–140. Raven, New York.
18. Edelman, I. S., and Marver, D. (1980). Mediating events in the action of aldosterone. *J. Steroid Biochem.* **12,** 219–224.
19. Farman, N., Kusch, M., and Edelman, I. S. (1978). Aldosterone receptor occupancy and sodium transport in the urinary bladder of *Bufo marinus*. *Am. J. Physiol.* **235,** C90–C96.
20. Fimognari, G. M., Porter, G. A., and Edelman, I. S. (1967). The role of the tricarboxylic acid cycle in the action of aldosterone on sodium transport. *Biochim. Biophys. Acta* **145,** 89–99.
21. Freytag, J. W., and Reynolds, J. A. (1981). Polypeptide molecular weights of the ($Na^+$,$K^+$)-ATPase from porcine kidney medulla. *Biochemistry* **20,** 7211–7214.
22. Frizzell, R. A., and Schultz, S. G. (1978). Effect of aldosterone on ion transport by rabbit colon in vitro. *J. Membr. Biol.* **39,** 1–26.
23. Garg, L. C., Knepper, M. A., and Burg, M. B. (1981). Mineralocorticoid effects on Na-K-ATPase in individual nephron segments. *Am. J. Physiol.* **240,** F536–F544.
24. Garrels, J. I. (1979). Two-dimensional gel electrophoresis and computer analysis of proteins synthesized by clonal cell lines. *J. Biol. Chem.* **254,** 7961–7977.
25. Geering, K., Girardet, M., Bron, C., Kraehenbühl, J-P., and Rossier, B. C. (1982). Hormonal regulation of ($Na^+$,$K^+$)-ATPase biosynthesis in the toad bladder: Effect of aldosterone and 3,5,3′-triiodo-L-thyronine. *J. Biol. Chem.* **257,** 10338–10343.
26. Geheb, M., Alvis, R., Hercker, E., and Cox, M. (1983). Mineralcorticoid-specificity of aldosterone-induced proteins in toad urinary bladder. *Biochem. J.* **214,** 29–35.
27. Geheb, M., Hercker, E., Singer, I., and Cox, M. (1981). Subcellular localization of aldosterone-induced proteins in toad urinary bladders. *Biochim. Biophys. Acta* **641,** 422–426.
28. Geheb, M., Huber, G., Hercker, E., and Cox, M. (1981). Aldosterone-induced proteins in toad urinary bladders. Identification and characterization using two-dimensional polyacrylamide gel electrophoresis. *J. Biol. Chem.* **256,** 11716–11723.
29. Girardet, M., Geering, K., Frantes, J. M., Geser, D., Rossier, B. C., Kraehenbühl, J. P., and Bron, C. (1981). Immunochemical evidence for a transmembrane orientation of both the ($Na^+$,$K^+$)-ATPase subunits. *Biochemistry* **20,** 6684–6691.
30. Goldsmith, M. R., Rattner, E. C., Koehler, M. D., Balikov, S. R., and Bock, S. C. (1979). Two-dimensional electrophoresis of small molecular weight proteins. *Anal. Biochem.* **99,** 33–40.
31. Handler, J. S., Perkins, F. M., and Johnson, J. P. (1980). Studies of renal cell function using cell culture techniques. *Am. J. Physiol.* **238,** F1–F9.
32. Handler, J. S., and Preston, A. S. (1976). Study of enzymes regulating vasopressin-stimulated cyclic AMP metabolism in separated mitochondria-rich and granular epithelial cells of toad urinary bladders. *J. Membr. Biol.* **26,** 43–50.
33. Handler, J. S., Preston, A. S., and Orloff, J. (1972). Effect of ADH, aldosterone, oubain and amiloride on toad bladder epithelial cells. *Am. J. Physiol.* **222,** 1071–1074.
34. Handler, J. S., Preston, A. S., Perkins, F. M., Matsumura, M., Johnson, J. P., and Watlington, C. O. (1981). The effect of adrenal steroid hormones on epithelia formed in culture by A6 cells. *Ann. N.Y. Acad. Sci.* **372,** 442–454.
35. Hendler, E. D., Toretti, J., Kupor, L., and Epstein, F. H. (1972). Effects of adrenalectomy and hormone replacement on Na-K-ATPase in renal tissue. *Am. J. Physiol.* **222,** 754–760.

36. Hill, J. H., Cortas, N., and Walser, M. (1973). Aldosterone action and sodium- and potassium-activated adenosine triphosphatase in toad bladder. *J. Clin. Invest.* **52,** 185–189.
37. Horster, M., Schmid, H., and Schmidt, U. (1980). Aldosterone *in vitro* restores nephron Na-K-ATPase of distal segments from adrenalectomized rabbits. *Pfluegers Arch.* **384,** 203–206.
38. Johnson, J. P., and Green, S. W. (1981). Aldosterone stimulates $Na^+$ transport without affecting citrate synthase activity in cultured cells. *Biochim. Biophys. Acta* **647,** 293–296.
39. Jørgensen, P. L. (1972). The role of aldosterone in the regulation of ($Na^+$-$K^+$)-ATPase in rat kidney. *J. Steroid Biochem.* **3,** 181–191.
40. Kinne, R., and Kirsten, R. (1968). Der einfluss von aldosteron auf die aktivitat mitochondrialer und cytoplasmatischer enzyme in der rattenniere. *Pfluegers Arch.* **300,** 244–254.
41. Kirsten, R., and Kirsten, E. (1972). Redox state of pyridine nucleotides in renal response to aldosterone. *Am. J. Physiol.* **223,** 229–235.
42. Kirsten, E., Kirsten, R., Leaf, A., and Sharp, G. W. G. (1968). Increased activity of enzymes of the tricarboxylic acid cycle in response to aldosterone in the toad bladder. *Pfluegers Archiv.* **300,** 213–225.
43. Kirsten, E., Kirsten, R., and Sharp. G. W. G. (1970). Effects of sodium transport stimulating substances on enzyme activities in the toad bladder. *Pfluegers Arch.* **316,** 26–33.
44. Knox, W. H., and Sen, A. D. (1974). Mechanism of action of aldosterone with particular reference to (Na + K)-ATPase. *Ann. N.Y. Acad. Sci.* **242,** 471–488.
45. Kusch, M., Farman, N., and Edelman, I. S. (1978). Binding of aldosterone to cytoplasmic and nuclear receptors of the urinary bladder epithelium of *Bufo marinus. Am. J. Physiol.* **235,** C82–C90.
46. Law. P. Y., and Edelman, I. S. (1978). Effect of aldosterone on the incorporation of amino acids into renal medullary proteins. *J. Membr. Biol.* **41,** 15–40.
47. Law, P. Y., and Edelman, I. S. (1978). Induction of citrate synthase by aldosterone in the rat kidney. *J. Membr. Biol.* **41,** 41–64.
48. Leaf, A., and MacKnight, A. D. C. (1972). The site of the aldosterone induced stimulation of sodium transport. *J. Steroid Biochem.* **3,** 237–245.
49. Lewis, S. A., Eaton, D. C., and Diamond, J. M. (1976). The mechanism of $Na^+$ transport by rabbit urinary bladder. *J. Membr. Biol.* **28,** 41–70.
50. MacKnight, A. D. C., DiBona, D. R., and Leaf, A. (1980). Sodium transport across toad urinary bladder. A model "tight" epithelium. *Physiol. Rev.* **60,** 616–715.
51. Marver, D., and Petty, K. J. (1981). Acute aldosterone-dependent increases in rabbit cortical collecting tubule citrate synthase activity are not sensitive to amiloride. *Annu. Meet. Am. Soc. Nephrol., 14th,* Washington, D.C., 155A.
52. Marver, D., and Schwartz, M. J. (1980). Identification of mineralocorticoid target sites in the isolated rabbit cortical nephron. *Proc. Natl. Acad. Sci. U.S.A.* **77,** 3672–3676.
53. Moolenaar, W. H., Mummery, C. L., Van der Saag, P. T., and DeLaat, S. W. (1981). Rapid ionic events and the initiation of growth in serum-stimulated neuroblastoma cells. *Cell* **23,** 789–798.
54. Nagel, W., and Crabbe, J. (1980). Mechanism of action of aldosterone on active $Na^+$ transport across toad skin. *Pfluegers Arch.* **385,** 181–187.
55. O'Farrell, P. H. (1975). High resolution two-dimensional electrophoresis of proteins. *J. Biol. Chem.* **250,** 4407–4021.
56. O'Farrell, P. Z., Goodman, H. M., and O'Farrell, P. H. (1977). High resolution two-dimensional electrophoresis of basic as well as acidic proteins. *Cell* **12,** 1133–1141.
57. Palmer, L. G., Edelman, I. S., and Lindemann, B. (1980). Current-voltage analysis of apical sodium transport in toad urinary bladder: Effects of inhibitors of transport and metabolism. *J. Membr. Biol.* **57,** 59–71.

58. Palmer, L. G., Li, J. H-Y., Lindemann, B., and Edelman, I. S. (1982). Aldosterone control of the density of sodium channels in the toad urinary bladder. *J. Membr. Biol.* **64,** 91–102.
59. Peterson, G. L., and Hokin, L. E. (1981). Molecular weight and stoichiometry of the sodium- and potassium activated adenosine triphosphatase subunits. *J. Biol. Chem.* **256,** 3751–3761.
60. Petty, K. J., Kokko, J. P., and Marver, D. (1981). Secondary effect of aldosterone on Na-K ATPase activity in the rabbit cortical collecting tubule. *J. Clin. Invest.* **68,** 1514–1521.
61. Reich, I. M., Skipski, I. A., and Scott, W. (1981). Mechanisms of hormonal modulation of ion transport in the toad's urinary bladder. Subcellular localization of aldosterone-induced proteins. *Biochim. Biophys. Acta* **676,** 379–385.
62. Rodriguez, H. J., Sinha, S. K., Starling, J., and Klahr, S. (1981). Regulation of renal $Na^+$-$K^+$-ATPase in the rat by adrenal steroids. *Am. J. Physiol.* **241,** F186–F195.
63. Rossier, B. C., Claire, M., Rafestin-Oblin, M. E., Geering, K., Gäggeler, H. P., and Corvol, P. (1983). Binding and antimineralocorticoid activities of spirolactones in toad bladder. *Am. J. Physiol.* **244,** C24–C31.
64. Rossier, B. C., Gäggeler, H. P., and Rossier, M. (1978). Effects of 3′ deoxyadenosine and actinomycin D on RNA synthesis in toad bladder: Analysis of response to aldosterone. *J. Membr. Biol.* **41,** 149–166.
65. Rossier, M., Rossier, B. C., Pfeiffer, J., and Kraehenbühl, J. P. (1979). Isolation and separation of toad bladder epithelial cells. *J. Membr. Biol.* **48,** 141–166.
66. Rossier, B. C., Wilce, P. A., and Edelman, I. S. (1974). Kinetics of RNA labelling in toad bladder epithelium: Effects of aldosterone and related steroids. *Proc. Natl. Acad. Sci. U.S.A.* **71,** 3101–3105.
67. Rossier, B. C., Wilce, P. A., and Edelman, I. S. (1977). Spironolactone antagonism of aldosterone action on $Na^+$ transport and RNA metabolism in toad bladder epithelium. *J. Membr. Biol.* **32,** 177–194.
68. Rossier, B. C., Wilce, P. A., Inciardi, F., Yoshimura, F. K., and Edelman, I. S. (1977). Effects of 3′ deoxycytidine on RNA synthesis in toad bladder: Analysis of response to aldosterone. *Am. J. Physiol.* **232,** C174–C179.
69. Rozengurt, E., Gelherter, T. D., Legg, A., and Pettican, P. (1981). Mellitin stimulates Na entry, Na-K pump activity and DNA synthesis in quiescent cultures of mouse cells. *Cell* **23,** 781–788.
70. Saito, T., and Essig, A. (1973). Effect of aldosterone on active and passive conductance and $E_{Na}$ in the toad bladder. *J. Membr. Biol.* **13,** 1–18.
71. Saito, T., Essig, A., and Caplan, S. R. (1973). The effect of aldosterone on the energetics of sodium transport in the frog skin. *Biochim. Biophys. Acta* **318,** 371–382.
72. Schmidt, U., Schmid, J., Schmid, H., and Dubach, U. C. (1975). Sodium and potassium-activated ATPase. A possible target of aldosterone. *J. Clin. Invest.* **55,** 655–660.
73. Scott, W. N., Reich, I. M., Brown, J. A., Jr., and Yang, C-P. H. (1978). Comparison of toad bladder aldosterone-induced proteins and proteins synthesized *in vitro* using aldosterone-induced messenger RNA as a template. *J. Membr. Biol.* **40** (Special Issue), 213–220.
74. Scott, W. N., Reich, I. M., and Goodman, D. B. P. (1979). Inhibition of fatty acid synthesis prevents the incorporation of aldosterone-induced proteins into membranes. *J. Biol. Chem.* **254,** 4957–4959.
75. Scott, W. N., and Sapirstein, V. S. (1975). Identification of aldosterone-induced proteins in the toad's urinary bladder. *Proc. Natl. Acad. Sci. U.S.A.* **72,** 4056–4060.
76. Scott, W. N., Sapirstein, V. S., and Yoder, M. J. (1974). Partition of tissue functions in epithelia: Localization of enzymes in mitochondria-rich cells of toad urinary bladder. *Science* **184,** 797–800.
77. Sharp, G. W. G., and Leaf, A. (1966). Mechanism of action of aldosterone. *Physiol. Rev.* **46,** 593–633.

78. Siegel, B., and Civan, M. M. (1976). Aldosterone and insulin effects on driving force of $Na^+$ pump in toad bladder. *Am. J. Physiol.* **230,** 1603–1608.
79. Smith, J. B., and Rozengurt, E. (1978). Serum stimulates the $Na^+$,$K^+$ pump in quiescent fibroblasts by increasing $Na^+$ entry. *Proc. Natl. Acad. Sci. U.S.A.* **75,** 5560–5564.
80. Spires, D., and Weiner, M. W. (1980). Use of an uncoupling agent to distinguish between direct stimulation of metabolism and direct stimulation of transport: Investigation of antidiuretic hormone and aldosterone. *J. Pharmacol. Exp. Therap.* **214,** 507–515.
81. Spooner, P. M., and Edelman, I. S. (1975). Further studies on the effect of aldosterone on electrical resistance of toad bladders. *Biochim. Biophys. Acta* **406,** 304–314.
82. Spooner, P. M., and Edelman, I. S. (1976). Effects of aldosterone on $Na^+$ transport in the toad bladder. I. Glycolysis and lactate production under aerobic conditions. *Biochim. Biophys. Acta* **444,** 653–662.
83. Spooner, P. M., and Edelman, I. S. (1976). Effects of aldosterone on $Na^+$ transport in the toad bladder. II. The anaerobic response. *Biochim. Biophys. Acta* **444,** 663–673.
84. Wilce, P. A. (1981). Metabolism of Poly(A)(+)RNA in toad bladder epithelial cells. *J. Membr. Biol.* **62,** 163–167.
85. Wilce, P. A., Rossier, B. C., and Edelman, I. S. (1976). Actions of aldosterone on polyadenylated ribonucleic acid and $Na^+$ transport in the toad bladder. *Biochemistry* **15,** 4279–4285.
86. Wilce, P. A., Rossier, B. C., and Edelman, I. S. (1976). Actions of aldosterone on rRNA and $Na^+$ transport in the toad bladder. *Biochemistry* **15,** 4286–4292.
87. Wu, J-Y., and Yang, J. T. (1970). Physicochemical characterization of citrate synthase and its subunits. *J. Biol. Chem.* **245,** 212–218.

# Chapter 17

# Development of an Isolation Procedure for the Brush Border Membrane of an Electrically Tight Epithelium: Rabbit Distal Colon

*MICHAEL C. GUSTIN* AND DAVID B. P. GOODMAN†*

**Department of Biochemistry*
*University of Wisconsin*
*Madison, Wisconsin*
*and*
*†Department of Pathology and Medicine*
*Hospital of the University of Pennsylvania*
*University of Pennsylvania School of Medicine*
*Philadelphia, Pennsylvania*



## I. THE NEED TO ISOLATE BRUSH BORDER MEMBRANE FROM TIGHT EPITHELIA

Epithelial tissues with high electrical resistance, so-called "tight" epithelia, have been used frequently for electrophysiological studies of transepithelial ion transport (Macknight *et al.*, 1980). These studies point to the apical or luminal plasma membrane as the principal permeability barrier or controller of sodium permeability. Additionally, in several tight epithelia, evidence indicates that the luminal membrane is the site of action of hormones that alter transepithelial ion and water permeability and transport (Frizzel and Schultz, 1978; Kachadorian *et*




ISBN 0-12-153320-4

*al.*, 1977). Direct study of the permeability and transport properties of this luminal plasma membrane in tight epithelia has been hampered by lack of a technique(s) to isolate this cellular membrane in sufficient purity and quantity. This has occurred despite considerable success in preparing luminal membranes from electrically "leaky" epithelia such as renal proximal tubule and small intestine (Murer and Kinne, 1980).

In these leaky epithelia, early preparations of luminal or brush border plasma membrane relied upon the relative stability of the brush border complex during cellular homogenization as compared to the relative ease of fragmentation of the basolateral plasma membrane (Berger and Sacktor, 1970; Miller and Crane, 1961). Thus, large brush border membrane fragments were obtained which differed sufficiently in density to make separation by density gradient centrifugation feasible. More recently the differential precipitation by divalent cations of brush border membranes, as opposed to basolateral plasma membrane and intracellular organelles and subcellular membranes, has been employed to rapidly isolate this portion of the plasma membrane (Booth and Kenny, 1974; Evers *et al.*, 1978; Kessler *et al.*, 1978).

Two critical problems were faced in choosing the appropriate tight epithelium to attempt isolation of the luminal membrane: cellular heterogeneity of the transport epithelial cell layer, a distinct disadvantage of the frog skin, and the availability of large amounts of starting material, a severely limiting factor in both the toad and rabbit urinary bladder. Consequently we chose the rabbit distal colon. The epithelial layer of this tissue consists of two cell types, the predominant transport epithelial cell and the mucus-secreting cell found in the mucosal crypts. Additionally, large amounts of starting material are readily available. Below, we review our experience in developing an isolation procedure for the brush border membrane of the rabbit distal colon and highlight some unique properties of the membrane that could be demonstrated only after isolation of the membrane in high purity (Gustin and Goodman, 1981, 1982).

## II. APPROACH TO PROBLEMS OF BRUSH BORDER MEMBRANE ISOLATION IN RABBIT DESCENDING COLON

The apical membrane of the rabbit descending colon epithelium has a brush border-type structure similar to that found in the small intestine (Forstner *et al.*, 1968). In order to isolate the apical membrane of the rabbit descending colon associated with the brush border-type structure, the procedure outline in Fig. 1 was developed (Gustin and Goodman, 1981). This finished procedure incorporates the solutions to two fundamental problems that hindered the successful isolation of apical membranes from the colonic mucosa: (1) mucus and (2) clean fractionation of colonic cell homogenates.

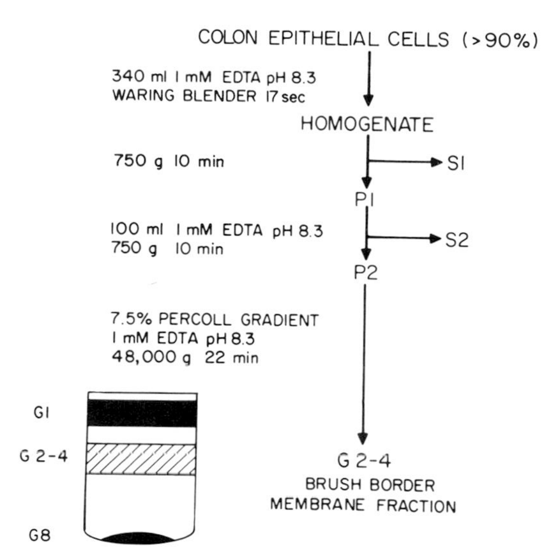


FIG. 1. Purification protocol for isolation of brush border membrane from rabbit descending colon. P, S, and G represent pellet, supernatant, and gradient fractions, respectively. Routinely, two animals are used for each preparation. Reproduced with permission from Gustin and Goodman (1981).

In contrast to the small intestine for which mucosal scrapings have been a successful starting point for brush border membrane isolation, the presence of large amounts of mucus residing in the crypt cells of the colonic mucosa interferes in the fractionation of whole mucosa homogenates. To avoid this problem, the procedure developed uses isolated surface epithelial cells as starting material from which apical membranes are isolated. In the rabbit distal colon this surface epithelial cell population contains less than 10% mucus-secreting cells. This makes the rabbit colon a particularly advantageous tissue compared to colonic epithelia from other species, particularly the rat.

Clean fractionation of the isolated epithelial cell homogenate was also a problem. Homogenization of the collected epithelial cells using a variety of procedures revealed the following: first, the cells were resistant to shear force; second, cytoplasmic structures were maintained, in large part, after disruption of the cell membrane; and third, aggregation of particulate material (cells, organelles, etc.) during homogenization and fractionation was commonly encountered. By trial and error, a homogenization strategy was developed that dealt satisfactorily with each problem.

The problem of the cells being resistant to shear force was solved by employing a hypotonic homogenization medium containing EDTA and a high-shear-force procedure involving a high-speed, 15- to 17-second homogenization using a Waring blender. The result was 100% cell breakage. This allowed brush

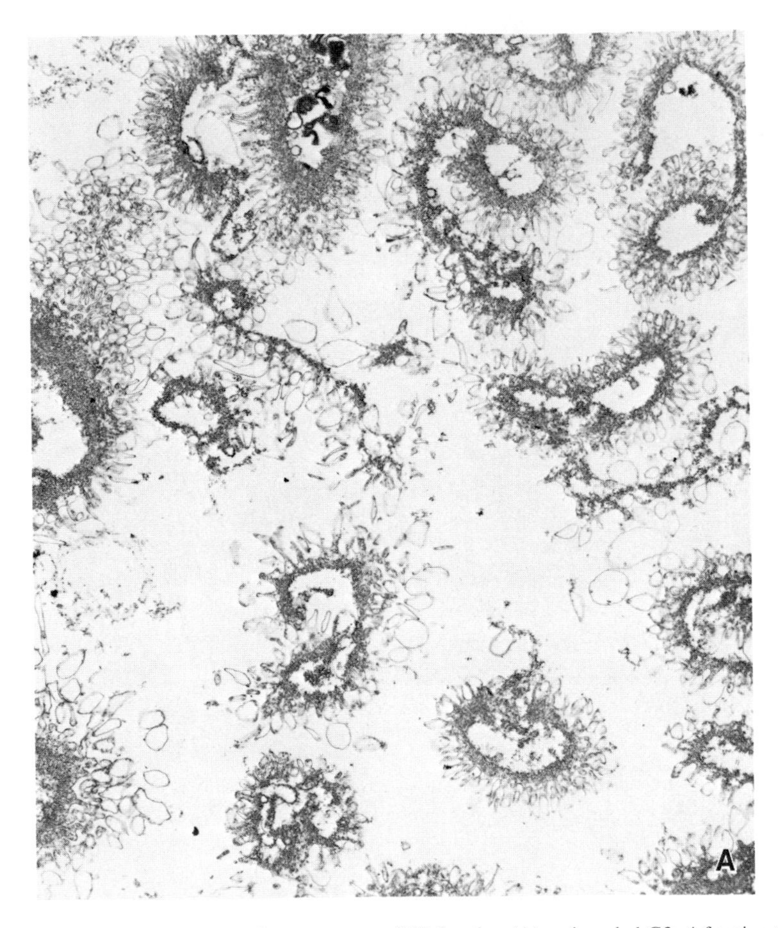


FIG. 2. Electron microscopic appearance of P2 fraction (A) and pooled G2–4 fraction (B). (A) Fraction contains primarily intact brush borders with attached pieces of basolateral membrane. ×4500. (B) Fraction contains primarily apical membranes in both vesicular and open forms. Occa-

borders to be sedimented by low-speed centrifugation without contamination by intact cells.

A more difficult problem was cytoplasmic dissolution. Although numerous brush borders were seen after homogenization in buffered (pH 7.4) medium, a large network of filaments and mitochondria was still attached to the cytoplasmic surface of the brush border. What essentially remained, therefore, was a nearly intact top one-third of the epithelial cell. Of the different approaches explored, two resulted in successful elimination of attached cytoplasmic structures. Repeated sonication in a bath-type sonicator resulted in a low yield of brush borders cleared of cytoplasmic structures. A higher yield was obtained by employing a high-pH (8.3), freshly prepared, unbuffered homogenization medium. After ho-

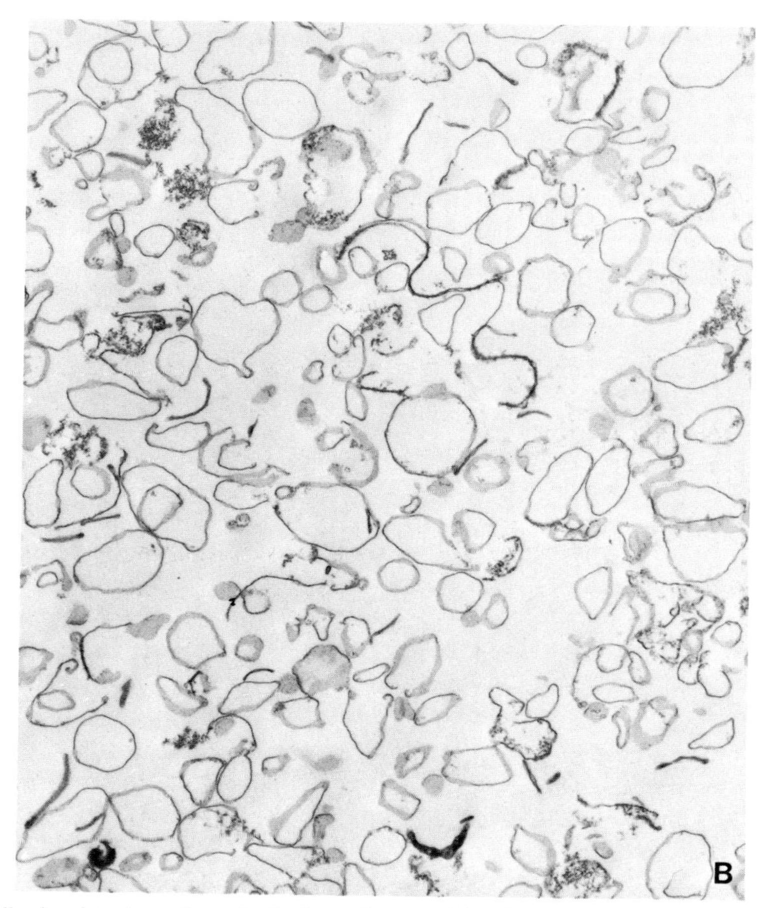


sionally, basolateral membrane in the form of synapsed junctional complexes was also observed. Some filamentous aggregates derived from the terminal web were also occasionally found in this pooled fraction. ×10,000. Reproduced with permission from Gustin and Goodman (1981).

mogenization and washing in a 1 m*M* disodium EDTA (pH 8.3) medium, virtually all of the brush borders obtained were free of gross cytoplasmic contamination.

The problem of aggregation of particulate material during homogenization and fractionation was solved by the use of (1) EDTA, (2) high dilution of the homogenate (approximately 100 mg of protein per 340 ml), and (3) low ionic strength.

Hence, the homogenization as outlined in Fig. 1 produced 100% cell breakage and intact cytoplasm-free brush borders and prevented aggregation which would interfere with subsequent fractionation of the homogenate. A washed pellet obtained after two successive sedimentations at low speed contained primarily brush borders (Fig. 2A) with occasional nuclear ghosts and unidentifiable filamentous material. To remove the latter contaminants a number of density gradient materials were tested. The most successful procedure was Percoll density centrifugation.

The use of a Percoll density gradient resulted in the achievement of two objectives: separation of nuclear ghosts from the apical membrane, and near quantitative removal of filamentous material, including filaments of the terminal web underlying the apical membrane. Nuclear ghosts and filamentous material formed an upper band on the Percoll gradient, this band actually being excluded from the Percoll phase. A small volume of 1–2 ml separated this upper band from a lower refractile band just within the Percoll gradient. Primarily, membranes in both vesicular and open forms made up this lower band (Fig. 2B). Vesicular space measurements with two types of radioactive tracers, a membrane-permeable and a membrane-impermeable tracer, confirmed that some of the membranes of this lower band form vesicles with a total space of between 4 and 10 μl/mg protein.

Although the origin of most of the membrane observed in electron micrographs of the lower band could not be determined, some membrane fragments, synapsed together, appear to be derived from regions of intercellular junctions. However, because of the distribution of apical membrane marker enzyme detailed below, most of the membrane is presumed to be of apical origin. Strengthening this notion, an increase in EDTA concentrations from 1 to 5 m*M* changed

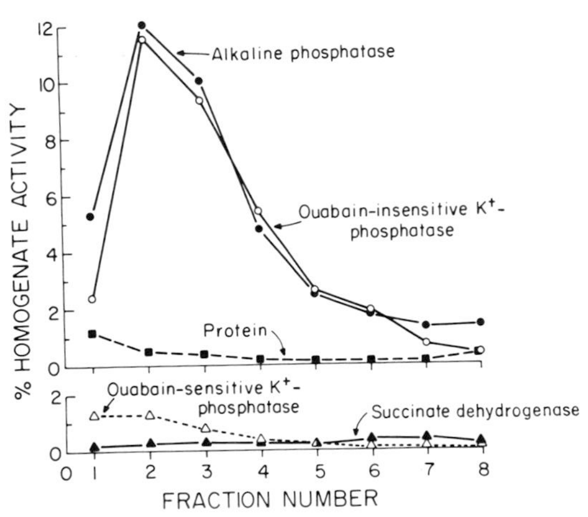


FIG. 3. Marker enzyme activity and protein profile of fractions obtained from the 7.5% Percoll gradient. Fraction 1, the least dense fraction, contained the entire first visible band. A second band within the Percoll gradient was contained in fractions 2, 3, and 4. Note the copurification of ouabain-insensitive $K^+$-phosphatase and alkaline phosphatase, an apical membrane marker. Reproduced with permission from Gustin and Goodman (1981).

TABLE I

MARKER ENZYME ANALYSIS OF SELECTED FRACTIONS OBTAINED DURING PURIFICATION OF RABBIT COLON BRUSH BORDER MEMBRANE: SUMMARY OF APICAL MEMBRANE PREPARATION[a]

| | Homogenate | Pellet 2 | Gradient pool 2–4 |
|---|---|---|---|
| Protein (mg) | 131.2 ± 4.1 | 8.7 ± 2.3 | 1.6 ± 0.1<br>(99.8 ± 2.9%) |
| Ouabain-sensitive | | | |
| K⁺-Phosphatase | | | |
| Total activity | 132.5 ± 19.8 | 9.9 ± 1.8 | 3.0 ± 1.0 |
| Specific activity | 1.0 ± 0.2 | 1.2 ± 0.6 | 1.9 ± 0.6 |
| Relative specific activity | 1.0 | 1.2 ± 0.4 | 1.9 ± 0.3<br>(105.7 ± 5.8%) |
| Alkaline phosphatase | | | |
| Total activity | 1.20 ± 0.08 | 0.49 ± 0.10 | 0.29 ± 0.04 |
| Specific activity | 0.01 ± 0.001 | 0.06 ± 0.01 | 0.19 ± 0.04 |
| Relative specific activity | 1.0 | 6.3 ± 0.6 | 20.2 ± 2.4<br>(101.3 ± 2.1%) |
| Ouabain-insensitive | | | |
| K⁺-Phosphatase | | | |
| Total activity | 75.5 ± 11.9 | 43.1 ± 11.3 | 19.8 ± 3.7 |
| Specific activity | 0.6 ± 0.1 | 4.9 ± 0.3 | 12.5 ± 1.3 |
| Relative specific activity | 1.0 | 8.7 ± 1.4 | 21.4 ± 1.5<br>(83.1 ± 9.6%) |
| Succinate dehydrogenase | | | |
| Total activity | 378.3 ± 125.4 | 7.1 ± 3.3 | 2.4 ± 1.2 |
| Specific activity | 2.9 ± 1.0 | 0.9 ± 0.6 | 1.6 ± 0.8 |
| Relative specific activity | 1.0 | 0.3 ± 0.2 | 0.5 ± 0.2<br>(92.4 ± 15.2%) |

[a] Units for enzyme activity are the following: total activity, μmol $hr^{-1}$ total volume; specific activity, μmol $hr^{-1}$ mg $protein^{-1}$; relative specific activity, specific activity in the fraction per specific activity in the homogenate. Recovery (values in parentheses) refers to the percentage recovery of enzyme activity over all fractions relative to total homogenate activity. All data are expressed as the mean (of three preparations) ± standard deviation. Alkaline phosphatase and ouabain-insensitive $K^+$-phosphatase had nearly the same relative specific activity in gradient fractions 2–4. Reproduced with permission from Gustin and Goodman (1981).

the membrane band and apical membrane marker enzyme activity to a higher density position lower in the Percoll gradient. Microscopic examination of this material revealed intact brush borders. Electron microscopic examination of brush borders in the lower ionic strength media (1 m*M* EDTA) revealed extensive loss of microvillar core filaments and blebbing of microvillar membrane to form large right-side-out vesicles.

The band containing apical membranes was collected in fractions G2–G4 of the Percoll gradient (Fig. 3). These pooled fractions were found to be highly enriched in alkaline phosphatase activity, with 25% of the total homogenate activity. This represented an approximate 20-fold increase in specific activity over the homogenate value (or in relative specific activity; Table I). Alkaline phosphatase had been shown by Vengesa and Hopfer (1979) to be a marker for the rabbit colon epithelial cell apical membrane. Succinate dehydrogenase, a mitochondrial enzyme marker, was distributed at a higher density position in the Percoll gradient (peak at fraction 6; see Fig. 3) and had less than 1% of its total homogenate activity in fractions G2–G4 (Table I). Other markers tested—cytochrome oxidase (inner mitochondrial membrane), monoamine oxidase (outer mitochondrial membrane), NADPH–cytochrome *c* reductase (endoplasmic reticulum and Golgi apparatus), acid phosphatase (lysosomes), and DNA (nucleus)—had less than 1% of their total homogenate activities represented in fractions G2–G4.

One significant contaminant of fractions G2–G4 appeared to be basolateral membrane, on the basis of the microscopic evidence, junctional membrane areas in Fig. 2B, and the activity of the basolateral enzyme marker, ouabain-sensitive $K^+$-activated *p*-nitrophenylphosphatase ($K^+$-phosphatase). Yields of this enzyme in fractions G2–G4 averaged 2% of the homogenate value (Table I). The peak of basolateral membrane marker enzyme activity was, however, generally observed to be one fraction higher in the Percoll gradient than the fraction containing the highest activity of alkaline phosphatase (Fig. 3). If sonication was employed in place of the low-ionic-strength procedure, contamination of the brush border membrane fraction by basolateral membrane was significantly reduced, but the yield of brush border membrane was also significantly diminished.

## III. OUABAIN-INSENSITIVE $K^+$-DEPENDENT PHOSPHOHYDROLASE

In addition to alkaline phosphatase and the ouabain-sensitive $K^+$-phosphatase, a third membrane-bound phosphohydrolase (phosphatase) could be detected in homogenates of rabbit colon epithelial cells. This enzyme was uncovered when it was found that not all of the $K^+$-dependent phosphatase activity could be inhibited by the addition of 1 m*M* ouabain to the assay medium. This enzyme, a ouabain-insensitive $K^+$-phosphatase, represented 37% of the total $K^+$-phosphatase activity of the homogenate (Table I). Furthermore, using the fractionation scheme described above, the enzyme copurified with alkaline phosphatase, the enzyme marker for the apical membrane.

In Fig. 3, the distribution of these two enzymes in a 7.5% Percoll gradient, as well as that of ouabain-sensitive $K^+$-phosphatase and succinate dehydrogenase, is compared. The codistribution of the ouabain-insensitive $K^+$-phosphatase and alkaline phosphatase is apparent. The other two enzyme markers are distributed very differently—ouabain-sensitive $K^+$-phosphatase activity is maximal in fractions G1 and G2, and succinate dehydrogenase activity is maximal in fractions G6 and G7. As with alkaline phosphatase, the ouabain-insensitive $K^+$-phosphatase activity comigrated with brush borders when the latter formed a higher density band in a 5 m*M* EDTA–Percoll gradient.

## IV. CRITICAL EVALUATION OF ISOLATION TECHNIQUE—FUTURE UTILITY

A technique has been developed for the isolation of the apical brush border membrane of the rabbit descending colon. This procedure is simple, reproducible, and rapid, requiring only 6 hours to complete. Based on recovery of alkaline phosphatase activity, ~25% of the total apical brush border membrane in the crude homogenate is obtained in the final membrane fraction (fractions G2–G4 of the Percoll gradient). This recovery of alkaline phosphatase activity represents a 20-fold enrichment in enzyme specific activity relative to the homogenate value.

The final membrane fraction is derived, via density gradient centrifugation, from a washed, low-speed pellet (P2) composed almost exclusively of intact brush border (Fig. 2A). Hence, contamination of the final membrane fraction by other subcellular fractions is minimal, a fact confirmed by measurements of marker enzyme activities. One exception appears to be contamination by ouabain-sensitive $K^+$-phosphatase, a marker for the basolateral membrane. This contamination is observed in the washed, low-speed pellet (P2), and is retained after density gradient centrifugation (Table I).

There are several possible explanations for this contamination by ouabain-sensitive $K^+$-phosphatase. First, the basolateral membrane, containing this enzyme, may pellet as sheets. However, extended sheets of membrane, not including brush borders, were not observed in electron micrographs of P2, making this an unlikely explanation for the observed contamination. Second, basolateral membrane fragments may be trapped in the low-speed pellet (P2). Two observations argue against this possibility: first, repeated washings of P2 in 1 m*M* disodium EDTA (pH 8.3) did not reduce the recovery of ouabain-sensitive $K^+$-phosphatase activity, and second, other enzyme markers had low activity in P2; this would not be expected for a nonspecific trapping phenomenon. A third possible explanation concerns the position of the shear plane across the colon epithelial cell. The shear plane that generated the brush border from the intact cell was generally below the outer luminal point of cell-to-cell contact (Fig. 2A). Hence, "brush borders" were not composed exclusively of the apical membrane domain, but included small portions of the basolateral membrane as well. At least part of the ouabain-sensitive $K^+$-phosphatase recovered in the final membrane fraction was derived from these membranes. A fourth possible explanation is that the barrier between the basolateral and apical membrane domains, the zonula occludens, was probably disrupted during the preparation of isolated epithelial cells in EDTA (Ceccarelli *et al.*, 1978), thus permitting the mixing of mobile membrane components from the two domains. Movement of basolateral enzyme into the apical membrane domain could therefore account for part of the observed contamination of the final membrane fraction with ouabain-sensitive $K^+$-phosphatase activity.

Using the procedure discussed above, approximately 2–4 mg of highly purified luminal membrane protein can be isolated from the distal colon of two New Zealand rabbits (2 kg). This preparation can be used for direct study of the permeability properties of the luminal membrane. Meaningful comparisons between data obtained from electrophysiological studies of intact tissue (Thompson *et al.*, 1982a,b) and this preparation should be forthcoming. Additionally, experiments to directly assess the factors controlling luminal membrane permeability can now be carried out.

ACKNOWLEDGMENT

The research described in this review was supported by USPHS Grant AM 28183.

REFERENCES

Berger, S. J., and Sacktor, B. (1970). Isolation and biochemical characterization of brush border membrane vesicles. *J. Cell Biol.* **47,** 637–645.

Booth, A. G., and Kenny A. J. (1974). A rapid method for the preparation of microvilli from rabbit kidney. *Biochem. J.* **142,** 575–581.

Ceccarelli, B., Hurlbut, W. P., DeCamelli, P., and Meldolesi, J. (1978). The effect of extracellular calcium on the topological organization of the plasmalemma in two secretory systems. *Ann. N.Y. Acad. Sci.* **307,** 653–655.

Evers, C., Haase, W., Murer, H., and Kinne, R. (1978). Properties of brush border vesicles from rat kidney cortex by calcium precipitation membrane. *Biochemistry* **1,** 203–219.

Forstner, G. G., Sabesin, S. M., and Isselbacher, K. J. (1968). Rat intestinal microvillus membrane purification and biochemical characterization. *Biochem. J.* **106,** 381–390.

Frizzel, R. A., and Schultz, S. G. (1978). Effect of aldosterone on ion transport by rabbit colon *in vitro*. *J. Membr. Biol.* **39,** 1–26.

Gustin, M. C., and Goodman, D. B. P. (1981). Isolation of brush-border membrane from the rabbit descending colon epithelium. Partial characterization of a unique $K^+$-activated ATPase. *J. Biol. Chem.* **256,** 10651–10656.

Gustin, M. C., and Goodman, D. B. P. (1982). Characterization of the phosphorylated intermediate of the $K^+$-ouabain-insensitive ATPase of the rabbit colon brush-border membrane. *J. Biol. Chem.* **257,** 9629–9633.

Kachadorian, W. A., Levine, S. D., Wade, J. B., DiScala, V. A., and Hays, R. M. (1977). Relationship of aggregated intramembranous particles to water permeability in vasopressin-treated toad urinary bladder. *J. Clin. Invest.* **59,** 576–581.

Kessler, M., Acuto, O., Storelli, C., Murer, H., Muller, M., and Semenza, G. (1978). A modified procedure for the rapid preparation of efficiently transporting vesicles from small intestinal brush border membranes. Their use in investigating some properties of D-glucose and choline transport systems. *Biochim. Biophys. Acta* **506,** 136–154.

MacKnight, A. D. C., DiBona, D. R., and Leaf, A. (1980). Sodium transport across toad urinary bladder: A model "tight" epithelium. *Physiol. Rev.* **60,** 615–715.

Miller, D., and Crane, R. K. (1961). The digestive function of the epithelium of the small intestine. II. Localization of disaccharide hydrolysis in the isolated brush border portion of intestinal epithelial cells. *Biochim. Biophys. Acta* **52,** 293–298.

Murer, H., and Kinne, R. (1980). The use of isolated membrane vesicles to study epithelial transport processes. *J. Membr. Biol.* **55,** 81–95.

Thompson, S. M., Suzuki, Y., and Schultz, S. G. (1982a). The electrophysiology of rabbit descending colon. I. Instantaneous transepithelial current-voltage relations and the current-voltage relations of the Na-entry mechanism. *J. Membr. Biol.* **66,** 41–54.

Thompson, S. M., Suzuki, Y., and Schultz, S. G. (1982b). The electrophysiology of the rabbit descending colon. II. Current-voltage relations of the apical membrane, the basolateral membrane, and the parallel pathways. *J. Membr. Biol.* **66,** 55–61.

Vengesa, P. B., and Hopfer, U. (1979). Cytochemical localization of alkaline phosphatase and $Na^+$ pump sites in adult rat colon. *J. Histochem. Cytochem.* **27,** 1231–1235.

# Index

## A

## B



## E



## F



## I

## M



## N



## P



## R



## S

# Contents of Previous Volumes

## Volume 6



## Volume 7



## Volume 8



## Volume 9

**Volume 10**



**Volume 11**

**Cell Surface Glycoproteins: Structure, Biosynthesis, and Biological Functions**



**Volume 12**

**Carriers and Membrane Transport Proteins**



**Volume 13**

**Cellular Mechanisms of Renal Tubular Ion Transport**

PART I: ION ACTIVITY AND ELEMENTAL COMPOSITION OF INTRAEPITHELIAL COMPARTMENTS

## Volume 14

## Carriers and Membrane Transport Proteins



## Volume 15

## Molecular Mechanisms of Photoreceptor Transduction

PART I: THE ROD PHYSIOLOGICAL RESPONSE



PART II: THE CYCLIC NUCLEOTIDE ENZYMATIC CASCADE AND CALCIUM ION

## Volume 16

## Electrogenic Ion Pumps

**Volume 17**

**Membrane Lipids of Prokaryotes**



**Volume 18**

PART I. ADENYLATE CYCLASE-RELATED RECEPTORS



PART II. RECEPTORS NOT INVOLVING ADENYLATE CYCLASE

## Volume 19

PART IX. Na,K-ATPase AND POSITIVE INOTROPY; ENDOGENOUS GLYCOSIDES

## PART X. PHYSIOLOGY AND PATHOPHYSIOLOGY OF THE Na/K PUMP